VOLUME ONE HUNDRED AND SIXTY FOUR

INTERNATIONAL REVIEW OF NEUROBIOLOGY

Sex and Gender Differences in Neurological Disease

Edited by

ELENA MORO
Grenoble Alpes University, CHU of Grenoble, Division of Neurology, Grenoble Institute of Neurosciences, Grenoble, France

GENNARINA ARABIA
Magna Graecia University, Movement Disorders Center, Neurology Unit, Catanzaro, Italy

MARIA CARMELA TARTAGLIA
Memory Clinic, Krembil Brain Institute, University Health Network; Canadian Concussion Centre, Krembil Brain Institute, University Health Network; Tanz Centre for Research in Neurodegenerative Diseases, University of Toronto, Toronto, Canada

MARIA TERESA FERRETTI
Women's Brain Project, Guntershausen, Switzerland

ELSEVIER

ACADEMIC PRESS
An imprint of Elsevier

Academic Press is an imprint of Elsevier
50 Hampshire Street, 5th Floor, Cambridge, MA 02139, United States
525 B Street, Suite 1650, San Diego, CA 92101, United States
The Boulevard, Langford Lane, Kidlington, Oxford OX5 1GB, United Kingdom
125 London Wall, London, EC2Y 5AS, United Kingdom

First edition 2022

ISBN: 978-0-323-91589-2
ISSN: 0074-7742

For information on all Academic Press publications
visit our website at https://www.elsevier.com/books-and-journals

Publisher: Zoe Kruze
Acquisitions Editor: Sam Mahfoudh
Developmental Editor: Federico Paulo Mendoza
Production Project Manager: Abdulla Sait
Cover Designer: Matthew Limbert

Typeset by STRAIVE, India

Contents

Contributors

Diana Aguiar de Sousa
Stroke Center, Lisbon Central University Hospital; CEEM and Institute of Anatomy, Faculty of Medicine, University of Lisbon, Lisbon, Portugal

Gennarina Arabia
Magna Graecia University, Movement Disorders Center, Neurology Unit, Catanzaro, Italy

Eva Becher
Sex- and Gender-Sensitive Medicine Unit, Medical Faculty OWL, Bielefeld University, Bielefeld, Germany

Esther Bui
Krembil Brain Institute, University Health Network, University of Toronto, Toronto, Canada

Cheryl Carcel
The George Institute for Global Health, Faculty of Medicine, University of New South Wales; Sydney School of Public Health, Sydney Medical School, The University of Sydney, Sydney, NSW, Australia

Valeria Caso
Stroke Unit, Santa Maria della Misericordia Hospital University of Perugia, Perugia, Italy

Samaneh Chaychi
Memory Clinic, Toronto; Canadian Concussion Centre, Krembil Brain Institute, University Health Network, Toronto, ON, Canada

Carly Davenport
Tanz Centre for Research in Neurodegenerative Diseases, University of Toronto, Toronto, ON, Canada

Karen D. Davis
Krembil Brain Institute, Krembil Research Institute, University Health Network; Institute of Medical Science; Department of Surgery, University of Toronto, Toronto, Canada

Antonio De Martino
Magna Graecia University, Movement Disorders Center, Neurology Unit, Catanzaro, Italy

Lara Diem
Department of Neurology, Inselspital, Bern University Hospital and University of Bern, Freiburgstrasse, Bern, Switzerland

Maria Teresa Ferretti
Women's Brain Project, Guntershausen, Switzerland

Helly Hammer
Department of Neurology, Inselspital, Bern University Hospital and University of Bern, Freiburgstrasse, Bern, Switzerland

Robert Hoepner
Department of Neurology, Inselspital, Bern University Hospital and University of Bern, Freiburgstrasse, Bern, Switzerland

Lauren Hophing
Krembil Brain Institute, University Health Network, University of Toronto, Toronto, Canada

M. Florencia Iulita
Sant Pau Memory Unit, Department of Neurology, Hospital de la Santa Creu i Sant Pau, Biomedical Research Institute Sant Pau, Universitat Autònoma de Barcelona, Barcelona; Center of Biomedical Investigation Network for Neurodegenerative Diseases (CIBERNED), Madrid, Spain; Women's Brain Project, Guntershausen, Switzerland

Paulina Kyriakopoulos
Division of Neurology, Department of Medicine, The Ottawa Hospital, Ottawa, Canada

Elena R. Lebedeva
Department of Neurology, The Ural State Medical University; International Headache Centre "Europe-Asia", Yekaterinburg, Russia

Elena Moro
Grenoble Alpes University, CHU of Grenoble, Division of Neurology, Grenoble Institute of Neurosciences, Grenoble, France

Sabine Oertelt-Prigione
Sex- and Gender-Sensitive Medicine Unit, Medical Faculty OWL, Bielefeld University, Bielefeld, Germany; Gender Unit, Department of Primary and Community Care, Radboud University Medical Center, Nijmegen, Netherlands

Natalie R. Osborne
Krembil Brain Institute, Krembil Research Institute, University Health Network; Institute of Medical Science, University of Toronto, Toronto, Canada

Max Pistor
Department of Neurology, Inselspital, Bern University Hospital and University of Bern, Freiburgstrasse, Bern, Switzerland

Jana Remlinger
Department of Neurology, Inselspital, Bern University Hospital and University of Bern, Freiburgstrasse; Department of Biomedical Research and Graduate School for Cellular and Biomedical Sciences, University of Bern, Bern, Switzerland

Anke Salmen
Department of Neurology, Inselspital, Bern University Hospital and University of Bern, Freiburgstrasse, Bern, Switzerland

Patrick Salwierz
Tanz Centre for Research in Neurodegenerative Diseases, University of Toronto, Toronto, ON, Canada

Else Charlotte Sandset
Department of Neurology, Oslo University Hospital; The Norwegian Air Ambulance Foundation, Oslo, Norway

Hermona Soreq
The Edmond & Lily Safra Center for Brain Sciences; The Alexander Silberman Institute of Life Sciences, The Hebrew University of Jerusalem, Jerusalem, Israel

Vishaal Sumra
Tanz Centre for Research in Neurodegenerative Diseases, University of Toronto, Toronto, ON, Canada

Maria Carmela Tartaglia
Memory Clinic, Krembil Brain Institute, University Health Network; Tanz Centre for Research in Neurodegenerative Diseases, University of Toronto; Canadian Concussion Centre, Krembil Brain Institute, University Health Network, Toronto, ON, Canada

Yonat Tzur
The Alexander Silberman Institute of Life Sciences, The Hebrew University of Jerusalem, Jerusalem, Israel

Eve Valera
Department of Psychiatry, Harvard Medical School, Boston; Department of Psychiatry, Massachusetts General Hospital, Charlestown, MA, United States

Katarzyna Winek
The Edmond & Lily Safra Center for Brain Sciences; The Alexander Silberman Institute of Life Sciences, The Hebrew University of Jerusalem, Jerusalem, Israel

Introduction

Elena Moro[a,*], Gennarina Arabia[b], Maria Carmela Tartaglia[c,d,f], and Maria Teresa Ferretti[e]

[a]Grenoble Alpes University, CHU of Grenoble, Division of Neurology, Grenoble Institute of Neuroscience, Grenoble, France
[b]Magna Graecia University, Movement Disorders Center, Neurology Unit, Catanzaro, Italy
[c]Tanz Centre for Research in Neurodegenerative Diseases, University of Toronto, Toronto, ON, Canada
[d]Memory Clinic, Krembil Brain Institute, University Health Network, Toronto, ON, Canada
[e]Women's Brain Project, Guntershausen, Switzerland
[f]Canadian Concussion Centre, Krembil Brain Institute, University Health Network, Toronto, ON, Canada
[*]Corresponding author: e-mail address: emoro@chu-grenoble.fr; g.arabia@unicz.it; carmela.tartaglia@uhn.ca; mariateresa.ferretti@gmail.com

Abstract

The editors have the great pleasure and honor to introduce the main contents of this first of its kind book, specifically focused on reviewing and analyzing sex and gender differences in the most common neurological diseases. This volume is intended to increase knowledge and awareness of sex and gender disparities and, therefore, to improve diagnosis, management, and research in neurological diseases.

Introduction

The recognition of important differences between sex and gender in medicine, and their impact on epidemiology, signs and symptoms, diagnosis, response to treatment, overall management, and prognosis in many diseases have become an important field of research (Mauvais-Jarvis et al., 2020). Nevertheless, whereas these differences have been well acknowledged and studied in several disciplines, like cardiology, oncology, or endocrinology, sex and gender disparities have only recently started to be investigated in neurology (Hentzen et al., 2022).

Furthermore, very few Faculties of Medicine around the world have included education on sex and gender as part of their curricula. Similarly, only a few scientific international societies have developed committees or task forces to better investigate and promote knowledge and foster proper management of sex and gender differences. Among these, the European Academy of Neurology has been the first large international society to constitute in May 2020 a dedicated Task Force to promote identification, knowledge, education, and research in sex and gender differences in neurological diseases (currently developed as a Coordinating Panel;

https://www.ean.org/home/organisation/scientific-coordinating-panels/coordinating-panel-on-diversity-equity-and-inclusion-in-neurology).

The dearth of resources and knowledge on sex and gender issues related to neurological diseases has been the impetus for both the publisher and the editors to write this book as the first available resource on sex and gender differences in neurological diseases. We have made the choice to focus here on some of the most frequently encountered neurological disorders. Given the sheer magnitude of neurological diseases, one book is not sufficient, and a second book may be envisioned to further this endeavor of illustrating sex and gender differences in neurological diseases.

The journey starts with a general overview of the history and conceptualization of gender medicine. The basic definitions of sex (biological differences) and gender (sociocultural construct) differences in medicine and how they are determinants of health are next. Chapter 1 describes the crucial events in the United States, Canada, and Europe that have over the past five decades led to the development of a cohesive structural framework for sex- and gender-sensitive medicine.

The reader will then learn more about biological underpinnings of sex differences, with a particular focus on sex-specific genetic, transcriptomic, and epigenetic profiles of various brain cells (neurons, glia, and blood-brain barrier cells) and how these could mediate some of the clinical sex-specific phenotypes observed (Chapter 2).

Chapter 3 will update the reader about the most important sex and gender differences in neurovascular diseases. Indeed, stroke has several sex differences concerning risk factors and prognosis that need to be recognized and identified for prevention and management.

Chapter 4 is dedicated to movement disorders, where Parkinson's disease is the most recognized disease in which sex differences are many and relevant to customize treatment and overall care of people with PD.

The reader's journey will move into neuroinflammatory diseases, reviewing several known differences between sexes in multiple sclerosis but also pointing out that in several other demyelinating/inflammatory diseases much remains unknown (Chapter 5).

In Chapter 6, sex and gender differences across neurodegenerative and cerebrovascular diseases causing dementia are presented. Although some work has been done in Alzheimer's disease, there is much to learn in the other causes of dementia.

Chapter 7 on sex and gender disparities in epilepsy covers the interaction of hormones and epilepsy throughout a woman's lifespan as well as the management of epilepsy in pregnancy.

Sex and gender disparities in pain are analyzed in Chapter 8. Chronic pain particularly affects women, in whom different genes and proteins differentially modulate pain circuitries.

Sex and gender differences in headache are well known, but Chapter 9 illustrates the clinical differences in migraine types between sexes, the role of gender, and the strategies for management and treatment of migraine, according to sex and age differences.

At the end of the journey, the reader will reach Chapter 10. Sex and gender differences in concussion/mild brain injury are becoming increasingly known and this chapter will highlight the latest findings. As well, this chapter includes a section on the often-ignored brain injury ensuing from intimate partner violence.

This volume provides the first broad comprehensive overview of the main sex and gender differences in frequently encountered neurological diseases. We hope that this work will help the dissemination of knowledge and foster research in the important role of sex and gender disparities in neurology, bringing us one step closer to personalized medicine in neurological diseases.

References

Hentzen, N. B., Ferretti, M. T., Santuccione, A., Jaarsma, J., de Visser, M., & Moro, E. (2022). Mapping of European activities on the integration of sex and gender factors in neurology and neuroscience. *Journal of Neurology*. https://doi.org/10.1111/ene.15439.

Mauvais-Jarvis, F., Bairey, M. N., Barnes, P. J., Brinton, R. D., Carrero, J. J., DeMeo, D. L., et al. (2020). Sex and gender: Modifiers of health, disease, and medicine. *Lancet*, *396*(10250), 565–582.

CHAPTER ONE

History and development of sex- and gender sensitive medicine (SGSM)

Eva Becher[a] and Sabine Oertelt-Prigione[a,b,*]
[a]Sex- and Gender-Sensitive Medicine Unit, Medical Faculty OWL, Bielefeld University, Bielefeld, Germany
[b]Gender Unit, Department of Primary and Community Care, Radboud University Medical Center, Nijmegen, Netherlands
*Corresponding author: e-mail address: sabine.oertelt-prigione@radboudumc.nl

Contents

Abstract

Sex- and gender-sensitive medicine has evolved from a feminist approach into an innovative cross-cutting approach to doing medicine. In the present chapter we define what sex and gender are in the context of biomedical research and describe the history of the development of this scientific approach. Looking back at crucial events in the U.S.A., Canada and Europe, we will outline how a structural framework has been established, ready to be filled with clinical and applied knowledge and to change the practice of medicine for decades to come.

International Review of Neurobiology, Volume 164
ISSN 0074-7742
https://doi.org/10.1016/bs.irn.2022.06.008

Abbreviation

AD	Alzheimer's disease
BC	before Christ
BSRI	Bem Sex Role Inventory
CAWMSET	Commission on Advancement of Women and Minorities in Science
CIHR	Canadian Institute of Health Research
DFG	German Research Foundation
EASE	European Association of Science Editors
EIGE	European Institute for Gender Equality
ERA	European Research Area
ESMO	European Society for Medical Oncology
EU	European Union
FDA	Food and Drug Administration
FDASIA	Food and Drug Administration Safety and Innovation Act
FP6	Sixth Framework Program (of the EU)
GBA (+)	gender based analysis (+)
GEP	Gender equality plan
GH5050	Global Health 50/50
GRACE	Gender and Cultures of Equality—Consortium
ICH	International Conference on Harmonization of Technical Requirements for Registration of Pharmaceuticals for Human Use
ICMJE	International Committee of Medical Journal Editors
IGH	Institute of Gender and Health
IOM	Institute of Medicine
LERU	League of European Research Universities
MS	multiple sclerosis
NIH	National Institute of Health
NSF	National Science Foundation
OBOS	Our Bodies, Ourselves
ORWH	Office of Research on Women's Health
PAQ	Personal Attributes Questionnaire
R&I	Research & Innovation
SAB	sex (assignment) at birth
SABV	sex be considered as biological variable
SAGER	Sex and Gender Equity in Research
SGBA	sex and gender based analysis
SGSM	sex and gender sensitive medicine
UK	United Kingdom
US	United States
WBHI	Women's Brain Health Initiative
WBP	Women's Brain Project
WHAM	Women's Health Access Matters

1. What is sex- and gender-sensitive medicine

A definition of sex- and gender-sensitive medicine is essential, although not always simple. The reasons are twofold; on the one side, different regions of the world have prioritized different aspects of the sex/gender dyad over time and on the other side, biomedical research has an unfortunate history of conflating the two terms.

The different prioritization of different areas is exemplified, for example, in the strong focus of US-based research on "sex as a biological variable (SABV)" with only recent shifts toward more gender-inclusive approaches to research (Arnegard et al., 2020). The Canadian Institutes of Health Research (CIHR) have focused on a more holistic approach with the motto "every cell has a sex and everybody has a gender," which translates into a broader and more comprehensive operationalization of sex and gender in research (Johnson & Beaudet, 2013). The European Commission has instead chosen a primary focus on the "gender dimension," an overarching concept which includes a focus on sex and gender in research, but also highlights the impact of structural gender inequalities on the research process (Commission, E., 2014).

In addition to the different priorities in different regions of the world, the concepts of sex and gender have been oftentimes used interchangeably in biomedicine (Nieuwenhoven & Klinge, 2010). This conflation of the terminology is not a methodological choice, but rather a shortcoming due to lack of clear methodological standards and sometimes a genuine ignorance towards the different meaning of the two terms. Researchers have used these concepts in an inconsistent way in the scientific literature. An analysis of the use of the terms "sex" and "gender" in articles in two journals on gender-specific medicine showed—in spite of the focus on "gender" in the journal titles—that there was a consistent conflation of "sex" and "gender" in the articles (Annandale & Hammarstrom, 2011).

Consideration of sex and gender differences represents a first step towards individualized or precision medicine. The FDA defines precision medicine as treatments that are tailored to specific characteristics of individuals, such as a person's genetic makeup, or the genetic profile of an individual's tumor (FDA, 2022).

Therefore, the term "precision medicine" is not synonymous with "sex- and gender sensitive medicine," because it is focused on single biological components. Next Generation Sequencing rapidly identifies or sequences large sections of a person's genome and can contribute to advances in the clinical applications of precision medicine. However, interpersonal differences related to their social gender and other environmental factors are not part of the analysis. Many models in the context of precision medicine operationalize precision along the P1-P4 model, where the Ps stand for prediction, prevention, personalization and participation (Hood & Friend, 2011). While this context would allow also for the inclusion of gender, the current focus mostly lies on the impact of biological sex.

2. But then, what are sex and gender exactly?

"Sex" refers to biological differences between female, male and intersex subjects—human, animal or even at the cellular level. Sex is generally operationalized through what can be summarized as the 3 "Gs," i.e., genes, gonads and genitalia (Blackless et al., 2000). Practically, this means that sex can be defined by three means. The first option is the investigation of the karyotype, looking at the presence and makeup of the X and Y chromosomes. The second option is the determination of the serum levels of steroid hormones, such as estrogens, androgens, their precursors and metabolites. The last option is the determination of sex by phenotype, i.e. through examination of the genitals at birth, generally defined sex-assigned-at-birth (SAB).

"Gender" encompasses a whole range of identities and socially constructed roles and behaviors (Johnson, Greaves, & Repta, 2009). Importantly, gender depends on the historical and cultural context and varies over time. The definition of gender is affected by its conceptual construction, the power dimensions examined and active processes of performative femininities and masculinities in a given context (West & Zimmerman, 1987). Beyond that, sex and gender are interwoven in a continuous interaction process (Alex et al., 2012; Heise et al., 2019) The consideration of gender aspects in biomedicine is still limited. This might be due to the fact that the concept of gender was initially defined in the social sciences with a deconstructive methodology that rarely finds application in medicine. Only recently, attempts have been made to operationalize the concept with traditional—mostly quantitative—biomedical methodology, simplifying its incorporation in ongoing research.

Practically, gender in biomedical research is generally explored as one or more of the following dimensions: gender identity, gender expression, gender roles, gender relations and institutionalized gender (Tannenbaum, Greaves, & Graham, 2016). Gender identity defines the subjectively felt identity of an individual, such as being a man, a woman, a non-binary person and many more. Gender identity is not necessarily matching with biological sex assigned at birth (SAB). Gender expression describes the subjectively perceived matching of one's presentation to a specific gender. It depends on the culturally defined stereotypes associated to a specific gender. Gender roles and/or norms are socially constructed expectations and behaviors attributed to a specific gender identity or expression. Gender relations describe how gender influences the interaction between individuals in the private, professional and public space. Institutionalized gender represents the broader impact of gender on organizations, societal structures, laws and access to resources (Oertelt-Prigione, 2022).

Persons whose gender identity does not align with their SAB are described as "transgender." This term can describe individuals who identify as men, as women, as both or neither masculine or feminine. Little is known about the effect of gender-affirming hormone therapy and surgery on optimal laboratory test interpretation. In the current three G framework, both their genetic and hormonal makeup need to be taken into account when making therapeutic choices and the main focus needs to lie on the component most impactful to the proposed therapy. Further research and increasing social and clinical awareness of the specific needs in health care for transgender individuals are needed (Goldstein, Corneil, & Greene, 2017).

The prevalence of mental health disparities and psychological distress among trans people compared to the general population is high. The prevalence of clinical depression in trans adults is estimated to be over 50% compared with an estimated 30% lifetime prevalence among the US general population (Kessler et al., 2012; Reisner et al., 2016). Possession of gender-concordant identity documents might improve mental health among trans persons (Scheim, Perez-Brumer, & Bauer, 2020). The National Women's Health Network provides specific information on Trans Health Care (National Women's Health Network, n.d.). FDA hosted in 2021 two patient listening sessions on healthcare challenges und unmet medical needs of transgender adults and adolescents to gain a better understanding of the barriers to obtaining health care, areas of unmet medical need, and the most important goals of individuals who are contemplating or undergoing transition to their affirmed gender (FDA, 2022a, 2022b).

3. Aims and clinical relevance of SGSM

The goal of SGSM, sometimes imprecisely called gender medicine, is to ensure awareness and to remove gender-based access barriers to healthcare. Furthermore, SGSM aims at improving the diagnosis and treatment of diseases for all genders, and at incorporating these results into medical education. Last, the integration of sex- and gender-sensitive analysis into research improves medical science by promoting rigorous, reproducible and responsible science. Sex and gender analysis is critical to the interpretation, validation, reproducibility and generalizability of research findings. It leads to the generation of novel scientific questions and enhances equality by ensuring that research findings are applicable across the whole society (Tannenbaum et al., 2019).

Cardiology played a pioneer role in sex- and gender-sensitive medicine. For decades myocardial infarction was described as a male disease. Meanwhile it is known, that women are less likely than men to have typical angina and are more likely to have atypical or non-anginal pain (Canto et al., 2012). Often, the diagnosis of coronary heart disease in women is delayed, especially at younger ages or in patients with diabetes. Moreover, compared to men, women are still less likely to undergo cardiac monitoring, enzyme measurement, recovery in coronary care unit, coronary angiography, and revascularization (Chambers, Bagai, & Ivascu, 2007). Also, an underestimation of the possible risk for myocardial infarction by women was identified as reason for higher mortality rates compared to men (Vaccarino et al., 1995). Aspirin as a preventive measure is more effective in men compared to women (Ridker et al., 2005).

Recently, sex and gender related differences are described to play a crucial role also for the crosstalk between cardiovascular disease and brain disorders. Both organs share common risk factors, such as hypertension, diabetes, smoking or dyslipidemia and are both affected by systemic inflammation, atherosclerosis and dysfunction of the neuroendocrine system. Sex and gender differences modify these factors and the interaction between brain and heart over the lifespan (Rossi et al., 2022).

Sex and Gender differences affect a large range of acute and chronic diseases. For many diseases the inflammatory response of the body is crucial. Differences in the immune response between male and female patients might be responsible for differences seen in a broad range of diseases. The hypothalamic-pituitary-adrenal-gonadal axis, which exerts important control on the adaptive immune system, differs between males and females. Exercise, stress, and depression all downregulate immune function (Irwin et al., 1990).

As a rule, estrogenic hormones upregulate and androgenic hormones downregulate the cellular effectors of human adaptive immunity: lymphocytes, macrophages, and dendritic cells (Fish, 2008; Oertelt-Prigione, 2012a). The adaptive immune response varies during the menstrual cycle (Oertelt-Prigione, 2012b).

Also, neurological diseases like multiple sclerosis (MS), pain or Parkinson's disease show sex and gender related differences. MS is mostly prevalent in females compared to males with a ratio of 3:1 and an earlier onset of the disease (Gottgens et al., 2020; Voskuhl, 2011).

These are only a few examples illustrating the essential contribution SGSM plays for an effective high-quality health care for all human beings.

4. Ancient history

In Hippocratic times (5th century BC), ideas about women's health markedly differed from today's understanding. It was believed that women's bodies were prone to accumulating excessive fluid because of the porosity of their body. Excessive porosity could lead to high quantities of blood in their breasts, a condition that drove them to hysteria. According to Hippocrates, the uterus could detach itself from the abdomen and migrate to attack different organs causing various forms of illness (Lippi, Bianucci, & Donell, 2020). The term "hysteria" derives from the Greek word for uterus, hysteria, and embodies long-held gender stereotypes about the expression of less-typical somatic symptoms. This idea persisted for several centuries, leading to the belief in Medieval times that "the womb is the origin of all diseases" (Tasca et al., 2012).

The consequences of these ideas about women and other genders still persists today. Women and other genders still fight to get sufficient attention when describing their physical symptoms. Symptoms presented by women are more often interpreted as "anxiety" or "stress" or "somatization" increasing the risk of missed and delayed diagnoses (Dusenbery, 2017).

5. How it started

The development of SGSM would not have been possible without women in medicine. Until today, most scientists researching SGSM are women (Nielsen et al., 2017). Equal numbers of men and women currently graduate from medical school, yet only a small fraction of the medical leaders are women (Alwazzan & Al-Angari, 2020), possibly contributing to the paucity of awareness and lack of practical implementation of sex and gender aspects in medicine.

Often the names of the pioneers are unknown to a broad public. Maria Petroncini Ferretti was the first female surgeon in Florence in 1788 (Lippi et al., 2020) and Florence Nightingale (1820–1910) was the founder of modern nursing based on scientific education and training (Allen, 2020).

The British physician Elizabeth Blackwell (1821–1910) was the first woman to receive a MD degree from an American Medical School (Blackwell, 1970). In Germany Rahel Hirsch (1870–1953) was the first women to be appointed a professor in medicine. These women paved the way for education of a number of female doctors in the 20th century.

The roots of sex- and gender sensitive medicine (SGSM) traces back through more than 50 years to the feminist movement. Women identified the impact of structural discrimination, of not having the same opportunities to work and be politically active as limiting to their development as human beings fully participating in society. Simone de Beauvoir wrote in 1949:

> "Humanity is male and man does not define woman as such, but in comparison to himself: she is not seen as an autonomous being (…) He is the subject, he is the absolute, she is the other." *(Beauvoir & Ralph Ellison Collection (Library of Congress), 1953).*

The focus on women and health started with the Women's Health movement, which initially targeted health conditions specific for women, namely reproductive medicine and obstetrics. The movement then expanded to the different representation of symptoms and diseases in women and men affected by the same condition. The newly established discipline was called "Gender-Medicine" or "Gender-specific medicine" (Legato, 2003), which might have contributed to some of the conflation of sex and gender in medicine. In fact, "Gender-Medicine" does not solely focus on gender, but mostly on biological sex and should possibly obtain a name more in line with this reality. This is the reason why today the term "Sex- and Gender-Sensitive Medicine (SGSM)" is preferred. The term focuses on biological (sex-) differences, socially determined (gender-) differences and the interaction between the two.

6. Milestones of the establishment of sex- and gender-sensitive medicine

The development of SGSM as an academic research subject mainly took place in Northern America and Europe, while other regions of the world have focused primarily on community-based gender-sensitive research. Although the development and implementation process of SGSM is an ongoing effort, the involvement of different stakeholders (government, funding organizations, universities, peer-reviewed journals) was crucial in its promotion. At the EU level the legal efforts for gender parity represented first and essential steps (Council, 1995). Funding organization in the USA (Clayton & Collins, 2014), Canada (Johnson & Beaudet, 2013) and the EU (Commission, E., 2014) then mandated the inclusion of sex and/or gender in grant applications. A growing number of scientific journals started requesting a focus on sex and gender in research reporting (Clark & Horton, 2019; Gogos et al., 2019). Compulsory guidelines for publication, such as the SAGER (Heidari et al., 2016) and the International Committee of Medical Journal Editors (ICMJE) guidelines, include the requirement to report sex and gender disaggregated data, to explain how the study design takes sex and gender differences into account, to discuss the results in a sex and gender sensitive manner and to develop conclusions and specific public health recommendations for men and women. More recently, under the umbrella of the current funding framework "Horizon Europe," the European Commission is also requesting gender equality plans (GEP) from universities submitting grants for funding in an attempt to strengthen the visibility of the interaction between scientific structures and scientific content produced.

In the following section we will list historic milestones of the development of SGSM in a chronological order (Fig. 1).

6.1 1960–1970

The history begins in the 1960s and 1970s with the international women's health movement with a focus on physical and sexual self-determination. Women realized how the health care system failed them and were no longer willing to accept men deciding about her bodies. Women sought to increase control over their own bodies using contraception and other reproductive technologies (Bird & Sharman, 2014; Maschewsky-Schneider et al., 2001).

Politically, a number of laws and regulations supported these developments, which initially aimed more at equality and avoidance of discrimination and then specifically intervened in the field of medicine. This took place in the US, but also in Europe.

In the United States and in the UK civil rights acts prohibited discrimination based on sex.

6.2 1970–1980

In May 1969, twelve women aged 23 -39 years met at a Women's Liberation Conference at Emmanuel College in Boston, to discuss women's health issues. These women were not professionals, but college-educated as teachers or social workers and politically active in the feminist movement.

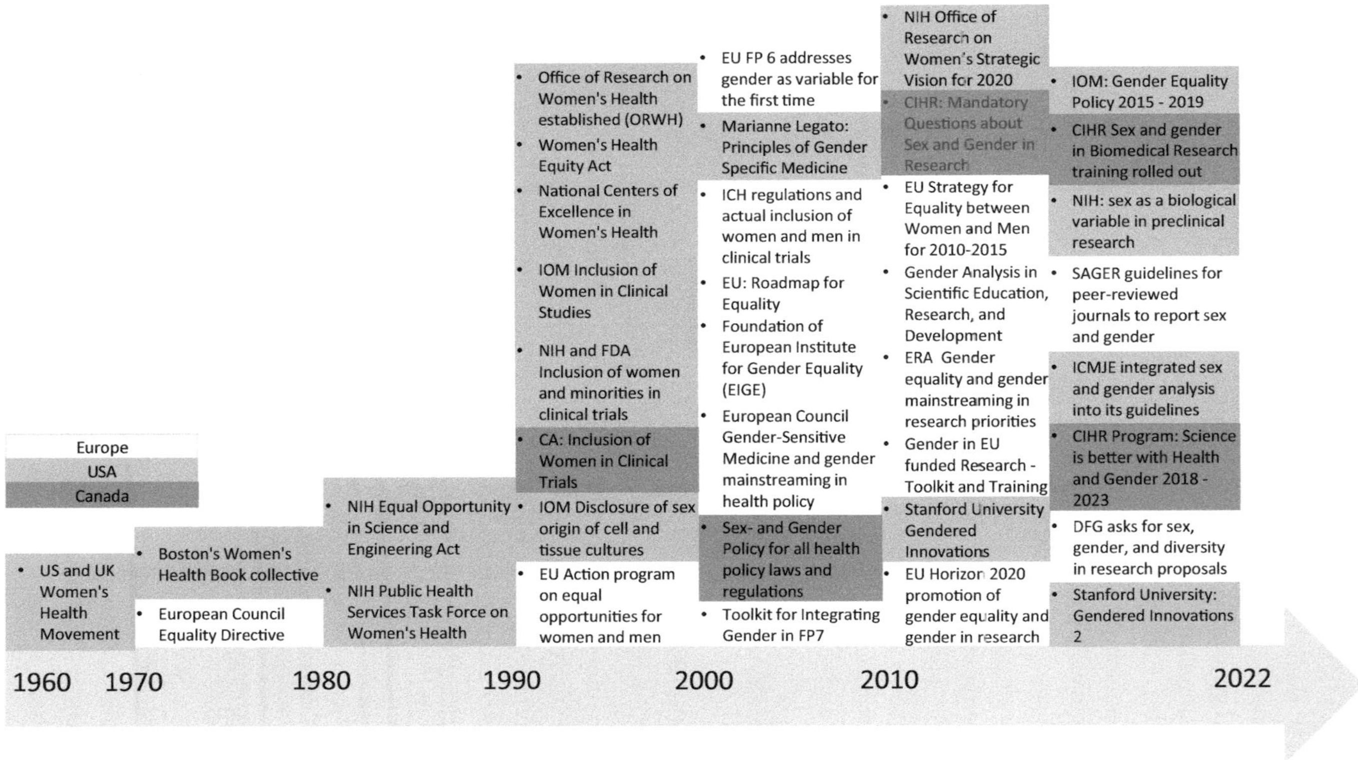

Fig. 1 Milestones in the implementation of sex- and gender-sensitive medicine.

During a workshop entitled "Women and their Bodies" they discovered that they had had similar experiences during pregnancy, labor, at gynecologists' offices and realized how little they knew about their own bodies. They decided to do their own research and share information about women's bodies and health. The Boston Women's Health Book Collective published a 193-page book named "Women and their Bodies." In 1971, the book was renamed to "Our Bodies, Ourselves." This will become synonymous for the Boston Women's Health Book Collective (www.ourbodiesourselves.org).

Meanwhile, in Europe, the European Council required member states to eliminate all discrimination on grounds of sex. This Directive includes administrative means to ensure equal pay for equal work (Council, E., 1975).

6.3 1980–1990

In 1980, the NIH brought the Equal Opportunity in Science and Engineering Act in place. This Act directs the National Institutes of Health (NIH) and National Science Foundation (NSF) to increase the participation of underrepresented groups in medicine, science, and engineering. Funding has been set aside to achieve these ends. In the eighties, the NIH established the Public Health Services Task Force on Women's Health. The task force concluded that a longstanding lack of research in women's health had compromised the quality of available information on diseases affecting women and the health care women receive (Merkatz, 1998).

In 1982, the U.S. National Science Foundation published the "Report on Women and Minorities in Science and Engineering" (National Science Foundation (U.S.), 1982), the first document offering explicit information about the underrepresentation of women in science.

6.4 1990–2000

In 1990, the Women's Health Equity Act (Congress 1990) was passed in the U.S. Congress. Subsequently, the NIH established the Office of Research on Women's Health (ORWH), the first national agency focusing on women's health issues. The task of the ORWH is to ensure that women's health is an integral part of the research at the NIH and in the scientific community.

In 1991, the NIH established the Women's Health Initiative (WHI) to address the most common causes of death, disability and impaired quality of life in postmenopausal women. Until today the WHI is one of the most comprehensive sources worldwide of information about the health of postmenopausal women.

After a lengthy period of exclusion from participation in clinical trials, in October 1992 the ORWH the Institute of Medicine (IOM) to establish the Committee on the Ethical and Legal Issues Relating to the inclusion of women in clinical studies. The committee had to consider ethical and legal implications of including women—particularly pregnant women and women of childbearing potential—in clinical studies, to examine known instances of litigation regarding injuries to research subjects and describe existing legal liabilities and protections and to provide practical advice on these issues.

As a consequence of this, in 1993, the NIH adopted the Health Revitalization Act (NIH, 2001), which legally established the inclusion of women and minorities in clinical trials. The Revitalization Act acknowledged sex differences in the physiology of men and women down to the molecular level (www.brighamandwomens.org) and acknowledged that differences in women's bodies were not limited to gynecology. The fact that the NIH had its first female director for the tenure 1991–1993 might have played a significant role in this progress: the cardiologist Dr. Bernadine Patricia Healy.

In 1994, the Women and Health Report from the IOM concluded: "*Women and men should be enrolled as participants in clinical studies in a manner that ensures that research yields scientifically generalizable results applicable to both genders.*" The report also recommended disclosing the sex origin of cell and tissue cultures used in research (Mastroianni, Faden, & Federman, 1994). Since 1995, the NIH requires grantees to include women and minorities in human subjects' research.

A little later, in its action program on equality between women and men, the European Council formalized the obligation to respect both sexes at all institutional levels (European Parliament, 1995).

In 1997, Health Canada (the equivalent of a "Ministry of Health") started requesting the inclusion of women in clinical trials.

6.5 2000–2010

In 2001, the landmark report of the US Institute of Medicine "*Exploring the biological contributions to human health: does sex matter?*" was published

(Exploring the biological contributions to human health, 2001). This laid out the basics of gender-specific medicine. The Report emphasizes that differences between males and females beyond the reproductive system exist and that they are relevant for health and illness. It recommends considering sex in research at the cellular level, since genes on the sex chromosomes might be expressed differently in male and female cells. The report recommends improving the knowledge base and research priorities for animal and cellular models to determine, when sex and gender differences exist and when they are relevant to biological functioning at different levels (cells, organs, organisms, societal beings). It should also address the issue of barriers for valid and productive research on sex and gender differences and develop strategies to overcome such barriers. In summary, the report concluded:

> *"Despite the progress made in focusing on women's health research and including women in clinical trials, such research will have limited value unless the underlying implications—that is, the actual differences between males and females that make such research so critical—are systematically studied and elucidated. Such research can enhance the basis for interpreting the results of separate studies with males and females, helping to clarify findings of no essential sex differences and suggesting mechanisms to be pursued when sex differences are found."*

In 2002, the EU initiated its sixth Research Framework Program (FP6), which for the first time addressed the topic of gender, both as a scientific issue and as a factor in research teams. The policy—which is still in effect today—pointed out three layers of impact of sex/gender on the research process. The three dimensions to address are: 1. Gender equality in scientific careers, 2. Gender balance in decision making and 3. Integration of gender dimension in to the content of research and innovation.

The clinical trial field also experienced a growing focus on sex and gender. In 2004, the International Conference on Harmonization of Technical Requirements for Registration of Pharmaceuticals for Human Use (ICH) published the first regulation calling for the inclusion of women and men in clinical trials. In the revised version of the document, published in 2009, the ICH also requires a differentiation of sex and gender in clinical trials (Singh, 2015).

In 2004, prof. Marianne Legato published the first academic textbook on SGSM, the "Principles of Gender specific Medicine," which is being updated for a fourth edition as we write this chapter. She described that the interest in SGSM was exploding internationally, not only in the US but also in other parts of the world like Japan, Israel, Canada, Sweden. Conferences about SGSM were being organized worldwide and centers for SGSM were being established. The process was initially focused on women's health and oftentimes civil society was part of the establishment of these centers. Over time women's specific medicine evolved into gender medicine, or SGSM today, "the science of how men and women differ in their normal functions and in the experience of disease." Regardless of the movement to increase a focus on SGSM and the establishment of structures, the transfer of knowledge into clinical practice was, however, still limited (Legato, 2006).

A progressive focus was also placed on publication dynamics. In 2006, prof. Reshma Jagsi and colleagues analyzed the frequencies of male and female first and last authors in original articles in six high impact peer reviewed journals (NEJM, J Am Med Ass, Ann Int Med, Ann Surg Obs Gyn, J Ped). The researchers found, that only 15.9% of the first authors and 10.3% of the senior authors were women (in 7249 original articles). The authors offer several possible explanations for the "gender gap" in medical publication, an essential one being the scarcity of women in the highest leadership positions in medical schools (Jagsi et al., 2006).

In 2006 the European Commission established in 2006 a "Roadmap for Equality between Women and Men." This Roadmap described priority areas in which EU policy must promote gender equality, including economic independence, work-life balance, and representation in decision-making. The roadmap also supported efforts to eliminate gender-based stereotypes and violence, and described strategies for achieving these ends.

Also in 2006, the European Parliament and Council Directive decided to establish a European Institute for Gender Equality (EIGE). Its responsibilities include collecting sex-disaggregated data, reflecting levels of equality between women and men, developing tools to eliminate gender discrimination, and increasing awareness of gender equality issues.

In 2008, the Council of Europe spoke out in favor of gender-sensitive medicine and the need for gender mainstreaming in health policy.

In 2009, Canada implemented the use of a Sex and Gender-based Analysis (SGBA) policy to review the gender equality of all health policy laws and regulations. SGBA is still in place today, but in an expanded version that takes into account the growing knowledge base about sex and gender in research.

In line with the Canadian efforts to streamline SGBA, in 2009, the EU also developed a Toolkit for Integrating Gender into EU-Funded Research for the Seventh Framework Program (FP7). Toolkits and training packages are available to give the European scientific research community practical tools and guidelines to promote the integration of gender into FP7 research priorities. It emphasizes that consideration of sex and gender aspects are relevant throughout the research cycle. The tool addresses not only gender in research but also participation of women and men in research (https://op.europa.eu/en/publication-detail/-/publication/c17a4eba-49ab-40f1-bb7b-bb6faaf8dec8).

6.6 2010–2022

In 2010, NIH Office of Research on Women's Health (ORWH) published its Strategic Vision for 2020. The report highlights how 20 years after establishment of the ORWH, significant progress has been achieved. Policies to ensure inclusion of women in NIH clinical research have been developed. The research on women's health and sex differences has increased. Programs to prepare researchers to conduct women's health research have been implemented. Interdisciplinary career development and sex differences research across the research continuum came into focus. The ORWH concluded that research on women's health has become a tangible reality (Pinn et al., 2010). The new goals that emerged from the strategic vision were the promotion of an increase in sex difference research in basic science studies, the incorporation of findings of sex/gender differences in the design and application of new technologies, medical devices and therapeutic drugs and the actualization of personalized prevention, diagnostics and therapeutics for girls and women. Further focus should be placed on the promotion of women's health in strategic alliances and partnerships, and on training and communication.

At the same time, the IOM also acknowledged substantial progress in the expansion of investment in women's health. However, in its report it also pointed out the inadequacy of translation of SGSM into practice. It identifies several barriers like inadequate attention to the social and environmental factors that, along with biological risk factors, influence health. The enforcement of policies requiring the representation of adequate numbers of women in clinical trials and the reporting of disaggregated results is insufficient. Poor communication of the results of women's health was identified as a problem. Accordingly, IOM gave clear recommendations to the US government agencies and other relevant organizations to sustain and strengthen their focus on women's health, including genetic, behavioral and social determinants of health over the life course. Furthermore, a goal should be to integrate women's health into all health research in such a way that differences between men and women are assessed in all health research.

It further recommended to conduct research on how to rapidly translate research findings on women's health into clinical practice (Wald & Wu, 2010).

The IOM recommended the adoption by the International Committee of Medical Journal Editors (ICMJE) of a guideline that requests sex-disaggregated analysis and reporting for all publications (2010). The ICMJE has since elaborated more specific guidance on sex and gender-sensitive reporting. The guideline recommends using the correct terms for sex and gender, reporting sex and/or gender of study participants, describing sex of animals or cells and explaining the methods to determine sex and gender. The recommendations are constantly updated (last in 2021) (http://www.icmje.org/recommendations/browse/manuscript-preparation/preparing-for-submission.html).

Since 2014 the Canadian Institute of Health Research (CIHR) Requires All Grant Applicants to Respond to Mandatory Questions about Sex and Gender in Research (Canadian Institute of Health Research, Gender, sex and health research guide; a tool for CIHR applicants, 2014. Available at http://www.cihr-irsc.gc.ca/e/32019.html).

The European Commission establishes a new strategy for equality between women and men for 2010-2015. The strategy for equality included measures to promote gender equality in economic independence, leadership, and other fields. It establishes measurable goals, such as 75% workforce participation for both, women and men, and at least 40% representation of each sex in European Commission expert groups. The Commission reaffirmed its "*commitment to making equality between women and men a reality*" by reinforcing the gender perspective in all EU policies and dedicating the necessary resources to achieve this goal.

In 2011, the United Nations (UN) published their Resolutions on Gender Analysis in Scientific Education, Research, and Development. The UN member states resolved thatc "gender-based analysis and gender impact assessments" should be integrated into science and technology research and development, and that "a gender perspective" should be incorporated into all science and technology curricula.

In 2012, the Food and Drug Administration Safety and Innovation Act (FDASIA) included a requirement (Section 907) that FDA study the availability of data on the participation of demographic subgroups (sex, age, race, and ethnicity) in clinical trials that support applications for new drugs, biologics, and devices.

The Canadian CIHR Institute of Gender and Health (IGH) published a Gender, Sex and Health Research Case Book ("What a difference sex and gender make," 2012 https://cihr-irsc.gc.ca/e/44734.html). The IGH is the world's first, and to that time the only, health research funding institute with a specific focus on gender, sex and health. Since then, the IGH promoted the integration of sex and gender in funding, ethics, publishing and policy in Canada.

Since 2013, the Gendered Innovations Project by Londa Schiebinger at Stanford University presents Methods, Terms and Checklists for sex and gender sensitive research in Science, Health and Medicine, Engineering and Environment. The project led to the publication: "Gendered Innovations: How Gender Analysis Contributes to Research" and a website with methods, checklists and case studies that is the reference for grantees and reviewers under the EU framework (Schiebinger, 2008).

The 8th EU Framework Program for Research and Innovation—Horizon 2020 (2014-2020) reiterates the relevance of the gender dimensions and wants to ensure "*the effective promotion of gender equality and the gender dimension in research and innovation content.*" Gender is a crosscutting issue and is mainstreamed in each of the different parts of the Work Program. The program re-defines the goals 1) to foster gender balance in research teams in order to promote participation of women 2) to ensure gender balance in decision making, at least 40% of the underrepresented sex should be reached in panels and advisory groups and 3) to integrate the gender dimension in research and innovation content to improve scientific quality and societal relevance of the produced knowledge, technology and/or innovation (Horizon 2020, official website of the European Union).

In 2014, the NIH announced policies requiring the inclusion of both sexes in NIH-funded biomedical research. Over half of the participants in NIH-funded clinical research at that time were women. The NIH policies demand that researchers report their plans to balance male and female cells and animals in their preclinical studies. Furthermore, the NIH planned to disseminate training on experimental design for NIH staff, trainees and grantees. In addition, the reviewers of the grant applications are now required to evaluate applicant's research plans regarding their inclusion of male and female animals and cells (Clayton & Collins, 2014).

The universities as essential stakeholders also start to take up their role in promoting the issue. The League of European Research Universities (LERU) analyses the role of gender and sex analysis in research and innovation (R&I), arguing that it needs to be better integrated into R&I funding, content and implementation process. It published twenty recommendations for stakeholders.

The European Association of Science Editors (EASE) established a Gender Policy Committee in 2012 and tasked it to develop a set of guidelines for reporting of sex and gender research "Sex and Gender Equity in Research" (SAGER). Part of this development process was an online survey among 716 journal editors, scientists and other members of the international publishing community. The average proportion of respondents who reported having sex and gender policies at their journals was 7%. Hereinafter, a panel of 13 experts from nine countries developed over a three-year period the guideline to provide a comprehensive procedure for reporting of sex and gender information in study design, data analysis, results and interpretation of findings. The SAGER guideline, published in 2016, is primarily for authors, but they are also useful for editors in their role as gatekeepers of science. Assessment of consideration of sex and gender in all parts of the manuscript should become an integral part of the editorial process (Heidari et al., 2016).

Since 2020 the German Research Foundation (DFG), a national agency, also asks for the integration of sex, gender, and diversity into research proposals. The agency provides a checklist for applicants on planning research projects(checkliste_en.pdf (dfg.de)) The DFG also introduced a new measure to promote gender equality in science and academia: a spokesperson belonging to an underrepresented gender at the management level in their specific field can apply for additional funds in conjunction with an establishment proposal or renewal proposal. The establishment of gender underrepresentation is based on the total number of professorships at all universities.

In 2021 an updated report of the Gendered Innovations project was published: *How Inclusive Analysis Contributes to Research and Innovation.* New case studies and methodological updates on the analysis of sex and gender in the following fields are included: health, artificial intelligence and robotics, energy, transport, marine science and climate change, urban planning, agriculture, fair taxation and venture funding, as well as COVID-19.

One of the most impactful innovations in the EU Framework program Horizon Europe is the introduction of a Gender Equality Plan (GEP) starting February 2022. For researchers to request funding under the umbrella of Horizon Europe, the applicant's organization needs to have a GEP in place. LERU highlights this by publishing a webpage that showcases the member universities' GEPs as examples for other research organizations (Gender Equality Plans for Horizon Europe | LERU).

7. The situation today

The timeline clearly demonstrates a growing interest and attention to the subject of SGSM. An increasing number of stakeholders is engaging in the provision of concepts, methods and structural support for a progressive institutionalization of the topic.

Groups outside of political institutions and universities emerge and aim to support and advocate for the topic, e.g., the Women's Brain Project (WBP). The WBP is a Swiss based non-profit organization, which studies sex and gender dimensions of health. It aims to contribute to achieve precision medicine (including biological and socio-cultural aspects) for sustainable and inclusive health care (Schumacher Dimech et al., 2021).

In the US, the Women's Health Access Matters (WHAM) group emphasizes the fact that women's health is an economic issue which cannot be ignored. WHAM is a non-profit organization working to increase awareness of and accelerate funding for women's health research. WHAM works across diseases that impact women differently and differentially—brain health, heart health, immune and autoimmune health and cancer. The WHAM Report shows the impact of accelerating sex and gender–based health research on women, their families, and the economy (https://thewhamreport.org/).

In Canada, the Women's Brain Health Initiative (WBHI) was funded to help protect women's brain health by focusing its resources on research to combat brain-aging disorders that disproportionately affect women. The initiative supports the creation of evidence-based preventative health education programs and provides everyday information, scientific information and caregiver information on brain health (https://womensbrainhealth.org/).

The UK registered charity organization Global Health 50/50 (GH5050) provides not only important sex-disaggregated data in global health but also dismantles gender inequality inside organizations and lack of gender sensitivity in health programs. GH5050 is an independent, evidence-driven initiative to advance action and accountability for gender equality in global health (https://globalhealth5050.org/).

Furthermore, the European Society for Medical Oncology (ESMO) set up a Gender Medicine Task Force. Its missions are to raise awareness of the presence of potential sex differences in biology and treatment outcomes of non-sex-related cancers and the impact of gender on access, quality of life and long-term consequences of tumor therapy. The task force will develop tools and materials for the society's educational programs. It is important, that precision medicine not only targets the treatment of the tumor, but also considers the host. Sex and gender are key host factors (Ozdemir et al., 2022).

The European Academy of Neurology established an own Task Force on Gender and Diversity issues in Neurology. This Task Force aims to increase awareness of sex and gender differences in neurological disorders, to improve the understanding of the different etiology, to recognize their impact on the practice of neurology and to promote gender and ethnic equality in care for neurological disorders. Further aims are the dissemination of knowledge and fostering health care systems changes.

These initiatives are an encouraging development, yet the translation of these efforts into practice will require constant monitoring and political support. The number of women scientists in the European region is progressively growing (Commission, 2021), the number of scientific publications considering sex and/or gender is increasing (Oertelt-Prigione et al., 2010) and the number of clinical trials including prevalence-adjusted numbers of female patients is growing (Steinberg et al., 2021). Nevertheless, we have not reached parity in the representation of women scientists, nor have we reaching a situation where the performance of SGSM is the norm. We are on our way, but much remains to be done.

The COVID-19 pandemic has exemplified how fragile these propositions can be under pressure. In a recent study we found that only 21 % of the registered COVID-19 studies explicitly planned to address sex/gender in the context of recruitment and only 4% planned to include sex/gender as an analytical variable. Although the inclusion of sex and gender improved in the published compared to the recorded trials, even in this group only 18% presented sex-disaggregated data in their results (Brady et al., 2021). Other authors have confirmed these numbers focusing on COVID-19 vaccines or pharmaceuticals (Vassallo et al., 2021). The pandemic also took a toll on women scientists' careers and productivity. The total number of publications produced by female authors decreased (Squazzoni et al., 2021), demonstrating how far societal patterns of gender inequality in care work still burden the careers of women scientists.

A comprehensive approach to knowledge translation will be needed to ensure a broad implementation of the envisioned strategy into practice. Current practitioners have to be trained in up-to-date methodology, clinicians will have to adapt their guidelines and practice and students will have to obtain medical training that "normalizes" a sex- and gender-sensitive approach to medicine.

The underlying reasons of resistance and barriers that need to be addressed have been summarized by professor Marianne Legato as follows: *The inevitable tendencies in the pursuit of gender medicine worldwide are*:

- "the effort to separate feminist angst (however justified and/or understandable) from the excitement of the pure science
- the nervousness of men's health advocates about the growing attention to women's biology
- *the need to expand a concentration on "women's health" into a broader study of the differences between males and females*"(Legato, 2004).

Inevitably, there is a variety of opinions, priorities and efforts in the pursuit of SGSM around the world. There is no doubt about the fact that any change in the way patients are treated has to be strongly evidence-based. However, feminisms and gender mainstreaming have played an essential role in the development of sex- and gender-sensitive medicine. Successful gender mainstreaming activities, such as the implementation of gender equality plans (GEPs) at the European Commission level, help to persuade and advocate for implementation of SGSM, as it is still mostly women who do the research in this field. Nevertheless, the interest and the target groups are expanding and a focus on contribution of SGSM to excellence, reproducibility and social impact are inviting an ever-growing group of diverse colleagues to join the effort. SGSM is not a "women's" topic, but an innovative approach to inclusive and person-centered care.

In her book "Doing Harm" (Dusenbery, 2017), Maya Dusenbery identified two major gaps responsible for the described inequities in health: the knowledge gap and the trust gap. The knowledge gap describes differences between men and women in symptoms, in prevalence, in effectiveness of medications and in the health-related behavior. Although, a growing body of knowledge is available, it is not systematically integrated into medical curricula and therefore not known to many practicing physicians. The trust gap, on the other hand, describes the lack of trust in the reporting of symptoms by a woman. The fact that being a woman is a risk factor for not being taken seriously and for having one's symptoms attributed to mental rather than somatic conditions.

Proper training will have to address both of these issues, and this will have to begin at the level of medical curricula (Ludwig et al., 2015). Furthermore, graduate and post-graduate training will need to be updated. Structural requirements will have to be enforced and monitored in order to guarantee their reliable uptake. Funding schemes will have to accommodate the missing information needed for adapting clinical guidelines and providing robust background data. Knowledge translation will have to involve the general public, turning the topic of SGSM into a normal request any patient could ask any doctor to illustrate.

8. Outlook

Considering sex- and gender in medical research throughout the entire research process (research question development, methodological design, analysis, interpretation and dissemination) leads to more reproducible, sound and socially-relevant results. Sex- and gender sensitive analysis is an essential component towards excellent and impactful care for all members of the population, and not just a selected group complying with a narrow research focus.

The growing number of policies to increase the uptake of SGSM and research practice is promising, but needs to be constantly monitored and supported. As the pandemic has shown, acute crises can revise years of progress. A broad stakeholder coalition has provided the basis for knowledge generation and dissemination, now the time to increase its production and guarantee its translation has come.

9. Key points/take home messages

- SGSM has experienced growing interest and support in the last decades
- Sex- and gender-sensitive medicine is obtaining growing institutional support through research funders, journals editors and policymakers, however, the practical translation of sex- and gender-sensitive knowledge into practice is still limited
- Medical curricula, graduate and postgraduate, as well as professional training needs to systematically include sex- and gender-sensitive knowledge to create a critical mass of informed professionals equipped to practice SGSM
- The implementation process of SGSM will need to be strategically monitored to avoid steps back due to unforeseeable competing priorities, as exemplified by the COVID-19 pandemic.

References

Alex, L., et al. (2012). Beyond a dichotomous view of the concepts of 'sex' and 'gender' focus group discussions among gender researchers at a medical faculty. *PLoS One*, *7*(11), e50275.

Allen, D. H. (2020). The Nightingale legacy. *Clinical Journal of Oncology Nursing*, *24*(5), 1–2.

Alwazzan, L., & Al-Angari, S. S. (2020). Women's leadership in academic medicine: A systematic review of extent, condition and interventions. *BMJ Open, 10*(1), e032232.

Annandale, E., & Hammarstrom, A. (2011). Constructing the 'gender-specific body': A critical discourse analysis of publications in the field of gender-specific medicine. *Health (London, England), 15*(6), 571–587.

Arnegard, M. E., et al. (2020). Sex as a biological variable: A 5-year progress report and call to action. *Journal of Women's Health (2002), 29*(6), 858–864.

Beauvoir, S. D., & Ralph Ellison Collection (Library of Congress). (1953). *The second sex. Vol. xxx* (1st American ed., p. 732). New York: Knopf. xiv.

Bird, C. E., & Sharman, Z. (2014). Gender-based analysis is essential to improving women's health and health care. *Womens Health Issues, 24*(2), e163–e164.

Blackless, M., et al. (2000). How sexually dimorphic are we? Review and synthesis. *American Journal of Human Biology, 12*(2), 151–166.

Blackwell, E. (1970). *Pioneer work in opening the medical profession to women: Autobiographical sketches. Vol. ix* (p. 264). New York: Source Book Press.

Brady, E., et al. (2021). Lack of consideration of sex and gender in COVID-19 clinical studies. *Nature Communications, 12*(1), 4015.

Canto, J. G., et al. (2012). Association of age and sex with myocardial infarction symptom presentation and in-hospital mortality. *JAMA, 307*(8), 813–822.

Chambers, T. A., Bagai, A., & Ivascu, N. (2007). Current trends in coronary artery disease in women. *Current Opinion in Anaesthesiology, 20*(1), 75–82.

Clark, J., & Horton, R. (2019). What is The Lancet doing about gender and diversity? *Lancet, 393*(10171), 508–510.

Clayton, J. A., & Collins, F. S. (2014). Policy: NIH to balance sex in cell and animal studies. *Nature, 509*(7500), 282–283.

Commission, E. (2021). *She figures handbook* (p. 2022). Available from: https://op.europa.eu/en/web/eu-law-and-publications/publication-detail/-/publication/67d5a207-4da1-11ec-91ac-01aa75ed71a1.

Commission, E. (2014). *Promoting gender equality in research and innovation*. Available from: https://ec.europa.eu/programmes/horizon2020/en/h2020-section/promoting-gender-equality-research-and-innovation.

Council, E. 1995; Available from: https://www.europarl.europa.eu/summits/mad1_en.htm#.

Council, E. (1975). *Council Directive 75/117/EEC of 10 February 1975 on the approximation of the laws of the Member States relating to the application of the principle of equal pay for men and women*. Available from: https://eur-lex.europa.eu/LexUriServ/LexUriServ.do?uri=CELEX:31975L0117:EN:HTML.

Dusenbery, M. (Ed.). (2017). *Doing harm: The truth about how bad medicine and lazy science leave women dismissed, misdiagnosed, and sick* (1st ed., p. 390). New York, NY: HarperOne.

Exploring the biological contributions to human health. (2001). Does sex matter? *Journal of Womens Health & Gender-Based Medicine, 10*(5), 433–439.

FDA. (2022). *Precision medicine*. Available from https://www.fda.gov/medical-devices/in-vitro-diagnostics/precision-medicine.

FDA. (2022a). *Healthcare challenges and unmet medical needs of transgender adults—FDA-requested listening session (session #1)*. Available from https://www.fda.gov/media/153932/download.

FDA. (2022b). *Healthcare challenges and unmet medical needs of transgender teens—FDA-requested listening session (session #2)*. Available from https://www.fda.gov/media/153933/download.

Fish, E. N. (2008). The X-files in immunity: Sex-based differences predispose immune responses. *Nature Reviews. Immunology, 8*(9), 737–744.

Gogos, A., et al. (2019). The importance of sex differences in pharmacology research. *British Journal of Pharmacology, 176*(21), 4087–4089.

Goldstein, Z., Corneil, T. A., & Greene, D. N. (2017). When gender identity doesn't equal sex recorded at birth: The role of the laboratory in providing effective healthcare to the transgender community. *Clinical Chemistry, 63*(8), 1342–1352.

Gottgens, I., et al. (2020). The impact of sex and gender on the multidisciplinary management of care for persons with Parkinson's disease. *Frontiers in Neurology, 11*, 576121.

Heidari, S., et al. (2016). Sex and gender equity in research: Rationale for the SAGER guidelines and recommended use. *Research Integrity and Peer Review, 1*, 2.

Heise, L., et al. (2019). Gender inequality and restrictive gender norms: Framing the challenges to health. *Lancet, 393*(10189), 2440–2454.

Hood, L., & Friend, S. H. (2011). Predictive, personalized, preventive, participatory (P4) cancer medicine. *Nature Reviews. Clinical Oncology, 8*(3), 184–187.

Irwin, M., et al. (1990). Reduction of immune function in life stress and depression. *Biological Psychiatry, 27*(1), 22–30.

Jagsi, R., et al. (2006). The "gender gap "in authorship of academic medical literature—A 35-year perspective. *The New England Journal of Medicine, 355*(3), 281 287.

Johnson, J. L., & Beaudet, A. (2013). Sex and gender reporting in health research: Why canada should be a leader. *Canadian Journal of Public Health, 104*(1), e80–e81.

Johnson, J. L., Greaves, L., & Repta, R. (2009). Better science with sex and gender: Facilitating the use of a sex and gender-based analysis in health research. *International Journal for Equity in Health, 8*, 14.

Kessler, R. C., et al. (2012). Twelve-month and lifetime prevalence and lifetime morbid risk of anxiety and mood disorders in the United States. *International Journal of Methods in Psychiatric Research, 21*(3), 169–184.

Legato, M. J. (2003). Beyond women's health the new discipline of gender-specific medicine. *The Medical Clinics of North America, 87*(5), 917–937. vii.

Legato, M. J. (2004). Gender-specific medicine: The view from Salzburg. *Gender Medicine, 1*(2), 61–63.

Legato, M. J. (2006). Rethinking gender-specific medicine. *Women's Health (London, England), 2*(5), 699–703.

Lippi, D., Bianucci, R., & Donell, S. (2020). Gender medicine: Its historical roots. *Postgraduate Medical Journal, 96*(1138), 480–486.

Ludwig, S., et al. (2015). A successful strategy to integrate sex and gender medicine into a newly developed medical curriculum. *Journal of Women's Health (2002), 24*(12), 996–1005.

Maschewsky-Schneider, U., et al. (2001). Women and sex-specific health research in the German Society for Social Medicine and Prevention. *Gesundheitswesen, 63*(Suppl 1), S89–S92.

Mastroianni, A. C., Faden, R., & Federman, D. (1994). Women and health research: A report from the Institute of Medicine. *Kennedy Institute of Ethics Journal, 4*(1), 55–62.

Merkatz, R. B. (1998). Inclusion of women in clinical trials: A historical overview of scientific, ethical, and legal issues. *Journal of Obstetric, Gynecologic, and Neonatal Nursing, 27*(1), 78–84.

National Science Foundation (U.S.). (1982). *Women and minorities in science and engineering* (p. 6). Washington, D.C.: National Science Foundation.

National Women's Health Network. Trans health care. Available from https://nwhn.org/issues/trans-health-care/, n.d.

Nielsen, M. W., et al. (2017). One and a half million medical papers reveal a link between author gender and attention to gender and sex analysis. *Nature Human Behaviour, 1*(11), 791–796.

Nieuwenhoven, L., & Klinge, I. (2010). Scientific excellence in applying sex- and gender-sensitive methods in biomedical and health research. *Journal of Women's Health (2002), 19*(2), 313–321.

Oertelt-Prigione, S. (2012a). The influence of sex and gender on the immune response. *Autoimmunity Reviews, 11*(6-7), A479–A485.

Oertelt-Prigione, S. (2012b). Immunology and the menstrual cycle. *Autoimmunity Reviews, 11*(6-7), A486–A492.

Oertelt-Prigione, S. (2022). The operationalization of gender in medicine. In M. Legato (Ed.), *Principles of gender-specific medicine*. New York: Elsevier.

Oertelt-Prigione, S., et al. (2010). Analysis of sex and gender-specific research reveals a common increase in publications and marked differences between disciplines. *BMC Medicine, 8*, 70.

Ozdemir, B. C., et al. (2022). Investigation of sex and gender differences in oncology gains momentum: ESMO announces the launch of a Gender Medicine Task Force. *Annals of Oncology, 33*(2), 126–128.

Pinn, V. W., et al. (2010). Public partnerships for a vision for women's health research in 2020. *Journal of Women's Health (2002), 19*(9), 1603–1607.

Reisner, S. L., et al. (2016). Social epidemiology of depression and anxiety by gender identity. *The Journal of Adolescent Health, 59*(2), 203–208.

Ridker, P. M., et al. (2005). A randomized trial of low-dose aspirin in the primary prevention of cardiovascular disease in women. *The New England Journal of Medicine, 352*(13), 1293–1304.

Rossi, A., et al. (2022). Heart-brain interactions in cardiac and brain diseases: Why sex matters. *European Heart Journal*. epub ahead of print.

Scheim, A. I., Perez-Brumer, A. G., & Bauer, G. R. (2020). Gender-concordant identity documents and mental health among transgender adults in the USA: A cross-sectional study. *The Lancet Public Health, 5*(4), e196–e203.

Schiebinger, L. L. (2008). *Gendered innovations in science and engineering* (p. 244). Stanford, Calif.: Stanford University Press. xii.

Schumacher Dimech, A., et al. (2021). The role of sex and gender differences in precision medicine: The work of the Women's Brain Project. *European Heart Journal, 42*(34), 3215–3217.

Singh, J. (2015). International conference on harmonization of technical requirements for registration of pharmaceuticals for human use. *The Journal of Pharmacy and Pharmacology, 6*(3), 185–187.

Squazzoni, F., et al. (2021). Gender gap in journal submissions and peer review during the first wave of the COVID-19 pandemic. A study on 2329 Elsevier journals. *PLoS One, 16*(10), e0257919.

Steinberg, J. R., et al. (2021). Analysis of female enrollment and participant sex by burden of disease in US clinical trials between 2000 and 2020. *JAMA Network Open, 4*(6), e2113749.

Tannenbaum, C., Greaves, L., & Graham, I. D. (2016). Why sex and gender matter in implementation research. *BMC Medical Research Methodology, 16*(1), 145.

Tannenbaum, C., et al. (2019). Sex and gender analysis improves science and engineering. *Nature, 575*(7781), 137–146.

Tasca, C., et al. (2012). Women and hysteria in the history of mental health. *Clinical Practice and Epidemiology in Mental Health, 8*, 110–119.

Vaccarino, V., et al. (1995). Sex differences in mortality after myocardial infarction. Is there evidence for an increased risk for women? *Circulation, 91*(6), 1861–1871.

Vassallo, A., et al. (2021). Sex and gender in COVID-19 vaccine research: Substantial evidence gaps remain. *Front Glob Womens Health, 2*, 761511.

Voskuhl, R. (2011). Sex differences in autoimmune diseases. *Biology of Sex Differences, 2*(1), 1.

Wald, C., & Wu, C. (2010). Biomedical research. of mice and women: The bias in animal models. *Science, 327*(5973), 1571–1572.

West, C., & Zimmerman, D. H. (1987). Doing gender. *Gender & Society, 1*(2), 125–151.

CHAPTER TWO

Biological underpinnings of sex differences in neurological disorders

Katarzyna Winek[a,b], Yonat Tzur[b], and Hermona Soreq[a,b,*]
[a]The Edmond & Lily Safra Center for Brain Sciences, The Hebrew University of Jerusalem, Jerusalem, Israel
[b]The Alexander Silberman Institute of Life Sciences, The Hebrew University of Jerusalem, Jerusalem, Israel
[*]Corresponding author: e-mail address: hermona.soreq@mail.huji.ac.il

Contents

Abstract

The importance of sex differences in neurological disorders has been increasingly acknowledged in recent clinical and basic research studies, but the complex biology and genetics underlying sex-linked biological heterogeneity and its brain-to-body impact remained incompletely understood. Men and women differ substantially in their susceptibility to certain neurological diseases, in the severity of symptoms, prognosis as well as the nature and efficacy of their response to treatments. The detailed mechanisms

International Review of Neurobiology, Volume 164
ISSN 0074-7742
https://doi.org/10.1016/bs.irn.2022.06.009

underlying these differences, especially at the molecular level, are being addressed in many studies but leave a lot to be further revealed. Here, we provide an overview of recent advances in our understanding of how sex differences in the brain and brain-body signaling contribute to neurological disorders and further present some future prospects entailed in terms of diagnostics and therapeutics.

1. How different are neurological diseases in males and females?—The brain perspective

Sex-specific differences in the incidence of neurological disorders are a widely known phenomenon. Women show a higher risk of multiple sclerosis (MS) than men, but men with this disease usually present more severe clinical picture. When compared to women, men further have a higher risk to develop Parkinson's disease (PD), whereas Alzheimer's disease (AD) is more common and progresses more rapidly in women (Clayton, 2016; Ferretti et al., 2018). The overall risk of stroke is higher in women than men, partly due to their longer life expectancy and stroke's association with hormonal status changing post-menopause; and female sex is linked with worse prognosis after a stroke (Clayton, 2016; Reeves et al., 2008). Fig. 1 illustrates the sex-bias in neurological disease, showing prevalence of neurological disorders (excluding infectious causes), as reported by the Global Burden of Disease Study in 2019 (Global Burden of Disease

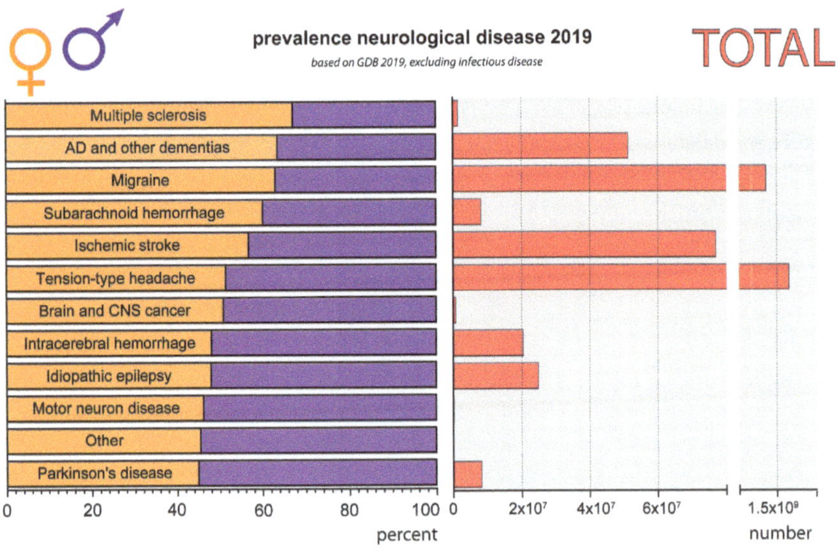

Fig. 1 Sex-bias in neurological disorders.

Collabortative Network, 2021). Multiple factors have already been established or suggested as contributors to this divergence, beginning with biological differences in brain features as linked to the cellular responses in the central nervous system (CNS), and commencing to systemic differences driven by hormonal and genetic makeup. Sex differences connected to the clinical aspects are further discussed in other chapters of this book, while in this part we focus on main mechanisms underlying the sex differences in neurological disorders.

Prevalence of neurological disorders in 2019, excluding infectious disease, as reported by the Global Burden of Disease Study (Global Burden of Disease Collabortative Network, 2021). Numbers indicate percentage of men and women (left panel) and total number of cases in both sexes (right panel).

At the global level, the debate on how different are brains between the sexes and whether we may actually consider sexual dimorphism in brain traits/specific regions is ongoing. Recent studies on this topic often spark vivid scientific discussions. This was certainly the case (Chekroud, Ward, Rosenberg, & Holmes, 2016; Del Giudice et al., 2016; Glezerman, 2016; Rosenblatta, 2016) with the proposal that brain features such as gray and white matter volume, connectivity as well as behavioral characteristics are not actually dimorphic, such that most humans will be somewhere on the "male-brain–female-brain-continuum," despite identification of sex-related differences (Joel et al., 2015). A more recent, comprehensive meta-analysis supporting this point of view shows that sex-specific differences in features such as the size of specific brain structures, connectome and lateralization are negligible. This report argues that the main factor driving the inequality between male and female brains is the bigger size of the brain in males (Eliot, Ahmed, Khan, & Patel, 2021). That study may foreseeably re-provoke a debate on the importance of sexual dimorphism and of studying sex-related differences in brain function (Galea, 2021). However, unlike the dubious structural features, functional and transcriptomic findings may demonstrate basic flaws in this concept, as is briefly detailed below. Notably, sex-specific effects have repeatedly been shown in the CNS molecular mechanisms and cellular functions, principally in all main cell populations of the brain.

1.1 Neurons

It is not surprising that the sex-related differences have been identified in neuronal cells. For instance, the distribution of steroid hormone receptors in hippocampal neurons differs sex-dependently. In both sexes, steroid membrane receptors are found on dendrites, but nuclear steroid receptors are located in GABAergic neurons in females and in pyramidal cells in males (McEwen & Milner, 2017); see also the section below *Sex hormones and the CNS*)). Moreover, sex-specific transcriptomic profiles have been shown in rodent hypothalamic neurons (Kim et al., 2019) and locus coeruleus cells (Mulvey et al., 2018). In the latter case, analysis of whole hindbrain or serotonergic neurons alone revealed negligible sex differences. However, focusing on noradrenergic cells, the authors identified 152 differentially expressed (DE) transcripts when comparing male and female mice. Examples include the Pterg3 transcript encoding the receptor for prostaglandin E2, which shows significant upregulation in females. Treatment with its agonist sulprostone reduces firing rates of locus coeruleus neurons in ex-vitro slice recordings and suppresses stress response in in-vivo tests of female animals (Mulvey et al., 2018).

Sex differences further exist in neurotransmitter systems including older studies reporting divergence in serotoninergic, dopaminergic and GABAergic transmission (for a review see (Cosgrove, Mazure, & Staley, 2007)), but also in glutamatergic (Wickens, Bangasser, & Briand, 2018) and cholinergic systems (Giacobini & Pepeu, 2018). For example, the synthesis rate of serotonin in the brains of men is higher than in women (Cosgrove et al., 2007; Nishizawa et al., 1997). When it comes to the GABAergic system, marked differences exist in $GABA_A$ receptor subunits in the superior temporal gyrus, with decreased expression in older men, compared to younger patients, when analyzing the effects of sex and age. In contradistinction, the expression of $GABA_B$ receptors remained unaffected (Pandya et al., 2019). The sex-linked difference in neurotransmitter systems affecting the levels of specific transmitters as well as the expression levels of respective receptors may be of great importance in pathological states—psychiatric diseases (Gray, Hyde, Deep-Soboslay, Kleinman, & Sodhi, 2015; Wickens et al., 2018) or neurodegenerative disorders such as AD, where for example the difference in the cholinergic system may contribute to sex-specific differences in treatment response (Ferretti et al., 2018; Giacobini & Pepeu, 2018).

1.2 Microglia

Although the previously postulated ratio of 10:1 glia cells to neurons seems to be overestimated, and although newer methods have shown that the actual proportion is with high probability just 1:1 (von Bartheld, Bahney, & Herculano-Houzel, 2016), glial cells play indispensable roles in the maintenance of the nervous system. Microglia are the specialized resident phagocytes as they participate in the immune surveillance of the CNS, providing protection in cases of infection and traumatic insults. Moreover, microglia have important functions in synapse formation (Colonna & Butovsky, 2017). Interestingly, experimental data showed that microglia in male mice are distributed at higher density in certain brain areas when compared to female mice, also depending on the age of the animals. Cell counting based on staining of the ionized calcium-binding adapter molecule 1 (Iba1—standard marker of microglia and macrophages) in five regions—cortex, hippocampus, amygdala, striatum and cerebellum, revealed that at the age of 3 weeks, male mice had more densely packed cells in the amygdala and lower microglial density in the hippocampus compared to females. In 13-weeks old mice, microglia in male animals (vs female mice) had higher density in the cortex, hippocampus and amygdala (Guneykaya et al., 2018). Further, microglia in males have bigger somata, higher antigen presenting abilities and more pronounced purinergic receptor capacities compared to cells of female origin (Guneykaya et al., 2018). Although microglia from males and females showed similar phagocytic activity in that study, another report did find sex-specific phagocytic differences, along with faster migration rate of male-originated microglia (Yanguas-Casás et al., 2018). Further, Staci Bilbo and coworkers showed that sex-related differences in microglia exist during development of the rat brain in a manner dependent on the brain region. For example, while females had more overall microglia at later embryonic stages and in the postnatal phase when compared to male rats, the numbers of microglia were higher in early development stages in males when inspecting the parietal cortex, hippocampus and amygdala. Interestingly, this phenomenon was not observed in the hypothalamic paraventricular nucleus (Schwarz, Sholar, & Bilbo, 2012), for a comprehensive review on sex differences in microglia, see (Hanamsagar & Bilbo, 2016). Additionally, microglia isolated from female mice were of more mature phenotype when compared to cells originating from male animals. This was measured by "microglial developmental index" (MDI), which identified genes up- and downregulated during development. In male cells, exposure to LPS increased the MDI and was associated with changes in morphology (Hanamsagar et al., 2017). Interestingly, the MDI can be also applied to human data and is higher in samples from certain pathological conditions such as AD and autism when compared to controls (Hanamsagar et al., 2017). Lastly, the MDI is not affected by the phase of the estrous cycle in female mice, which might be an important point when discussing limitations of the use of female animals in neurobiology research (see Section 6.2).

Sex-specific differences in other pathological states linked to microglial response, such as glioma, have also been observed in experimental and clinical studies. In a mouse glioma model, microglia from male animals showed higher expression of MHC II molecules when compared to female mice (Ochocka et al., 2021), similarly to the report of Guneykaya et al. in naïve animals (Guneykaya et al., 2018). Moreover, this effect was confirmed in a separate analysis of human brain samples (Ochocka et al., 2021). In neurosurgical studies, microglia isolated from human patients seem to differ between the sexes, although in the case of CNS disease including brain trauma, tumor, hemorrhage, hydrocephalus as compared to healthy control tissue, the pathology itself was identified as the major driver of transcript profile differences when comparing conditions (Young et al., 2021). Genes involved in the complement pathway and synaptic pruning which were DE in microglia isolated from men and women brains were more susceptible to be upregulated in female patients (Young et al., 2021). Interestingly, the incidence of the most aggressive brain tumor—glioblastoma multiforme is significantly higher in men (1.6:1 ratio). Recent analyses show, however, that women have better outcomes when undergoing standard glioblastoma multiforme treatment and that the tumors can be divided into sex-specific molecular subtypes where different pathways are crucial for survival (Yang et al., 2019). These findings are of great importance when thinking about designing personalized therapies.

1.3 Astrocytes

Astrocytes provide metabolic support to neurons, and are important regulators of synapse forming during development (Farhy-Tselnicker et al., 2021). Interestingly, sex seems to be an important variable in synaptogenesis supported by astrocytes, whereby astrocyte-secreted thrombospondin induces potent synapse formation in cortical neurons derived from male but not female rats, regulated by locally-produced estrogen ((Mazur et al., 2021), see also section on neurosteroids). A recent study identified developmental transcriptomic profiles of mouse cortical astroglia and showed that astroglia from male animals reach maturity faster than cells of female origin (Rurak et al., 2022). Furthermore, cortical astrocytes isolated from male and female mouse pups show sex-dependent differences in their response to the inflammatory impact of bacterial lipopolysaccharide (LPS), whereby LPS stimulation of astrocytes isolated from males leads to more pronounced upregulation of interleukin (IL)-6, TNF-α and IL-1β expression compared to cells derived from females. This sex-specific variation is driven by sex hormones, as the differences can be abolished by pretreatment of female mice with testosterone (Santos-Galindo, Acaz-Fonseca, Bellini, & Garcia-Segura, 2011).

Under pathological conditions such as experimental cerebral ischemia, both astrocytes and microglia exhibit sex-linked changes. Sex-specific effects can be found, for example in astrocyte calcium signaling (higher calcium peak frequency in the ipsilateral hemisphere in female animals compared to males), and expression of CD11b, generally higher in females (Morrison & Filosa, 2016), also in sham- control animals. These high basal CD11b levels did not change after cerebral ischemia in females, but the initially lower CD11b levels in male mice were increased in response to injury (Morrison & Filosa, 2016). Further support of sex-driven transcriptomic differences was provided by exploring the neuroprotective properties of female-originated microglia under hormonal signaling-independent states. Microglia isolated from female mice and transplanted into the brains of male animals provided protection of male brains in the experimental stroke model (Villa et al., 2018), further indicating context-independent sex-related differences.

1.4 Oligodendrocytes

Oligodendrocytes produce the myelin sheath protecting neurons and improving conductivity (Bradl & Lassmann, 2010). Rodent brain analyses comparing male and female animals revealed higher oligodendrocyte density and thicker myelin sheaths in males, and showed higher cellular turnover in females. Both these differences seem to arise from hormonal effects and are established during the first days after birth (Abi Ghanem et al., 2017; Cerghet et al., 2006). Properties of oligodendrocyte precursor cells (OPCs) are also sex-specific in rodent models. Specifically, female-derived rat OPCs showed increased proliferation and migration potential compared to male-isolated cells. Cells isolated from male animals had higher in-vitro differentiation rate and higher expression of myelin basic protein (MBP), all of these features were confirmed by in-vivo experiments.(Yasuda et al., 2020).

Recently, two groups investigated the transcriptomic features of AD (as compared to controls) in specific cells using single cell sequencing (Belonwu et al., 2022) in samples from the prefrontal cortex (Mathys et al., 2019) and entorhinal cortex (Grubman et al., 2019). Both identified effects specific for the cell subtype (neurons, astrocytes, oligodendrocytes, OPCs, microglia, endothelial cells and pericytes) and highlighted the possible involvement of pathways linked to oligodendrocytes and myelination in the pathogenesis of AD (Grubman et al., 2019; Mathys et al., 2019). Additionally, both analyses pointed out the impact of sex on differences in transcriptomic responses to AD. Further in-depth exploration of these datasets, focusing explicitly on sex differences, showed that more sex-specific effects are present in prefrontal cortex glial cells than in neurons (Belonwu et al., 2022). Analysis of the dataset derived from cells of the entorhinal cortex (Grubman et al., 2019) uncovered that many pathways are adversely regulated in males when compared to females (with most of those upregulated in males and downregulated in females). For example, the expression of genes involved in binding of amyloid beta and fibril formation, as well as demyelination/remyelination was downregulated in female and upregulated in male AD patients (Belonwu et al., 2022).

Adding to the neuronal and glial cell type-specific effects, sex-linked variability has been reported in the blood-brain barrier (BBB; (Weber & Clyne, 2021)) or the cerebral vasculature (Krause, Duckles, & Pelligrino, 2006), both of great importance for brain health and homeostasis. Furthermore, peripheral systems and variables are often subjected to the impact of sex-related differences which influence the susceptibility, course and prognosis of neurological disorders, adding another layer of complexity. Here, we provide two examples of such effects linked to the immune system and metabolism.

2. Peripheral sex-derived differences affecting the CNS

2.1 Sex differences in immune response

The most recent example of sex differences in immunity can be derived from studies of the COVID-19 response. Importantly, men had generally worse outcomes than women when infected with SARS-CoV-2 (Jun et al., 2021). Sex-specific differences in the immune reaction to the SARS-CoV-2 infection include significantly larger increases of plasma innate immune cytokines and potent activation of non-classical monocytes in male patients, whereas the immune response in female patients is linked with stronger activation of T lymphocytes (Takahashi et al., 2020). Notably, the impact of sex differences in the immune response is of relevance beyond infectious disease, and numerous sex-specific effects have been identified in recent years. Therefore, sex differences in immune response should be considered in experimental and clinical studies of neurological disease at large. Also, in physiological conditions the immune response differs between men and women. In particular, macrophages isolated from females show higher phagocytic capacities and activation levels, whereas male-originated cells produce higher levels of pro-inflammatory cytokines (Klein & Flanagan, 2016). Females further present higher counts of CD4+ and males of CD8+ T lymphocytes (for an extensive review see (Klein & Flanagan, 2016)).

The connection between the brain and the immune system has been experimentally demonstrated in the 1970s, when immunosuppression was shown to be achieved by behavioral conditioning (Ader & Cohen, 1975; Dantzer, 2018). Later discoveries identified that the brain-immune communication occurs over the hypothalamus-pituitary-adrenal glands (HPA) axis and the autonomic nervous system. It was further shown that specific immune cells express receptors for neurotransmitters, including for example the nicotinic α7 receptor, which is found on macrophages and mediates immunosuppressive actions upon binding of acetylcholine (Borovikova et al., 2000; Rosas-Ballina et al., 2008). Similarly, the CNS reacts to peripheral events triggering immune response, with the impact of this effect depending mostly on the action of cytokines secreted by immune cells (Dantzer, 2018). Finally, although the brain has long been considered as an immune-privileged organ, devoid of lymphatic vessels, recent studies have proven that such vessels actually exist in the dural sinuses (Louveau et al., 2015).

It is therefore not surprising that the immune system contributes to a plethora of neurological disorders (Rickenbach & Gericke, 2022), culminated by straight-forward immune-mediated diseases such as MS (Baecher-Allan, Kaskow, & Weiner, 2018), but featuring also stroke (Iadecola, Buckwalter, & Anrather, 2020) or AD (Bettcher, Tansey, Dorothée, & Heneka, 2021). Nevertheless, MS may serve as a classic example of a neurological disorder mediated by the immune system, which presents a strong sex bias in disease susceptibility. Whereas the specific trigger remains unknown, MS associates with activation of CD4+ T cells in the periphery and their migration to the brain, where they initiate local inflammatory reaction (in another hypothesis, the disease actually starts in the CNS (Baecher-Allan et al., 2018)). Cells involved in pathological processes leading to MS further include CD8+ T cells, Th17 cells, B cells, macrophages and microglia contributing to demyelination (Baecher-Allan et al., 2018). A fine balance between the pro- and anti-inflammatory responses seems to be crucial for the events linked to disease development. Interestingly, new sex-related mechanisms have been proposed in recent studies. For example, IL-33 has been suggested as contributing to MS progression, based on investigations in experimental autoimmune encephalomyelitis mice

(EAE, a rodent model of MS) (Russi, Ebel, Yang, & Brown, 2018). The authors showed that IL-33 secreted by mast cells induced Th2-mediated protective response involving type 2 innate lymphoid cells. Interestingly, expression of the Il33 gene in mast cells could be upregulated by testosterone, which could therefore convey disease protection in males (Russi et al., 2018).

In ischemic stroke, peripheral immune cells enter the brain due to increased permeability of the BBB and inflammatory reaction in the CNS (Iadecola et al., 2020). Given the protective nature of estrogens in stroke and cardiovascular disease (Hurn & Brass, 2003; Lisabeth & Bushnell, 2012), the risk of stroke in females increases significantly in the post-menopausal years. T cells provide a pronounced example of the sex-dependent involvement of post-stroke peripheral immune cells in CNS pathophysiological processes. T cells at large respond to cerebral ischemia (Cramer, Benakis, & Liesz, 2019; Iadecola et al., 2020), whereas generally IL-17-producing T cells have been linked with detrimental post-stroke effects and T regulatory cells associate with beneficial effects in experimental models (Iadecola et al., 2020). At least one human study showed that the Treg response after stroke was sex-specific, with limited anti-inflammatory action of Tregs in female patients (Yan et al., 2012). On top of that, males seem to have generally higher Treg counts than females (Klein & Flanagan, 2016). A summary of stroke-relevant sex differences in peripheral immune response can be found in (Spychala, Honarpisheh, & McCullough, 2017).

AD is another important example of the involvement of the immune system in neurological disease, which involves pathological processes in both the brain and periphery. Although the role of the peripheral immune system in AD is not fully elucidated (Bettcher et al., 2021; Cao & Zheng, 2018; Dai & Shen, 2021), it is generally agreed that sex differences in immune response may affect both the AD risk and its prognosis. A recent meta-transcriptomic analysis identified sex-specific signatures in both brain and blood from AD patients (Paranjpe et al., 2021). Further tests based on a machine learning algorithm indicated better prediction of AD when clinical features were combined with blood transcriptomic data. This effect was, however sex-specific and the algorithm performed better when trained with female data only, which was attributed to the faster progression of AD in women (Paranjpe et al., 2021).

Highlighting the role of sex differences in AD, some analyses have shown that the APOE ε4 genotype poses a greater risk of AD in females when compared to males (Altmann, Tian, Henderson, & Greicius, 2014).

Refining this concept, a recent meta-analysis demonstrated that this effect is actually present only in females aged 65 – 75 years (Neu et al., 2017). Further, female carriers of this mutation feature much earlier loss of amyloid beta-specific CD4+ T cells than male patients (Begum et al., 2014). Moreover, despite the primary focus on the brain and microglial responses, several experimental studies have provided a link between peripheral immune responses and pathophysiological processes in AD, pursuing the search of relevant cellular populations and detailed molecular mechanisms (reviewed in (Cao & Zheng, 2018)). Recent reports show aggregation of IL-17-producing cells in the brain and meninges of transgenic mice with AD pathology (3xTg-AD), an accumulation which was linked with the onset of cognitive impairment but only occurred in female animals (Brigas et al., 2021). Additionally, early treatment with the anti-IL-17 antibody prevented cognitive decline if applied before the formation of amyloid plaques and neurofibrillary tangles. In comparison, late intervention could only delay cognitive impairment, indicating causal relevance to both cognition at large and to the rate of its decline (Brigas et al., 2021). Considering the sex-related differences in immune responses and AD pathology hence calls for detailed investigations of the involvement of peripheral immunity stratified by sex, especially since the peripheral immune system is of particular interest as a therapeutic target in AD (Novak et al., 2021).

2.2 Peripheral metabolic alterations in neurological diseases showing sex-related differences

Importantly, several neurological disorders such as stroke, AD and depression associate with impaired peripheral metabolism (Farooqui, Farooqui, Panza, & Frisardi, 2011). This indicates that sex differences in metabolic diseases may contribute to the development and progression of neurological disorders. Supporting this notion, men and women demonstrate profound differences in certain metabolic aspects, including glucose homeostasis, lipid metabolism, deposition and lipoprotein profiles, which are often attributed to the distinct profiles of secreted sex hormones (Mauvais-Jarvis, Arnold, & Reue, 2017; Mauvais-Jarvis, Clegg, & Hevener, 2013). Healthy premenopausal women have greater body fat mass, subcutaneous adipose tissue and high-density lipoprotein (HDL) cholesterol levels, whereas men tend to accumulate visceral adipose tissue and have greater levels of plasma triglycerides (Link & Reue, 2017; Palmisano, Zhu, Eckel, & Stafford, 2018; Power & Schulkin, 2008). Insulin signaling also differs between the sexes, as women are more insulin-sensitive and show higher capacity for insulin secretion than men (Kautzky-Willer et al., 2012; Nuutila et al., 1995). However, when glucose tolerance deteriorates towards type 2 diabetes, both sexes show similar levels of insulin resistance (Tura et al., 2018). These metabolic differences between men and women affect their predisposition to various metabolic disorders, some of which are linked to neurological diseases.

Ischemic stroke is twice more prevalent in older postmenopausal women and its outcome is often worse compared to that of older men. Several pathological conditions were identified as important risk factors for stroke, including diabetes, obesity and dyslipidemia, and sex differences are implicated in each of them (Bonkhoff, Karch, Weber, Wellmann, & Berger, 2021; Branyan & Sohrabji, 2020). A meta-analysis based on data from 64 cohort studies revealed that diabetic women had a 27% greater relative risk for stroke compared to diabetic men (Peters, Huxley, & Woodward, 2014). Almdal et al. investigated the independent effect of type 2 diabetes on stroke in a population-based study of over 13,000 men and women with 20 years of follow-up. This study showed that type 2 diabetes was linked with an increased risk of stroke (1.5- to 2-fold in men and 1.5- to 6.5-fold in women). However, this sex difference was limited to the age group 55–64 years (Almdal, Scharling, Jensen, & Vestergaard, 2004). Diabetes further associates with stroke outcome and mortality; doubling death rates among men, and tripling among women (Gnatiuc et al., 2018). Obesity is another major risk factor for stroke; overweight and abdominal obesity were associated with progressively increasing risk of stroke (Johnson, Onuma, Owolabi, & Sachdev, 2016; Strazzullo et al., 2010). Obesity is generally more prevalent in women than men (Sarma, Sockalingam, & Dash, 2021), and postmenopausal women experience changes in the adipose tissue distribution and accumulate more visceral fat. The UK Biobank cohort study, which included ~500,000 participants, revealed that overweight and obesity increased the risk of any type of stroke to a similar extent in women and men. However, a higher risk of ischemic stroke was associated with obesity in women (Peters, Carcel, Millett, & Woodward, 2020). These examples demonstrate how sex differences in peripheral metabolic aspects can contribute to the sex-related predisposition to stroke, and to its higher prevalence in postmenopausal women compared to men.

AD affects approximately 35 million people worldwide, and these numbers are predicted to triple by 2050. AD often coexists with comorbidities including obesity, diabetes, hypertension, and cardiovascular diseases (Tyagi et al., 2020). After the age of 65, the risk of AD is 1 in 6 for women and 1 in

11 for men, and nearly two-thirds of current AD cases are women (Zhao, Mao, Woody, & Brinton, 2016). To explore the female brain susceptibility to AD, Zhao et al. used custom-designed Taqman low-density arrays to study sex differences in brain aging profiles in female and male mice, and focused on changes in energy and amyloid metabolism which are most significantly affected in the preclinical development of AD. They discovered that female and male brains demonstrate profoundly dissimilar trajectories as they age and present great differences in gene expression profiles. Brain age-related changes occur much earlier in females than in males, and include a decreased bioenergetic function and increased amyloid dyshomeostasis. These early changes in female brains indicate the onset of a hypometabolic phenotype, contributing to their susceptibility to AD (Zhao et al., 2016). Metabolic-biological sex differences were also shown to play a role in affecting the risk for developing AD in men and women. For example, women have a higher risk of diabetic complications than men, including myocardial infarction and coronary heart disease, which are risk factors for AD (Kautzky-Willer, Harreiter, & Pacini, 2016). Therefore, even though a diagnosis of hypertension, high cholesterol, and diabetes in midlife has been associated with a greater risk of developing AD for both women and men, the risk of AD may be higher for women presenting these factors than for men (Nebel et al., 2018). Recent evidence from 1,857 participants enrolled in the population-based Mayo Clinic Study of Aging showed that the majority of cardiovascular diseases were more strongly linked to cognition in women. For example, coronary heart disease and other cardiovascular diseases were only found to be linked to global cognitive loss in women. Furthermore, diabetes, cholesterol, and coronary heart disease were solely linked with a declined language z score in women. Congestive heart failure, on the other hand, was solely linked with a reduction in the language z score in men (Huo et al., 2022). These data suggest that association between midlife cardiovascular conditions and midlife cognition differs by sex; therefore, although cardiovascular conditions and risk factors are more prevalent in men than women in midlife, the presence of these conditions in men and women has a different effect on cognitive decline. The APOE ε4 genotype is another strong risk factor for AD that also leads to many metabolic differences. APOE ε4 confers a greater AD risk in women and associates with AD biomarkers in a sex-dependent way, with larger risk estimates for women than for men at least in older age (Altmann et al., 2014; Neu et al., 2017). In addition, association analyses of 139 serum metabolites in 1,517 individuals from the AD Neuroimaging Initiative with AD biomarkers

revealed some metabolic effects specific to female APOε4 carriers, which suggests that females experience greater impairment of mitochondrial energy production than males (Arnold et al., 2020).

Considering the mechanistic aspects, there are three main modulators of all systems affected in a sex-specific manner: sex hormones, genetic makeup and gene expression. These factors are tightly interlinked, since steroid hormones affect gene expression (see below) and because all cells primarily depend on their genetic information and epigenetic modulators.

3. Sex hormones and the CNS

Sex hormones belong to the class of steroid hormones (main female hormones: estrogen and progesterone, main male hormones: testosterone and dihydrotestosterone) and their production underlies the control of the HPA axis. The hypothalamus produces gonadotropin releasing hormone (GnRH) to facilitate synthesis of gonadotropins in the pituitary gland: luteinizing hormone (LH) and follicle-stimulating hormone (FSH). LH and FSH jointly stimulate target tissues including the adrenal glands and gonads to produce sex-specific hormones. This axis operates in a mode of a negative feedback loop—high levels of sex hormones will inhibit the production of GnRH, FSH and LH (Wierman, 2007).

Notably, studies from as early as the 1980s show that the brain, spinal cord and even the peripheral nervous system may all locally produce steroid hormones ("neurosteroids" (Baulieu & Robel, 1990; Mellon & Griffin, 2002)) from their precursor cholesterol, or by conversion of circulating steroid hormones (Mellon & Griffin, 2002). Neurosteroids can be secreted by both neurons and glial cells. Examples of endogenous neurosteroid hormones include allopregnanolone (produced from progesterone), tetrahydrodeoxycorticosterone (derived from deoxycorticosterone) and androstanediol (produced from testosterone); synthetic neurosteroids comprise alphaxalone and ganaxolone (Compagnone & Mellon, 2000; Mellon & Griffin, 2002).

Neuroactive steroids, meaning sex hormones produced in the periphery and neurosteroids produced in the nervous system (Paul & Purdy, 1992), can act through their nuclear receptors and additionally bind to the membrane neurotransmitter receptors, which explains their postulated involvement in many brain functions (Marrocco & McEwen, 2016). Classical steroid receptors are located intracellularly and mediate genomic actions linked to the modulation of gene expression. In the brain, estrogen receptors (ERα/β), progesterone receptors (PRA/B) and androgen receptors can be found e.g. in the amygdala, hypothalamus and hippocampus (Barth, Villringer, & Sacher, 2015; Cunningham, Lumia, & McGinnis, 2012). Steroid hormones binding to the nonclassical membrane receptors, including neurotransmitter receptors ($GABA_A$, NMDA, serotonin or dopamine) mediate rapid, nongenomic effects (Barth et al., 2015). Modulation of neurotransmitter signaling by sex hormones can be involved in cognitive processes, such as executive function (NMDA receptors in the prefrontal cortex and hippocampus), regulation of mood (via serotonin and $GABA_A$ receptors), anxiety ($GABA_A$ receptors), motor coordination, decision making, reward and motivation (e.g. dopamine receptors (Barth et al., 2015)). Importantly, these effects often depend on the nature and dose of the specific hormone, length of exposure and the molecular composition of the receptor involved. For example, the pentameric $GABA_A$ receptor can be formed by variable combinations of 19 different subunits (6α, 3β, 3γ, 3ρ, δ, ε, θ and π (Sigel & Steinmann, 2012)). High progesterone concentrations are linked to anxiolytic actions, but prolonged exposure to allopregnanolone leads to increased anxiety, which correlates with expression of the α4 $GABA_A$ subunit in the hippocampus (Gulinello & Smith, 2003; M. Wang, 2011). Moreover, steroid sex hormones in both males and females might be causally involved in adult neurogenesis, influencing the susceptibility to cognitive impairment and mood disorders (as shown in experimental models and reviewed in (Yagi & Galea, 2019)) For example, experiments in female meadow voles and rats have shown that estradiol affects cell proliferation in the hippocampal dentate gyrus, increasing the rate of new cell generation in case of short-term treatment or triggering under long-term treatment (48 h) inhibition of cell proliferation (Ormerod, Lee, & Galea, 2003). In male rats, testosterone increased survival of cells in the hippocampal granule cell layer in the dentate gyrus (Spritzer & Galea, 2007) and further studies have shown that this process likely occurs thanks to cooperation with neurons from other regions in the hippocampus (e.g. the CA3 layer), since neuronal cells in the dentate gyrus do not express androgen receptors (Hamson et al., 2013).

The impact of steroid hormones on neurogenesis may further link with age and reproductive history (Mahmoud, Wainwright, & Galea, 2016). Specifically, estrogens are well known for their neuroprotective actions and protection against neurological disorders such as stroke or AD (Garcia-Segura, Azcoitia, & DonCarlos, 2001). Supporting evidence from experimental and clinical studies is, however, not free from controversies. The beneficial effects of estrogen span its influence on systemic variables linked to certain neurological diseases, such as balancing endothelial function, which may limit the risk of stroke (Hurn & Brass, 2003). Estrogens induce transcription of diverse genes involved in cellular apoptosis or structural proteins, including nerve growth factor, brain-derived neurotrophic factor (BDNF) and insulin-like growth factor (IGF1) (Behl, 2002). Estrogen also displays antioxidant properties and can activate survival routes such as the MAPK pathway (Behl, 2002). Correspondingly, recent studies using specific knockout mouse models indicate that experimental cerebral ischemia initiates the release of astrocytes-induced estrogen which is essential in inducing astrocytes activation and neuroprotective actions of microglia (Wang et al., 2020).

Based on the beneficial effects of estrogen and the exacerbated postmenopausal risk of cardiovascular, neurological and other disorders, exogenous supplementation of sex hormones has been proposed as a potentially preventive and therapeutic measure (Rossouw et al., 1995). Combination of estrogen and progesterone (or estrogen alone in hysterectomized women) has been suggested as hormonal replacement therapy (HRT) to alleviate menopause symptoms and complications. Surprisingly, large studies within The Women Health Initiative (WHI) involving more than 68,000 participants aged 50–79 (Anderson et al., 1998) showed that HRT actually increases the risk of cardiovascular events, blood clotting, breast cancer and dementia, depending on the starting age of treatment (Manson et al., 2013; Rossouw et al., 2002). The Women Health Initiative Memory Study (WHIMS), including a subgroup of patients from the WHI trial, all above 65 years old, showed that HRT was linked with significantly increased risk of dementia (Craig, Maki, & Murphy, 2005). However, HRT initiated in younger WHI participants (50–54 years old), showed no detrimental effects on cognitive health (Espeland et al., 2017). Therefore, when it comes to HRT prescription, age as well as individual health limitations and risk factors should be always into consideration.

Given the complexity of the hormonal effects on the CNS, more research is urgently needed to fully elucidate the role of hormones in the pathophysiology of neurological disorders. Additionally, recent years brought a constantly growing body of evidence indicating the involvement of genetic components in sex differences of neurological disease. This provided another level of regulatory mechanisms, highlighted as a graphical summary of sex-biased variables and systems influencing neurological disorders (Fig. 2).

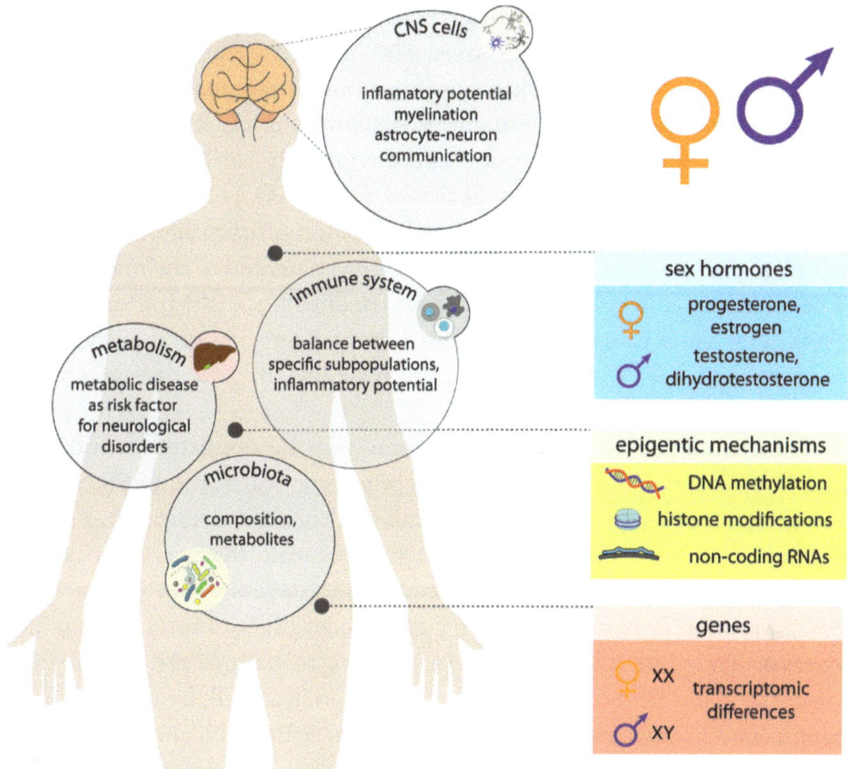

Fig. 2 The importance of sex differences in the brain and periphery for neurological diseases. General summary of factors and systems potentially influencing susceptibility, symptoms, prognosis and outcome of neurological disorders, identified in experimental and clinical studies and discussed in this manuscript.

4. Genetics of sex differences and their impact on the brain

Genetically, sex in humans is determined by the sex chromosomes (allosomes = X and Y), with XX being female and XY—male. Female sex is therefore homogametic, meaning harboring morphologically identical sex chromosome sets; and males are heterogametic. In case of allosomal aneuploidies (see below), the male sex is determined by the presence of the Y chromosome (Simchovitz-Gesher & Soreq, 2020), but it also should be mentioned here that ultimately there are more variables, such as hormonal status influencing the individual phenotype (not included in this review). Both allosomes developed from autosomes, but X and Y differ substantially in their size, structure and number of genes carried on them. The X chromosome includes over 800 protein-coding genes (Schurz et al., 2019), whereas the Y chromosome, which developed ca. 180 million years ago in a common therian ancestor (Cortez et al., 2014), underwent substantial decay during evolution and covers only ca. 78 protein-coding genes in modern *Homo sapiens* (Skaletsky et al., 2003). Importantly, it has been suggested that the extent of Y chromosome shrinkage is connected to both preservation of homologous X–Y gene pairs and to the maintenance of genes essential in sexual development (Bellott et al., 2014; Cortez et al., 2014). This view provides insights into the greater than previously anticipated role of the Y chromosome in male health and disease (Bellott et al., 2014).

The central role of the Y chromosome in male sex determination, has been well characterized. The Y chromosome carries the SRY gene (sex-determining region Y protein, called also testis-determining factor, TDF), which operates as the main switch directing the precursor cells to develop into testicular cells during embryonic development (Berta et al., 1990). Females, in comparison, carry two copies of chromosome X, which is heavily loaded with genetic information. To re-balance the genetic load, one of the X chromosomes becomes inactivated in all cells by condensing its material into heterochromatin (Fang, Disteche, & Berletch, 2019). This process is initiated by a long noncoding RNA, Xist - X-inactive specific transcript, which is turned on in the inactivated chromosome (Richler, Soreq, & Wahrman, 1992) Recent research shows that some cells express Xist from both chromosomes at an early stage, later determining which chromosome will maintain Xist expression permanently (Pacini et al., 2021).

Having two copies of the X chromosome may provide a certain protection from disorders linked to mutations of genes found on this chromosome, which are generally less pronounced in females. Interestingly, not all regions of the X chromosome become inactivated, leading to lower than predicted 50% incidence of symptomatic X-linked disorders in females when compared to males (Migeon, 2020). Around 15% of the X chromosome genes avoid this inactivation (Carrel & Willard, 2005), with some that escape inactivation in all females, and others that do it only in certain individuals (Shvetsova et al., 2018). Since the inactivation occurs in a random manner, females are actually mosaics, with a different degree of mosaicism, depending on which (paternal or maternal) X chromosome became inactivated in a given cell collection/tissue (Migeon, 2020), usually with a preferential ratio of either maternal or paternal X chromosome with sustained activity. This phenomenon, known as "skewed X inactivation" is commonly found in the general population (Shvetsova et al., 2018). There are, however, some X-linked diseases with unconventional clinical presentation, with heterozygotic females showing more severe symptoms. One such example is ephrin-B1 (EFNB1) deficiency resulting in a cranio-frontonasal syndrome. In males, the missing ephrin is substituted by other variants of this protein. Importantly, this is not possible in heterozygotes due to yet unknown mechanisms (Migeon, 2020). In other disorders that are lethal to males, such as Rett syndrome (see below), females are the only clinically symptomatic group (Amir et al., 1999).

Recent studies show that X chromosome genes are critically involved in brain development and function, and that the X chromosome is particularly enriched for brain-expressed genes (Mallard et al., 2021). Correspondingly, X-linked intellectual disabilities are generally more common in males due to their single copy of the X chromosome. An exemption might be again the Rett syndrome, linked to mutations in the MECP2 gene encoding methyl CpG binding protein 2, which binds to methylated DNA and is thereby involved in regulating gene expression. The MECP2 protein is expressed in all tissues, but shows highest abundance in the brain (Sandweiss, Brandt, & Zoghbi, 2020). Patients with Rett syndrome suffer from gradual loss of motor skills, speech impairments, seizures, autism, ataxia and microcephaly (Amir et al., 1999). Interestingly, duplications in the MECP2 gene sequence lead to neurodevelopmental disorders sharing several clinical features with Rett syndrome (Sandweiss et al., 2020).

In comparison, fragile X syndrome is the most common inherited intellectual disability affecting both sexes. This disorder is caused by anticipated expansion of a CGG trinucleotide in the FMR1 gene, where over 200 repeats induce the disease (Hayward & Usdin, 2021). This leads to methylation of this DNA fragment and its silencing ((Moore, Le, & Fan, 2013), see also *Epigenetics of sex differences* below). Lower repeat numbers are linked to the fragile X-associated tremor/ataxia syndrome (FXTAS) and fragile X-associated primary ovarian insufficiency (FXPOI) in women and pose an anticipated risk for >200 repeats in the next generation (Hayward & Usdin, 2021). Furthermore, aneuploidies in sex chromosomes are linked to a spectrum of neurological symptoms, intellectual disabilities and behavioral alterations. Aneuploidies are most commonly caused by nondisjunction events and although they belong to the common genetic disorders, they often remain undiagnosed (Hong & Reiss, 2014; Skuse, Printzlau, & Wolstencroft, 2018). The most frequent aneuploidies comprise: 45, X0 = Turner syndrome; 47, XXY = Klinefelter syndrome (1:600 births); 47, XYY and 47, XXX (Skuse et al., 2018).

All of the above indicates that it is difficult to make a clear distinction between the sex-related consequences of genotype and hormone differences, driven in fact by the genotype (Arnold & Chen, 2009; Ngun, Ghahramani, Sánchez, Bocklandt, & Vilain, 2011). To approach this issue experimentally, specific animal models have been proposed, such as "the four core genotypes model." In this model, the offspring of XX and XY^{-}Sry mice (with Sry gene inserted into an autosome) having the following genotypes: XX, XY^{-}, XY^{-}Sry and XXSry may help to distinguish between hormonally and genetically-driven differences linked to the allosomes. Mice with the XX and XY^{-} genotypes develop ovaries and XY^{-}Sry and XXSry mice develop testes, demonstrating additionally that the presence of the Sry gene is the cause, even if not necessarily located on the sex chromosome. This model also allows exploring the impacts of maternal or paternal X imprinting (Arnold & Chen, 2009). Initial investigations using the four core genotypes model highlighted the dominant role of sex hormones in sexual differentiation, and identified several behavioral and cellular phenotypes associated with neuronal numbers in certain brain areas. Large effects attributed to sex chromosomes have been identified more recently, including differences in avoidance behavior (more pronounced in females) or aggression (XY females were more aggressive than XX animals) (Arnold & Chen, 2009), which further raise sociological issues.

Interestingly, sex-associated differences in gene expression are not limited to the sex chromosomes. A new study taking advantage of the human Genotype Tissue Expression (GTEx (Lonsdale et al., 2013)) dataset and investigating 29 tissues identified sex-dependent transcription factors and their target genes, creating gene regulatory networks (Lopes-Ramos et al., 2020). Surprisingly, however, sex-stratified DE genes were not enriched in motifs of sex hormone receptors, indicating that other regulators of the observed heterogeneities might be involved. In this context, an in-depth analysis of regulatory networks in basal ganglia found that most transcription factors had different sex-dependent targets, which may indicate their involvement in sex-biased neurological disorders such as AD or PD (Lopes-Ramos et al., 2020). Further, yet more extensive investigations of the GTEx dataset demonstrated sex-related transcriptomic differences in 44 tissues. Although those sex-specific effects were very small, they were broadly present and generally specific for a given tissue (Oliva et al., 2020). Another genetic survey, comprising 12 tissues (adipose tissue, brain, colon, heart, liver, lung, muscle, skin, spleen, thyroid, adrenal and pituitary glands) in five mammalian species focused on the evolutionary aspect of sex differences in gene expression. That survey showed that the majority of those differences emerged late in evolution, further advising caution when transferring findings in experimental animal models to humans, especially when it comes to sex-related differences (Naqvi et al., 2019), which is also of importance in case of brain-expressed non-coding genes (Barbash, Shifman, & Soreq, 2014; Hanan et al., 2020)

Importantly, sex-specific differences in the incidence of neurological disorders such as PD may in fact be driven by genetic factors. A recent report indicated that SRY is significantly upregulated in both in-vivo and cell culture models of PD and that reduced nigrostriatal SRY expression leads to improvement of motor function and decrease in degeneration of dopaminergic neurons (Lee et al., 2019). Altogether, the impact of sex hormones on gene expression complicates the dissociation of the effects of sex-specific gene expression and hormonal actions (Lopes-Ramos et al., 2020; Van Nas et al., 2009). The fact that males are hemizygous for the X-linked genes, technical issues related to data analyses and experimental design and often not justified fears that sex-stratified analyses may decrease the statistical power, have generally led to frequent exclusion of allosomal genes from genetic screening studies (Ferretti & Santuccione Chadha, 2021; Wise, Gyi, & Manolio, 2013). This issue is very problematic considering e.g. recent reports connecting X-linked genes to the progression and symptomology in AD, a disease known to have a certain sex-bias (Davis et al., 2021; Ferretti & Santuccione Chadha, 2021). Similar effects may be present in other pathological states.

5. Epigenetics of sex differences

Epigenetics refers to regulatory mechanisms controlling gene expression without altering the genetic code (Wu & Morris, 2001). Examples here include posttranscriptional regulation by noncoding RNAs (such as microRNAs binding in their main mechanisms of action to the 3′ untranslated region (3′UTR) of a transcript leading to its degradation or transfer RNA fragments acting in a similar way), histone modifications and DNA methylation.

Epigenetic processes play significant roles in the fine-tuning of gene expression and may be influenced by sex. The topic of sex differences in epigenetics is, however, relatively new and therefore not fully explored. DNA methylation was one of the first identified epigenetic regulators of gene expression. In this process, the commonly affected base is cytosine, undergoing a chemical modification to 5-methylcytosine by transfer of the methyl group, which often leads to transcriptional repression (Li & Zhang, 2014). Interestingly, the brain has one of the highest contents of methylated DNA sequences when compared to other organs, and proper functioning of the DNA methylation mechanism is critical in neuronal development (Moore et al., 2013). For example, in the X-linked Rett syndrome discussed before, the expanded CGG triplet undergoes methylation, leading to silencing of the MECP2 gene, which is indispensable for correct neuronal growth and synaptic function (Moore et al., 2013).

Numerous studies showing differences in autosomal DNA methylation have been published, but comparing different datasets reveals relatively inconsistent methylation profiles (Gatev et al., 2021). A recent, stringent approach focusing on regions with specific methylation patterns, analyzed data from whole blood samples of 3795 individuals (males and females) and identified 179 autosomal sex-associated co-methylated genomic regions. Importantly, the methylation status of these regions was consistent and showed sex-linked profiles over the lifespan (Gatev et al., 2021). Other studies have reported sex-dependent changes in DNA methylation profiles associated with age (McCartney et al., 2019; Yusipov et al., 2020) or metabolism (García-Calzón, Perfilyev, de Mello, Pihlajamäki, & Ling, 2018).

Another layer of epigenetic control of gene expression is provided by histone modifications. DNA is wrapped around histone complexes forming nucleosomes, and composing chromatin. Histones can be modified by several processes, including methylation, phosphorylation, adenylation or ubiquitination impacting chromatin structure and therefore transcription processes (Bannister & Kouzarides, 2011). Recent clinical studies using positron emission tomography with tracer for histone deacetylases (HDAC) found age- and sex-related changes in HDACs expression in the cerebral white matter (Gilbert et al., 2019). A comprehensive summary of sex-related differences in histone modifications (and other epigenetic processes) in neurological disorders has been provided by Qureshi and Mehler (Qureshi & Mehler, 2010).

Further aspect of epigenetic regulation involves sex-linked variations in non-coding RNAs participating in the post-transcriptional control over gene expression. These variations may contribute to male/female differences in pathophysiological processes of the CNS. One of the best-known families of non-coding RNAs are the microRNAs (miRNAs). Mature miRNAs are small RNA molecules, 21–25 nucleotides long which in their main mechanism of action bind to the complementary sequence in the 3′-UTR of mRNA transcripts, leading to translational repression (Bartel, 2009). Interestingly, the X chromosome is particularly rich in miRNA sequences and contains ca. 10% of all miRNA genes. This feature is conserved evolutionarily, but limited to mammals (Di Palo et al., 2020). In contrary, to date, only two exclusively Y chromosome-derived miRNAs are registered in the miRBase (hsa-miR-9985 and hsa-miR-12120) (Kozomara, Birgaoanu, & Griffiths-Jones, 2019). Unfortunately, the experimental evidence on these miRNAs is lacking and more research would be needed to validate these molecules and identify their roles.

Importantly, the sex-related differences in miRNA expression may be linked to brain pathology. Male and female wildtype mice and transgenic animals with tau pathology (P201S) showed distinct miRNA expression patterns. Microglia-specific decreases in mature miRNAs in Dicer-depleted animals led to sex-specific effects with greater impact on the transcriptome in male mice (Kodama et al., 2019). Another example may be proved by a neuronal knockout of the Atrx gene causing sex-specific increases of miR-137 expression in the hippocampus and selectively associates with learning and memory impairments in male mice (Tamming et al., 2020). This is reminiscent of the ATRX mutations in humans, which cause the α-thalassemia X-linked mental retardation (ATR-X) syndrome associated with brain pathology and intellectual disability. The ATRX protein is involved in chromatin remodeling and its loss leads to increased levels of miR-137, which impact the target genes involved in the regulation of cognitive processes. Changes in another epigenetic mechanism—histone methylation status were proposed as the main driver of increased miR-137 expression (Tamming et al., 2020).

Unfortunately, clinical studies of sex differences in miRNA profiles in neurological disease are rather scarce, despite the sex-bias in the involvement of miRNAs in numerous brain disorders. Further, hoping for their future diagnostic utility, human studies focus mostly on miRNA profiles in biofluids. An analysis of human miRNA expression generated from publicly available datasets and including brain, colorectal mucosa, whole blood and cord blood identified sex-related miRNA patterns in all tissues (73 female-biased miRNAs and 163 male-biased miRNAs) (Cui et al., 2018). Functional enrichment tests linked female-biased miRNAs with processes such as apoptosis, inflammation or cell cycle; and male-biased miRNAs with pathways connected to e.g. circadian rhythm, histone modifications and cell reprograming (Cui et al., 2018). When it comes to mechanisms underlying different miRNA levels in males and females, it has been postulated that those may be driven by sex hormones or escapee genes located on the X chromosome (Di Palo et al., 2020; Sharma & Eghbali, 2014). Advances in the recent years have suggested more candidate modulators of neurological disorders in experimental and clinical settings (e.g. (Lobentanzer, Hanin, Klein, & Soreq, 2019)), providing further interesting directions for the development of these new research areas, as is highlighted in Fig. 2.

6. Recently identified factors influencing neurological disease with sex-specific signatures

6.1 The gut microbiota

Accumulating evidence highlights the involvement of bacterial commensals in physiological and pathological processes of the host. The gut microbiota is the largest bacterial commensal community which resides within the intestines. Ample experimental studies reveal influences of these bacteria on host metabolism (Cani & Delzenne, 2009), immunity (Hooper, Littman, & Macpherson, 2012) and even BBB permeability (Braniste et al., 2014) or microglial function (Erny et al., 2015). Numerous reports further show that gut microbiota may influence the individual susceptibility to neurological disorders such as stroke or neurodegenerative processes or modulate the outcome after disease onset (Cryan, O'Riordan, Sandhu, Peterson, & Dinan, 2020).

Until recently, the differences in microbiota composition between males and females have not been investigated in detail. While a defined, sex-specific composition is not fully established, it has been reported that sex might have some impact on the microbial makeup, mostly as an influence of sex hormones in experimental systems (Org et al., 2016). In humans, the effects of biological sex also involve the composition of microbiota, showing increased microbial diversity and the presence of metabolically beneficial species in women before menopause (Zhang et al., 2021). Together, all these findings further characterize age-related changes in the microbiome which may affect susceptibility to age- and sex-related disorders.

Additionally, some very interesting insights are derived from experimental studies in germ-free mice. Germ-free status implies delivery over cesarean section in sterile conditions and protection from any exposure to microorganisms throughout life. Recent investigations showed that when non-obese diabetic mice (NOD mice) are raised under germ-free conditions, they show no sex-bias in disease incidence; in contrary, when grown with intact microbiota, female mice are more prone to diabetes than males

(Yurkovetskiy et al., 2013)). Moreover, another study in this model showed that colonization of female animals with male-derived microbiota increased testosterone levels and protected female recipients from disease development (Markle et al., 2013). The authors argue that in humans as well, the gut microbiota may impact sex-bias in certain disorders, including neurological diseases such as MS (Markle et al., 2013). Involvement of microbial commensals in MS autoimmune responses was studied using microbiota transplants from twins discordant for the disease into mice genetically prone to MS. Microbiota transplant from an MS patient resulted in higher autoimmunity when compared to transplant from a healthy twin donor (Berer et al., 2017). Finally, involvement of the microbiome in neurodegenerative disorders has been postulated in recent years (Needham, Kaddurah-Daouk, & Mazmanian, 2020) and was shown to be affected by sex-related factors. For example, transgenic amyloidosis (APPPS1-21) male mice only showed decreased accumulation of β-amyloid after microbiome-shift induced by antibiotic treatment, and this decline was accompanied by reduced phagocytic activity of microglia (Dodiya et al., 2019). Importantly, the causal involvement of microbiota was confirmed by recolonization experiments, where restoring the microbial makeup found in APPPS1-21 animals aggravated the amyloid pathology (Dodiya et al., 2019).

6.2 Sex differences in experimental systems and clinical trials

The ultimate goals of characterizing sex-associated differences in neurological disorders are undoubtedly aimed at the development of better, personalized therapies and identification of novel biomarkers that might be sex-specific. Since the basis for that aim leans on solid basic research, it is crucial to underscore the urgent need of inclusion of both sexes in experimental studies (Rich-Edwards, Kaiser, Chen, Manson, & Goldstein, 2018; Shansky & Murphy, 2021) and the importance of analysis of men and women data, which is strongly supported and advised by agencies such as the National Institutes of Health, issuing the "NIH Revitalization Act of 1993 Public Law 103-43" requesting inclusion of women and minorities in clinical research (National Institutes of Health, 1994). Interestingly, the trend was different in the second half of the 20th century (National Institutes of Health, 2021). Regrettable decisions of introducing thalidomide to the market without proper assessment of teratogenic potential (except of the USA, where it did not receive the FDA approval), led to numerous cases of birth defects (Kim & Scialli, 2011). This has motivated the FDA to recommend exclusion of women, who might become pregnant, from phase I and II clinical drug studies and resulted in broad omission of women in trials (National Institutes of Health, 2021). Fig. 3 shows some of the landmarks in discoveries of factors contributing to sex differences and policies on studies involving both sexes.

The issue of inclusion of both sexes in research studies, reaches back even to in-vitro studies where the genetic cellular makeup should be reported. Notably, recent research from our group showed divergent miRNA response to neurokine stimulation in neuroblastoma cells of male (LA-N-5) and female (LA-N-2) origin. These differences may drive the male/female variance in neuronal cholinergic differentiation, which is of major importance since impaired cholinergic signaling is involved in many neuropsychiatric disorders (Lobentanzer et al., 2019). Nowadays, due to the availability of numerous cell lines, models and systems, sex-specific differences may be included in in-vitro investigations in more complicated setups, such as e.g. studies of

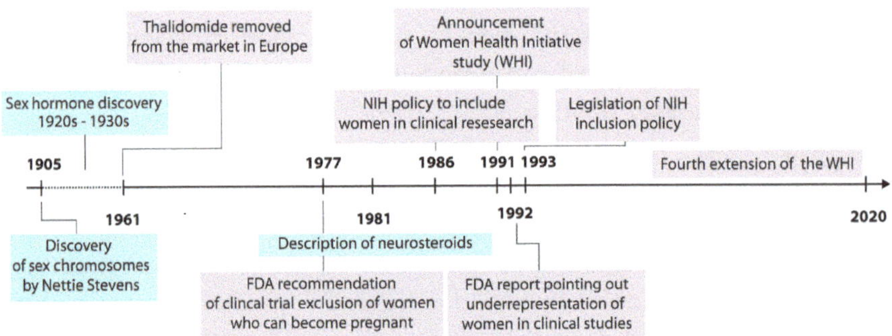

Fig. 3 Timeline of selected discoveries and policies on studies including both sexes. Selected relevant landmark biological discoveries of factors contributing to sex-associated differences and policies/studies concerning inclusion of both sexes to clinical research. *Based on Baulieu, E. E., Robel, P., & Schumacher, M. (2001). Neurosteroids: Beginning of the story.* International Review of Neurobiology, 46, *1–32; Food and Drug Administration. (2021). Understanding sex differences at FDA. Retrieved from https://www.fda.gov/science-research/womens-health-research/understanding-sex-differences-fda; Furman, B. L. S., Metzger, D. C. H., Darolti, I., Wright, A. E., Sandkam, B. A., et al. (2020). Sex chromosome evolution: So many exceptions to the rules.* Genome Biology and Evolution, 12, *750–763; National Institutes of Health. (2021).* NIH inclusion outreach toolkit: How to engage, recruit, and retain women in clinical research. History of Women's Participation in Clinical Research. *Retrieved from https://orwh.od.nih.gov/toolkit/recruitment/history#7%0A; Stevens, N. (1905). Studies in spermatogenesis Washington D.C.: Carnegie Institution of Washington.; Tata, J. R. (2005). One hundred years of hormones.* EMBO Reports, 6, *490; Women's Health Initiative. (2021). About WHI, Women's Health Initiative. Retrieved from https://www.whi.org/page/about-whi.*

the BBB, an important player in neurodegenerative disease (Weber & Clyne, 2021). Particularly problematic are in-vivo studies with limited n-numbers, where often one (mostly male) sex is chosen to reduce heterogeneity; a decision which is actually based on wrong assumptions (see below).

Two leading theories of the possible origin of sex-related variabilities profoundly impact the design of experimental studies: (1) According to the evolutionary view, heterogametic animals (males in mammals, XY) should be more heterogenous when it comes to their biological characteristics; (2) The hormonal perspective highlights the influences of "estrus-mediated variability" in females (Zajitschek et al., 2020). An unfortunate example of this type of sex bias in experimental research may be numerous studies which had been performed in stroke models, where researchers aimed to exclude the protective effects of estrogens and the phase of the estrus cycle. Nevertheless, recognizing all of the limitations of experimental systems, animals of both sexes should be included in in-vivo studies. A recent study addressed these concerns by analyzing over 29,000 mice and 218 biological features and found that the actual heterogeneity strongly depends on the trait. Males are more heterogenous when it comes to physical features, whereas female animals differ more in immunological parameters (Zajitschek et al., 2020). Other analyses show that the female animals are not more variable than male mice in multiple traits including e.g. behavioral and molecular data (Beery, 2018; Prendergast, Onishi, & Zucker, 2014; Shansky & Murphy, 2021). Therefore, the effects of the estrus cycle in mice and rats seem to be largely overestimated in the experimental design (Prendergast et al., 2014).

Based on all of the above, development of new research tools is urgently needed to broaden the appreciation of the importance of sex differences, which is essential for further research advances (Mauvais-Jarvis et al., 2017). Regrettably, clinical studies are still not flawless when it comes to inclusion of both sexes. Another recent example from the field of stroke shows significant underrepresentation of women in stroke trials (Carcel et al., 2021). This might be surprising considering that the lifetime risk of stroke is higher in female patients. In other cases, such as AD studies, although the percentage of enrolled women is higher than that of men participants, the numbers still do not reflect the proportions of female and male patients present in the general population (Martinkova et al., 2021). Interestingly, females are overrepresented for example in trials focusing on psychiatric diseases, trauma or nutritional and metabolic disorders (Steinberg et al., 2021).

Optimized experimental setup, including both sexes and using appropriate statistical analysis e.g. with factorial designs (Beery, 2018) may help to address some of the above-mentioned issues. A solution for experimental research might be including both sexes in the first study checking for differences between male and female animals, and based on their presence/absence continuation with a mixed cohort or choice of one sex for further tests (Shansky & Murphy, 2021).

Importantly, sex-stratified analyses should be designed and performed carefully, which unfortunately is not always the case. Due to the lack of training of researchers for such analyses, sex-linked effects are often analyzed separately in the two sexes, leading to the conclusion that a certain intervention has significantly different outcomes in males compared to females, which is statistically wrong (since the sexes have not been compared directly) (Garcia-Sifuentes & Maney, 2021). A recent report highlights this problem, demonstrating its occurrence in >70% of 147 analyzed manuscripts published in 2019 from different areas of biological sciences (such as behavioral science, immunology or neuroscience) (Garcia-Sifuentes & Maney, 2021). This study further shows that identifying sex-associated differences is not straightforward, and points out that there might be some overreporting of sex-linked effects that were identified due to improper statistical design. Publication bias excluding studies with negative results from being successfully published further increases the complexity of proper reporting of sex-related differences, as many of the negative studies are not shared with the scientific community.

6.3 Sex does not equal gender

At the social sciences level, it is important to note that sex and gender are not one and the same term, and that both sex and gender have to be considered in research and medical practice. Unfortunately, the term "sex" has often been confused and is commonly used interchangeably with "gender," which actually covers social, behavioral and cultural factors (Rich-Edwards et al., 2018). According to the definition of the World Health Organization: "Gender refers to the characteristics of women, men, girls and boys that are socially constructed. This includes norms, behaviours and roles associated with being a woman, man, girl or boy, as well as relationships with each other. As a social construct, gender varies from society to society and can change over time. (...) Gender interacts with but is different from sex, which refers to the different biological and physiological characteristics of females,

males and intersex persons, such as chromosomes, hormones and reproductive organs." (World Health Organisation, 2022). Despite the fact that recent years have put the research on sex and gender differences into the spotlight, misreporting of results could be noted even in 2019. In the already discussed analysis by Garcia-Sifuentes and Maney, gender was reported as a variable in nonhuman mammals in 9% of 147 analyzed papers and the terms "sex" and "gender" have been used interchangeably in 19% of them (Garcia-Sifuentes & Maney, 2021).

Gender is yet another important variable in health and disease, since it may influence both individual behavior (including e.g. nutritional habits and risk exposure) and social interactions (Mauvais-Jarvis et al., 2020). Ultimately, gender may also impact epigenetic modifications, however testing this hypothesis is quite challenging (Cortes, Cisternas, & Forger, 2019). Moreover, needs of transgender patients and slightly different profile when it comes to disease risk have to be considered in clinical practice (von Vaupel-Klein & Walsh, 2021). It also should not be forgotten that transgender individuals undergoing transition are treated with hormonal therapies that might affect several pharmacological parameters modifying drug action (summarized in (Cirrincione & Huang, 2021). Therefore, both males and females as well as transgender individuals may further be subjected to antagonistic pleiotropy, affecting their health and wellbeing in a manner dependent on their age, sex and gender.

7. Conclusion

Sex differences in the CNS initiate at the cellular and molecular levels, and experimental studies have reported sex-bias in main brain-resident cell populations, including neurons, microglia, astrocytes and oligodendrocytes. The question to which extent this divergence contributes to the development of neurological disorders should be addressed in further research, ideally including samples from human patients of the two sexes that are analyzed and compared. Moreover, peripheral factors such as the immune system, metabolism or gut microbiota may contribute to the susceptibility, course and prognosis of neurological disease and sex-related differences that have been observed in these systems. Therefore, future studies of CNS disorders should not focus solely on the brain, but try to reveal the broader picture, including brain-body communication pathways. Most importantly, both sexes should be included and properly reported and compared in experimental and clinical investigations, and both sex and gender should be accounted for in clinical research. Elucidating the mechanisms underlying sex differences in individuals living with neurological disease is essential for achieving successful development of personalized and more effective therapies.

Main points

1. Men and women vary in their susceptibility to neurological disorders and sex-related differences are driven by genetic, epigenetic and hormonal factors.
2. Sex differences affect the brain as well as the peripheral systems such as the immune system or metabolism, playing important roles in the pathogenesis of neurological disorders.
3. The field investigating the impact of sex differences in pathological conditions is rapidly evolving and new studies should be performed in both sexes and using proper statistical design.

Acknowledgment

KW is a Shimon Peres Post-Doctoral Fellow at the ELSC and HS is supported by the Israel Science Foundation Grant no. 1016/18 and the Israel Science Foundation Precision Medicine Partnership (IPMP) Grant no. 3213/19.

References

Abi Ghanem, C., Degerny, C., Hussain, R., Liere, P., Pianos, A., Tourpin, S., et al. (2017). Long-lasting masculinizing effects of postnatal androgens on myelin governed by the brain androgen receptor. *PLoS Genetics*, *13*. https://doi.org/10.1371/journal.pgen.1007049.

Ader, R., & Cohen, N. (1975). Behaviorally conditioned immunosuppression. *Psychosomatic Medicine*, *37*, 333–340.

Almdal, T., Scharling, H., Jensen, J. S., & Vestergaard, H. (2004). The independent effect of type 2 diabetes mellitus on ischemic heart disease, stroke, and death: A population-based study of 13,000 men and women with 20 years of follow-up. *Archives of Internal Medicine*, *164*, 1422–1426.

Altmann, A., Tian, L., Henderson, V. W., & Greicius, M. D. (2014). Sex modifies the APOE-related risk of developing Alzheimer disease. *Annals of Neurology*, *75*, 563–573.

Amir, R. E., Van Den Veyver, I. B., Wan, M., Tran, C. Q., Francke, U., & Zoghbi, H. Y. (1999). Rett syndrome is caused by mutations in X-linked MECP2, encoding methyl-CpG-binding protein 2. *Nature Genetics*, *23*, 185–188.

Anderson, G., Cummings, S., Freedman, L. S., Furberg, C., Henderson, M., Johnson, S. R., et al. (1998). Design of the Women's Health Initiative clinical trial and observational study. The Women's Health Initiative Study Group. *Controlled Clinical Trials*, *19*, 61–109.

Arnold, A. P., & Chen, X. (2009). What does the "four core genotypes" mouse model tell us about sex differences in the brain and other tissues? *Frontiers in Neuroendocrinology*, *30*, 1–9.

Arnold, M., Nho, K., Kueider-Paisley, A., Massaro, T., Huynh, K., Brauner, B., et al. (2020). Sex and APOE ε4 genotype modify the Alzheimer's disease serum metabolome. *Nature Communications*, *11*. https://doi.org/10.1038/S41467-020-14959-W.

Baecher-Allan, C., Kaskow, B. J., & Weiner, H. L. (2018). Multiple sclerosis: Mechanisms and immunotherapy. *Neuron*, *97*, 742–768.

Bannister, A. J., & Kouzarides, T. (2011). Regulation of chromatin by histone modifications. *Cell Research*, *21*(3), 381–395.

Barbash, S., Shifman, S., & Soreq, H. (2014). Global coevolution of human microRNAs and their target genes. *Molecular Biology and Evolution*, *31*, 1237–1247.

Bartel, D. P. (2009). MicroRNAs: Target recognition and regulatory functions. *Cell*, *136*, 215–233.

Barth, C., Villringer, A., & Sacher, J. (2015). Sex hormones affect neurotransmitters and shape the adult female brain during hormonal transition periods. *Frontiers in Neuroscience*, *9*. https://doi.org/10.3389/FNINS.2015.00037/ABSTRACT.

Baulieu, E. E., & Robel, P. (1990). Neurosteroids: A new brain function? *The Journal of Steroid Biochemistry and Molecular Biology*, *37*, 395–403.

Beery, A. K. (2018). Inclusion of females does not increase variability in rodent research studies. *Current Opinion in Behavioral Sciences*, *23*, 143.

Begum, A. N., Cunha, C., Sidhu, H., Alkam, T., Scolnick, J., Rosario, E. R., et al. (2014). Women with the Alzheimer's risk marker ApoE4 lose Aβ-specific $CD4^+$ T cells 10-20 years before men. *Translational Psychiatry*, *4*(7). https://doi.org/10.1038/TP.2014.51.

Behl, C. (2002). Oestrogen as a neuroprotective hormone. *Nature Reviews Neuroscience*, *3*, 433–442.

Bellott, D. W., Hughes, J. F., Skaletsky, H., Brown, L. G., Pyntikova, T., Cho, T.-J., et al. (2014). Mammalian Y chromosomes retain widely expressed dosage-sensitive regulators. *Nature*, *508*, 494–499.

Belonwu, S. A., Li, Y., Bunis, D., Rao, A. A., Solsberg, C. W., Tang, A., et al. (2022). Sex-stratified single-cell RNA-Seq analysis identifies sex-specific and cell type-specific transcriptional responses in Alzheimer's disease across two brain regions. *Molecular Neurobiology*, *59*, 276–293.

Berer, K., Gerdes, L. A., Cekanaviciute, E., Jia, X., Xiao, L., Xia, Z., et al. (2017). Gut microbiota from multiple sclerosis patients enables spontaneous autoimmune encephalomyelitis in mice. *Proceedings of the National Academy of Sciences of the United States of America*, *114*, 10719–10724.

Berta, P., Hawkins, J. B., Sinclair, A. H., Taylor, A., Griffiths, B. L., Goodfellow, P. N., et al. (1990). Genetic evidence equating SRY and the testis-determining factor. *Nature*, *348*, 448–450.

Bettcher, B. M., Tansey, M. G., Dorothée, G., & Heneka, M. T. (2021). Peripheral and central immune system crosstalk in Alzheimer disease—A research prospectus. *Nature Reviews. Neurology*, *17*, 689–701.

Bonkhoff, A. K., Karch, A., Weber, R., Wellmann, J., & Berger, K. (2021). Female stroke: Sex differences in acute treatment and early outcomes of acute ischemic stroke. *Stroke*, *52*, 406–415.

Borovikova, L. V., Ivanova, S., Zhang, M., Yang, H., Botchkina, G. I., Watkins, L. R., et al. (2000). Vagus nerve stimulation attenuates the systemic inflammatory response to endotoxin. *Nature*, *405*, 458–462.

Bradl, M., & Lassmann, H. (2010). Oligodendrocytes: Biology and pathology. *Acta Neuropathologica*, *119*, 37.

Braniste, V., Al-Asmakh, M., Kowal, C., Anuar, F., Abbaspour, A., Toth, M., et al. (2014). The gut microbiota influences blood-brain barrier permeability in mice. *Science Translational Medicine*, *6*, 263ra158.

Branyan, T. E., & Sohrabji, F. (2020). Sex differences in stroke co-morbidities. *Experimental Neurology, 332*. https://doi.org/10.1016/J.EXPNEUROL.2020.113384.
Brigas, H. C., Ribeiro, M., Coelho, J. E., Gomes, R., Gomez-Murcia, V., Carvalho, K., et al. (2021). IL-17 triggers the onset of cognitive and synaptic deficits in early stages of Alzheimer's disease. *Cell Reports, 36*, 109574.
Cani, P. D., & Delzenne, N. M. (2009). Interplay between obesity and associated metabolic disorders: New insights into the gut microbiota. *Current Opinion in Pharmacology, 9*, 737–743.
Cao, W., & Zheng, H. (2018). Peripheral immune system in aging and Alzheimer's disease 11 medical and health sciences 1109 neurosciences. *Molecular Neurodegeneration, 13*, 1–17.
Carcel, C., Harris, K., Peters, S. A., Sandset, E. C., Balicki, G., Bushnell, C. D., et al. (2021). Representation of women in stroke clinical trials: A review of 281 trials involving more than 500,000 participants. *Neurology*. https://doi.org/10.1212/WNL.0000000000012767.
Carrel, L., & Willard, H. F. (2005). X-inactivation profile reveals extensive variability in X-linked gene expression in females. *Nature, 434*, 400–404.
Cerghet, M., Skoff, R. P., Bessert, D., Zhang, Z., Mullins, C., & Ghandour, M. S. (2006). Proliferation and death of oligodendrocytes and myelin proteins are differentially regulated in male and female rodents. *Journal of Neuroscience, 26*, 1439–1447.
Chekroud, A. M., Ward, E. J., Rosenberg, M. D., & Holmes, A. J. (2016). Patterns in the human brain mosaic discriminate males from females. *Proceedings of the National Academy of Sciences of the United States of America, 113*, E1968.
Cirrincione, L. R., & Huang, K. J. (2021). Sex and gender differences in clinical pharmacology: Implications for transgender medicine. *Clinical Pharmacology & Therapeutics, 110*, 897–908.
Clayton, J. A. (2016). Sex influences in neurological disorders: Case studies and perspectives. *Dialogues in Clinical Neuroscience, 18*, 357.
Colonna, M., & Butovsky, O. (2017). Microglia function in the central nervous system during health and neurodegeneration. *Annual Review of Immunology, 35*, 441–468. https://doi.org/10.1146/annurev-immunol-051116-052358.
Compagnone, N. A., & Mellon, S. H. (2000). Neurosteroids: Biosynthesis and function of these novel neuromodulators. *Frontiers in Neuroendocrinology, 21*, 1–56.
Cortes, L. R., Cisternas, C. D., & Forger, N. G. (2019). Does gender leave an epigenetic imprint on the brain? *Frontiers in Neuroscience, 13*. https://doi.org/10.3389/FNINS.2019.00173.
Cortez, D., Marin, R., Toledo-Flores, D., Froidevaux, L., Liechti, A., Waters, P. D., et al. (2014). Origins and functional evolution of Y chromosomes across mammals. *Nature, 508*, 488–493.
Cosgrove, K. P., Mazure, C. M., & Staley, J. K. (2007). Evolving knowledge of sex differences in brain structure, function and chemistry. *Biological Psychiatry, 62*, 847.
Craig, M. C., Maki, P. M., & Murphy, D. G. (2005). The women's health initiative memory study: Findings and implications for treatment. *The Lancet Neurology, 4*, 190–194.
Cramer, J. V., Benakis, C., & Liesz, A. (2019). T cells in the post-ischemic brain: Troopers or paramedics? *Journal of Neuroimmunology, 326*, 33–37. Elsevier B.V.
Cryan, J. F., O'Riordan, K. J., Sandhu, K., Peterson, V., & Dinan, T. G. (2020). The gut microbiome in neurological disorders. *The Lancet Neurology, 19*, 179–194.
Cui, C., Yang, W., Shi, J., Zhou, Y., Yang, J., Cui, Q., et al. (2018). Identification and analysis of human sex-biased microRNAs. *Genomics, Proteomics & Bioinformatics, 16*, 200–211.
Cunningham, R. L., Lumia, A. R., & McGinnis, M. Y. (2012). Androgen receptors, sex behavior, and aggression. *Neuroendocrinology, 96*, 131–140.
Dai, L., & Shen, Y. (2021). Insights into T-cell dysfunction in Alzheimer's disease. *Aging Cell, e13511*.
Dantzer, R. (2018). Neuroimmune interactions: From the brain to the immune system and vice versa. *Physiological Reviews, 98*, 477–504.

Davis, E. J., Solsberg, C. W., White, C. C., Miñones-Moyano, E., Sirota, M., Chibnik, L., et al. (2021). Sex-specific association of the X chromosome with cognitive change and Tau pathology in aging and Alzheimer disease. *JAMA Neurology*, *78*, 1249–1254.

del Giudice, M., Lippa, R. A., Puts, D. A., Bailey, D. H., Bailey, J. M., & Schmitt, D. P. (2016). Joel et al.'s method systematically fails to detect large, consistent sex differences. *Proceedings of the National Academy of Sciences of the United States of America*, *113*, E1965.

Di Palo, A., Siniscalchi, C., Salerno, M., Russo, A., Gravholt, C. H., & Potenza, N. (2020). What microRNAs could tell us about the human X chromosome. *Cellular and Molecular Life Sciences*, 77, 4069–4080.

Dodiya, H. B., Kuntz, T., Shaik, S. M., Baufeld, C., Leibowitz, J., Zhang, X., et al. (2019). Sex-specific effects of microbiome perturbations on cerebral Aβ amyloidosis and microglia phenotypes. *The Journal of Experimental Medicine*, *216*, 1542.

Eliot, L., Ahmed, A., Khan, H., & Patel, J. (2021). Dump the "dimorphism": Comprehensive synthesis of human brain studies reveals few male-female differences beyond size. *Neuroscience & Biobehavioral Reviews*, *125*, 667–697.

Erny, D., Hrabe de Angelis, A. L., Jaitin, D., Wieghofer, P., Staszewski, O., David, E., et al. (2015). Host microbiota constantly control maturation and function of microglia in the CNS. *Nature Neuroscience*, *18*, 965–977.

Espeland, M. A., Rapp, S. R., Manson, J. A. E., Goveas, J. S., Shumaker, S. A., Hayden, K. M., et al. (2017). Long-term effects on cognitive trajectories of postmenopausal hormone therapy in two age groups. *The Journals of Gerontology. Series A, Biological Sciences and Medical Sciences*, 72, 838–845.

Fang, H., Disteche, C. M., & Berletch, J. B. (2019). X inactivation and escape: Epigenetic and structural features. *Frontiers in Cell and Developmental Biology*, 7, 219.

Farhy-Tselnicker, I., Boisvert, M. M., Liu, H., Dowling, C., Erikson, G. A., Blanco-Suarez, E., et al. (2021). Activity-dependent modulation of synapse-regulating genes in astrocytes. *ELife*, *10*. https://doi.org/10.7554/ELIFE.70514.

Farooqui, A. A., Farooqui, T., Panza, F., & Frisardi, V. (2011). Metabolic syndrome as a risk factor for neurological disorders. *Cellular and Molecular Life Sciences*, *69*, 741–762.

Ferretti, M. T., Iulita, M. F., Cavedo, E., Chiesa, P. A., Dimech, A. S., Chadha, A. S., et al. (2018). Sex differences in Alzheimer disease—The gateway to precision medicine. *Nature Reviews Neurology*, *14*, 457–469.

Ferretti, M. T., & Santuccione Chadha, A. (2021). The missing X factor in Alzheimer disease. *Nature Reviews. Neurology*, *17*, 727–728.

Galea, L. A. M. (2021). Chasing red herrings and wild geese: Sex differences versus sex dimorphism. *Frontiers in Neuroendocrinology*, *63*, 100940.

García-Calzón, S., Perfilyev, A., de Mello, V. D., Pihlajamäki, J., & Ling, C. (2018). Sex differences in the methylome and transcriptome of the human liver and circulating HDL-cholesterol levels. *The Journal of Clinical Endocrinology and Metabolism*, *103*, 4395.

Garcia-Segura, L. M., Azcoitia, I., & DonCarlos, L. L. (2001). Neuroprotection by estradiol. *Progress in Neurobiology*, *63*, 29–60.

Garcia-Sifuentes, Y., & Maney, D. L. (2021). Reporting and misreporting of sex differences in the biological sciences. *ELife*, *10*.

Gatev, E., Inkster, A. M., Negri, G. L., Konwar, C., Lussier, A. A., Skakkebaek, A., et al. (2021). Autosomal sex-associated co-methylated regions predict biological sex from DNA methylation. *Nucleic Acids Research*, *49*, 9097–9116.

Giacobini, E., & Pepeu, G. (2018). Sex and gender differences in the brain cholinergic system and in the response to therapy of Alzheimer disease with cholinesterase inhibitors. *Current Alzheimer Research*, *15*, 1077–1084.

Gilbert, T. M., Zürcher, N. R., Catanese, M. C., Tseng, C. E. J., Di Biase, M. A., Lyall, A. E., et al. (2019). Neuroepigenetic signatures of age and sex in the living human brain. *Nature Communications*, *10*.

Glezerman, M. (2016). Yes, there is a female and a male brain: Morphology versus functionality. *Proceedings of the National Academy of Sciences of the United States of America, 113*, E1971.

Global Burden of Disease Collabortative Network. (2021). *Global Burden of Disease Study 2019 (GBD 2019) Reference Life Table*. Seattle, United States of America: Institute for Health Metrics and Evaluation (IHME).

Gnatiuc, L., Herrington, W. G., Halsey, J., Tuomilehto, J., Fang, X., Kim, H. C., et al. (2018). Sex-specific relevance of diabetes to occlusive vascular and other mortality: a collaborative meta-analysis of individual data from 980 793 adults from 68 prospective studies. *The Lancet. Diabetes & Endocrinology, 6*, 538–546.

Gray, A. L., Hyde, T. M., Deep-Soboslay, A., Kleinman, J. E., & Sodhi, M. S. (2015). Sex differences in glutamate receptor gene expression in major depression and suicide. *Molecular Psychiatry, 20*, 1057–1068.

Grubman, A., Chew, G., Ouyang, J. F., Sun, G., Choo, X. Y., McLean, C., et al. (2019). A single-cell atlas of entorhinal cortex from individuals with Alzheimer's disease reveals cell-type-specific gene expression regulation. *Nature Neuroscience, 22*, 2087–2097.

Gulinello, M., & Smith, S. S. (2003). Anxiogenic effects of neurosteroid exposure: Sex differences and altered GABAA receptor pharmacology in adult rats. *Journal of Pharmacology and Experimental Therapeutics, 305*, 541–548.

Guneykaya, D., Ivanov, A., Hernandez, D. P., Haage, V., Wojtas, B., Meyer, N., et al. (2018). Transcriptional and translational differences of microglia from male and female brains. *Cell Reports, 24*, 2773–2783.e6.

Hamson, D. K., Wainwright, S. R., Taylor, J. R., Jones, B. A., Watson, N. V., & Galea, L. A. M. (2013). Androgens increase survival of adult-born neurons in the dentate gyrus by an androgen receptor-dependent mechanism in male rats. *Endocrinology, 154*, 3294–3304.

Hanamsagar, R., Alter, M. D., Block, C. S., Sullivan, H., Bolton, J. L., & Bilbo, S. D. (2017). Generation of a microglial developmental index in mice and in humans reveals a sex difference in maturation and immune reactivity. *Glia, 65*, 1504–1520.

Hanamsagar, R., & Bilbo, S. D. (2016). Sex differences in neurodevelopmental and neurodegenerative disorders: Focus on microglial function and neuroinflammation during development. *Journal of Steroid Biochemistry and Molecular Biology, 160*, 127–133. Pergamon.

Hanan, M., Simchovitz, A., Yayon, N., Vaknine, S., Cohen-Fultheim, R., Karmon, M., et al. (2020). A Parkinson's disease CircRNAs resource reveals a link between circSLC8A1 and oxidative stress. *EMBO Molecular Medicine, 12*.

Hayward, B. E., & Usdin, K. (2021). Mechanisms of genome instability in the Fragile X-related disorders. *Genes, 12*, 1633.

Hong, D. S., & Reiss, A. L. (2014). Cognitive and neurological aspects of sex chromosome aneuploidies. *The Lancet Neurology, 13*, 306–318.

Hooper, L. V., Littman, D. R., & Macpherson, A. J. (2012). Interactions between the microbiota and the immune system. *Science (New York, N.Y.), 336*, 1268.

Huo, N., Vemuri, P., Graff-Radford, J., Syrjanen, J., Machulda, M., Knopman, D. S., et al. (2022). Sex differences in the association between midlife cardiovascular conditions or risk factors with midlife cognitive decline. *Neurology, 98*, E623–E632.

Hurn, P. D., & Brass, L. M. (2003). Estrogen and stroke: A balanced analysis. *Stroke, 34*, 338–341.

Iadecola, C., Buckwalter, M. S., & Anrather, J. (2020). Immune responses to stroke: Mechanisms, modulation, and therapeutic potential. *The Journal of Clinical Investigation, 130*, 2777–2788.

Joel, D., Berman, Z., Tavor, I., Wexler, N., Gaber, O., Stein, Y., et al. (2015). Sex beyond the genitalia: The human brain mosaic. *Proceedings of the National Academy of Sciences of the United States of America, 112*, 15468–15473.

Johnson, W., Onuma, O., Owolabi, M., & Sachdev, S. (2016). Stroke: A global response is needed. *Bulletin of the World Health Organization*, *94*, 634.

Jun, T., Nirenberg, S., Weinberger, T., Sharma, N., Pujadas, E., Cordon-Cardo, C., et al. (2021). Analysis of sex-specific risk factors and clinical outcomes in COVID-19. *Communications Medicine*, *1*, 1–8.

Kautzky-Willer, A., Brazzale, A. R., Moro, E., Vrbíková, J., Bendlova, B., Sbrignadello, S., et al. (2012). Influence of increasing BMI on insulin sensitivity and secretion in normotolerant men and women of a wide age span. *Obesity (Silver Spring, Md.)*, *20*, 1966–1973.

Kautzky-Willer, A., Harreiter, J., & Pacini, G. (2016). Sex and gender differences in risk, pathophysiology and complications of type 2 diabetes mellitus. *Endocrine Reviews*, *37*, 278.

Kim, J. H., & Scialli, A. R. (2011). Thalidomide: The tragedy of birth defects and the effective treatment of disease. *Toxicological Sciences*, *122*, 1–6.

Kim, D. W., Yao, Z., Graybuck, L. T., Kim, T. K., Nguyen, T. N., Smith, K. A., et al. (2019). Multimodal analysis of cell types in a hypothalamic node controlling social behavior. *Cell*, *179*, 713–728.e17.

Klein, S. L., & Flanagan, K. L. (2016). Sex differences in immune responses. *Nature Reviews Immunology*, *16*, 626–638. Nature Publishing Group.

Kodama, L., Guzman, E., Etchegaray, J. I., Li, Y., Sayed, F. A., Zhou, L., et al. (2019). Microglial microRNAs mediate sex-specific responses to tau pathology. *Nature Neuroscience*, *23*, 167–171.

Kozomara, A., Birgaoanu, M., & Griffiths-Jones, S. (2019). miRBase: From microRNA sequences to function. *Nucleic Acids Research*, *47*, D155–D162.

Krause, D. N., Duckles, S. P., & Pelligrino, D. A. (2006). Influence of sex steroid hormones on cerebrovascular function. *Journal of Applied Physiology (Bethesda, Md. : 1985)*, *101*, 1252–1261.

Lee, J., Pinares-Garcia, P., Loke, H., Ham, S., Vilain, E., & Harley, V. R. (2019). Sex-specific neuroprotection by inhibition of the Y-chromosome gene, SRY, in experimental Parkinson's disease. *Proceedings of the National Academy of Sciences of the United States of America*, *116*, 16577–16582.

Li, E., & Zhang, Y. (2014). *DNA methylation in mammals* (p. 6). Cold Spring Harbor Perspectives in Biology.

Link, J. C., & Reue, K. (2017). Genetic basis for sex differences in obesity and lipid metabolism. *Annual Review of Nutrition*, *37*, 225–245.

Lisabeth, L., & Bushnell, C. (2012). Stroke risk in women: The role of menopause and hormone therapy. *The Lancet Neurology*, *11*, 82–91.

Lobentanzer, S., Hanin, G., Klein, J., & Soreq, H. (2019). Integrative transcriptomics reveals sexually dimorphic control of the cholinergic/neurokine interface in schizophrenia and bipolar disorder. *Cell Reports*, *29*, 764–777.e5.

Lonsdale, J., Thomas, J., Salvatore, M., Phillips, R., Lo, E., Shad, S., et al. (2013). The genotype-tissue expression (GTEx) project. *Nature Genetics*, *45*, 580–585.

Lopes-Ramos, C. M., Chen, C. Y., Kuijjer, M. L., Paulson, J. N., Sonawane, A. R., Fagny, M., et al. (2020). Sex differences in gene expression and regulatory networks across 29 human tissues. *Cell Reports*, *31*, 107795.

Louveau, A., Smirnov, I., Keyes, T. J., Eccles, J. D., Rouhani, S. J., Peske, J. D., et al. (2015). Structural and functional features of central nervous system lymphatic vessels. *Nature*, *7560*(523), 337–341.

Mahmoud, R., Wainwright, S. R., & Galea, L. A. M. (2016). Sex hormones and adult hippocampal neurogenesis: Regulation, implications, and potential mechanisms. *Frontiers in Neuroendocrinology*, *41*, 129–152.

Mallard, T. T., Liu, S., Seidlitz, J., Ma, Z., Moraczewski, D., Thomas, A., et al. (2021). X-chromosome influences on neuroanatomical variation in humans. *Nature Neuroscience, 24*, 1216–1224.

Manson, J. A. E., Chlebowski, R. T., Stefanick, M. L., Aragaki, A. K., Rossouw, J. E., Prentice, R. L., et al. (2013). The Women's Health initiative hormone therapy trials: Update and overview of health outcomes during the intervention and post-stopping phases. *JAMA : The Journal of the American Medical Association, 310*, 1353.

Markle, J. G. M., Frank, D. N., Mortin-Toth, S., Robertson, C. E., Feazel, L. M., Rolle-Kampczyk, U., et al. (2013). Sex differences in the gut microbiome drive hormone-dependent regulation of autoimmunity. *Science, 339*, 1084–1088.

Marrocco, J., & McEwen, B. S. (2016). Sex in the brain: Hormones and sex differences. *Dialogues in Clinical Neuroscience, 18*, 373.

Martinkova, J., Quevenco, F. C., Karcher, H., Ferrari, A., Sandset, E. C., Szoeke, C., et al. (2021). Proportion of women and reporting of outcomes by sex in clinical trials for Alzheimer disease: A systematic review and meta-analysis. *JAMA Network Open, 4*(9), e2124124. https://doi.org/10.1001/jamanetworkopen.2021.24124.

Mathys, H., Davila-Velderrain, J., Peng, Z., Gao, F., Mohammadi, S., Young, J. Z., et al. (2019). Single-cell transcriptomic analysis of Alzheimer's disease. *Nature, 570*(7761), 332–337.

Mauvais-Jarvis, F., Arnold, A. P., & Reue, K. (2017). A guide for the design of pre-clinical studies on sex differences in metabolism. *Cell Metabolism, 25*, 1216–1230.

Mauvais-Jarvis, F., Bairey Merz, N., Barnes, P. J., Brinton, R. D., Carrero, J. J., DeMeo, D. L., et al. (2020). Sex and gender: Modifiers of health, disease, and medicine. *The Lancet, 396*, 565–582. Lancet Publishing Group.

Mauvais-Jarvis, F., Clegg, D. J., & Hevener, A. L. (2013). The role of estrogens in control of energy balance and glucose homeostasis. *Endocrine Reviews, 34*, 309–338.

Mazur, A., Bills, E. H., Deschepper, K. M., Williamson, J. C., Henderson, B. J., & Risher, W. C. (2021). Astrocyte-derived thrombospondin induces cortical synaptogenesis in a sex-specific manner. *ENeuro, 8*. https://doi.org/10.1523/ENEURO.0014-21.2021.

McCartney, D. L., Zhang, F., Hillary, R. F., Zhang, Q., Stevenson, A. J., Walker, R. M., et al. (2019). An epigenome-wide association study of sex-specific chronological ageing. *Genome Medicine, 12*, 1–11.

McEwen, B. S., & Milner, T. A. (2017). Understanding the broad influence of sex hormones and sex differences in the brain. *Journal of Neuroscience Research, 95*, 24–39.

Mellon, S. H., & Griffin, L. D. (2002). Neurosteroids: Biochemistry and clinical significance. *Trends in Endocrinology & Metabolism, 13*, 35–43.

Migeon, B. R. (2020). X-linked diseases: Susceptible females. *Genetics in Medicine, 22*(7), 1156–1174.

Moore, L. D., Le, T., & Fan, G. (2013). DNA methylation and its basic function. *Neuropsychopharmacology : Official Publication of the American College of Neuropsychopharmacology, 38*, 23–38.

Morrison, H. W., & Filosa, J. A. (2016). Sex differences in astrocyte and microglia responses immediately following middle cerebral artery occlusion in adult mice. *Neuroscience, 339*, 85–99.

Mulvey, B., Bhatti, D. L., Gyawali, S., Lake, A. M., Kriaucionis, S., Ford, C. P., et al. (2018). Molecular and functional sex differences of noradrenergic neurons in the mouse locus coeruleus. *Cell Reports, 23*, 2225–2235.

Naqvi, S., Godfrey, A. K., Hughes, J. F., Goodheart, M. L., Mitchell, R. N., & Page, D. C. (2019). Conservation, acquisition, and functional impact of sex-biased gene expression in mammals. *Science (New York, N.Y.)*, 365. https://doi.org/10.1126/SCIENCE.AAW7317.

National Institutes of Health. (1994). *NIH Revitalization Act of 1993 Public Law 103-43*. Women and Health Research - NCBI Bookshelf. Retrieved May 15, 2022, from https://www.ncbi.nlm.nih.gov/books/NBK236531/.

National Institutes of Health. (2021). *NIH inclusion outreach toolkit: How to engage, recruit, and retain women in clinical research*. History of Women's Participation in Clinical Research. Retrieved from https://orwh.od.nih.gov/toolkit/recruitment/history#7%0A.

Nebel, R. A., Aggarwal, N. T., Barnes, L. L., Gallagher, A., Goldstein, J. M., Kantarci, K., et al. (2018). Understanding the impact of sex and gender in Alzheimer's disease: A call to action. *Alzheimer's & Dementia : The Journal of the Alzheimer's Association, 14*, 1171–1183.

Needham, B. D., Kaddurah-Daouk, R., & Mazmanian, S. K. (2020). Gut microbial molecules in behavioural and neurodegenerative conditions. *Nature Reviews Neuroscience, 12*(21), 717–731.

Neu, S. C., Pa, J., Kukull, W., Beekly, D., Kuzma, A., Gangadharan, P., et al. (2017). Apolipoprotein E genotype and sex risk factors for Alzheimer disease: A meta-analysis. *JAMA Neurology, 74*, 1178–1189.

Ngun, T. C., Ghahramani, N., Sánchez, F. J., Bocklandt, S., & Vilain, E. (2011). The genetics of sex differences in brain and behavior. *Frontiers in Neuroendocrinology, 32*, 227.

Nishizawa, S., Benkelfat, C., Young, S. N., Leyton, M., Mzengeza, S., De Montigny, C., et al. (1997). Differences between males and females in rates of serotonin synthesis in human brain. *Proceedings of the National Academy of Sciences of the United States of America, 94*, 5308–5313.

Novak, P., Kovacech, B., Katina, S., Schmidt, R., Scheltens, P., Kontsekova, E., et al. (2021). ADAMANT: A placebo-controlled randomized phase 2 study of AADvac1, an active immunotherapy against pathological tau in Alzheimer's disease. *Nature Aging, 6*(1), 521–534.

Nuutila, P., Knuuti, M. J., Mäki, M., Laine, H., Ruotsalainen, U., Teräs, M., et al. (1995). Gender and insulin sensitivity in the heart and in skeletal muscles. Studies using positron emission tomography. *Diabetes, 44*, 31–36.

Ochocka, N., Segit, P., Walentynowicz, K. A., Wojnicki, K., Cyranowski, S., Swatler, J., et al. (2021). Single-cell RNA sequencing reveals functional heterogeneity of glioma-associated brain macrophages. *Nature Communications, 12*.

Oliva, M., Muñoz-Aguirre, M., Kim-Hellmuth, S., Wucher, V., Gewirtz, A. D. H., Cotter, D. J., et al. (2020). The impact of sex on gene expression across human tissues. *Science (New York, N.Y.)*, 369.

Org, E., Mehrabian, M., Parks, B. W., Shipkova, P., Liu, X., Drake, T. A., et al. (2016). Sex differences and hormonal effects on gut microbiota composition in mice. *Gut Microbes, 7*, 313.

Ormerod, B. K., Lee, T. T. Y., & Galea, L. A. M. (2003). Estradiol initially enhances but subsequently suppresses (via adrenal steroids) granule cell proliferation in the dentate gyrus of adult female rats. *Journal of Neurobiology, 55*, 247–260.

Pacini, G., Dunkel, I., Mages, N., Mutzel, V., Timmermann, B., Marsico, A., et al. (2021). Integrated analysis of Xist upregulation and X-chromosome inactivation with single-cell and single-allele resolution. *Nature Communications, 12*(1), 1–17.

Palmisano, B. T., Zhu, L., Eckel, R. H., & Stafford, J. M. (2018). Sex differences in lipid and lipoprotein metabolism. *Molecular Metabolism, 15*, 45.

Pandya, M., Palpagama, T. H., Turner, C., Waldvogel, H. J., Faull, R. L., & Kwakowsky, A. (2019). Sex- and age-related changes in GABA signaling components in the human cortex. *Biology of Sex Differences, 10*(1), 1–16.

Paranjpe, M. D., Belonwu, S., Wang, J. K., Oskotsky, T., Gupta, A., Taubes, A., et al. (2021). Sex-specific cross tissue meta-analysis identifies immune dysregulation in women with Alzheimer's disease. *Frontiers in Aging Neuroscience, 13*, 622.

Paul, S. M., & Purdy, R. H. (1992). Neuroactive steroids. *FASEB Journal : Official Publication of the Federation of American Societies for Experimental Biology*, *6*, 2311–2322.
Peters, S. A. E., Carcel, C., Millett, E. R. C., & Woodward, M. (2020). Sex differences in the association between major risk factors and the risk of stroke in the UK Biobank cohort study. *Neurology*, *95*, e2715–e2726.
Peters, S. A. E., Huxley, R. R., & Woodward, M. (2014). Diabetes as a risk factor for stroke in women compared with men: A systematic review and meta-analysis of 64 cohorts, including 775 385 individuals and 12 539 strokes. *The Lancet*, *383*, 1973–1980.
Power, M. L., & Schulkin, J. (2008). Sex differences in fat storage, fat metabolism, and the health risks from obesity: Possible evolutionary origins. *The British Journal of Nutrition*, *99*, 931–940.
Prendergast, B. J., Onishi, K. G., & Zucker, I. (2014). Female mice liberated for inclusion in neuroscience and biomedical research. *Neuroscience & Biobehavioral Reviews*, *40*, 1–5.
Qureshi, I. A., & Mehler, M. F. (2010). Genetic and epigenetic underpinnings of sex differences in the brain and in neurological and psychiatric disease susceptibility. *Progress in Brain Research*, *186*, 77–95.
Reeves, M. J., Bushnell, C. D., Howard, G., Gargano, J. W., Duncan, P. W., Lynch, G., et al. (2008). Sex differences in stroke: Epidemiology, clinical presentation, medical care, and outcomes. *The Lancet Neurology*, *7*, 915–926.
Rich-Edwards, J. W., Kaiser, U. B., Chen, G. L., Manson, J. A. E., & Goldstein, J. M. (2018). Sex and gender differences research design for basic, clinical, and population studies: Essentials for investigators. *Endocrine Reviews*, *39*, 424–439.
Richler, C., Soreq, H., & Wahrman, J. (1992). X inactivation in mammalian testis is correlated with inactive X-specific transcription. *Nature Genetics*, *2*, 192–195.
Rickenbach, C., & Gericke, C. (2022). Specificity of adaptive immune responses in central nervous system health, aging and diseases. *Frontiers in Neuroscience*, *15*.
Rosas-Ballina, M., Ochani, M., Parrish, W. R., Ochani, K., Harris, Y. T., Huston, J. M., et al. (2008). Splenic nerve is required for cholinergic antiinflammatory pathway control of TNF in endotoxemia. *Proceedings of the National Academy of Sciences of the United States of America*, *105*, 11008–11013.
Rosenblatta, J. D. (2016). Multivariate revisit to "sex beyond the genitalia". *Proceedings of the National Academy of Sciences of the United States of America*, *113*, E1966–E1967.
Rossouw, J. E., Anderson, G. L., Prentice, R. L., LaCroix, A. Z., Kooperberg, C., Stefanick, M. L., et al. (2002). Risks and benefits of estrogen plus progestin in healthy postmenopausal women: Principal results From the Women's Health Initiative randomized controlled trial. *Journal of the American Medical Association*, *288*, 321–333.
Rossouw, J. E., Finnegan, L. P., Harlan, W. R., Pinn, V. W., Clifford, C., & McGowan, J. A. (1995). The evolution of the Women's Health Initiative: Perspectives from the NIH. *Journal of the American Medical Women's Association*, *1972*(50), 50–55.
Rurak, G. M., Simard, S., Freitas-Andrade, M., Green, J. R., Coppola, G., Lacoste, B., et al. (2022). Sex differences in developmental patterns of neocortical astroglia: A mouse translatome database. *Cell Reports*, *38*.
Russi, A. E., Ebel, M. E., Yang, Y., & Brown, M. A. (2018). Male-specific IL-33 expression regulates sex-dimorphic EAE susceptibility. *Proceedings of the National Academy of Sciences of the United States of America*, *115*, E1520–E1529.
Sandweiss, A. J., Brandt, V. L., & Zoghbi, H. Y. (2020). Advances in understanding of Rett syndrome and MECP2 duplication syndrome: Prospects for future therapies. *The Lancet Neurology*, *19*, 689–698.
Santos-Galindo, M., Acaz-Fonseca, E., Bellini, M. J., & Garcia-Segura, L. M. (2011). Sex differences in the inflammatory response of primary astrocytes to lipopolysaccharide. *Biology of Sex Differences*, *2*, 1–11.

Sarma, S., Sockalingam, S., & Dash, S. (2021). Obesity as a multisystem disease: Trends in obesity rates and obesity-related complications. *Diabetes, Obesity & Metabolism, 23*(Suppl. 1), 3–16.
Schurz, H., Salie, M., Tromp, G., Hoal, E. G., Kinnear, C. J., & Möller, M. (2019). The X chromosome and sex-specific effects in infectious disease susceptibility. *Human Genomics, 13*, 1–12.
Schwarz, J. M., Sholar, P. W., & Bilbo, S. D. (2012). Sex differences in microglial colonization of the developing rat brain. *Journal of Neurochemistry, 120*, 948–963.
Shansky, R. M., & Murphy, A. Z. (2021). Considering sex as a biological variable will require a global shift in science culture. *Nature Neuroscience, 24*, 457–464.
Sharma, S., & Eghbali, M. (2014). Influence of sex differences on microRNA gene regulation in disease. *Biology of Sex Differences, 5*, 3.
Shvetsova, E., Sofronova, A., Monajemi, R., Gagalova, K., Draisma, H. H. M., White, S. J., et al. (2018). Skewed X-inactivation is common in the general female population. *European Journal of Human Genetics, 27*(3), 455–465.
Sigel, E., & Steinmann, M. E. (2012). Structure, function, and modulation of GABAA receptors. *Journal of Biological Chemistry, 287*, 40224–40231.
Simchovitz-Gesher, A., & Soreq, H. (2020). Pharmaceutical implications of sex-related RNA divergence in psychiatric disorders. *Trends in Pharmacological Sciences, 41*, 840–850.
Skaletsky, H., Kuroda-Kawaguchl, T., Minx, P. J., Cordum, H. S., Hlllier, L. D., Brown, L. G., et al. (2003). The male-specific region of the human Y chromosome is a mosaic of discrete sequence classes. *Nature, 423*, 825–837.
Skuse, D., Printzlau, F., & Wolstencroft, J. (2018). Sex chromosome aneuploidies. *Handbook of Clinical Neurology, 147*, 355–376.
Spritzer, M. D., & Galea, L. A. M. (2007). Testosterone and dihydrotestosterone, but not estradiol, enhance survival of new hippocampal neurons in adult male rats. *Developmental Neurobiology, 67*, 1321–1333.
Spychala, M. S., Honarpisheh, P., & McCullough, L. D. (2017). Sex differences in neuroinflammation and neuroprotection in ischemic stroke. *Journal of Neuroscience Research, 95*, 462–471.
Steinberg, J. R., Turner, B. E., Weeks, B. T., Magnani, C. J., Wong, B. O., Rodriguez, F., et al. (2021). Analysis of female enrollment and participant sex by burden of disease in US clinical trials between 2000 and 2020. *JAMA Network Open, 4*, e2113749.
Strazzullo, P., D'Elia, L., Cairella, G., Garbagnati, F., Cappuccio, F. P., & Scalfi, L. (2010). Excess body weight and incidence of stroke: Meta-analysis of prospective studies with 2 million participants. *Stroke, 41*.
Takahashi, T., Ellingson, M. K., Wong, P., Israelow, B., Lucas, C., Klein, J., et al. (2020). Sex differences in immune responses that underlie COVID-19 disease outcomes. *Nature, 588*(7837), 315–320.
Tamming, R. J., Dumeaux, V., Jiang, Y., Shafiq, S., Langlois, L., Ellegood, J., et al. (2020). Atrx deletion in neurons leads to sexually dimorphic dysregulation of miR-137 and spatial learning and memory deficits. *Cell Reports, 31*, 107838.
Tura, A., Pacini, G., Moro, E., Vrbíková, J., Bendlová, B., & Kautzky-Willer, A. (2018). Sex- and age-related differences of metabolic parameters in impaired glucose metabolism and type 2 diabetes compared to normal glucose tolerance. *Diabetes Research and Clinical Practice, 146*, 67–75.
Tyagi, A., Mirita, C., Taher, N., Shah, I., Moeller, E., Tyagi, A., et al. (2020). Metabolic syndrome exacerbates amyloid pathology in a comorbid Alzheimer's mouse model. *Biochimica et Biophysica Acta. Molecular Basis of Disease, 1866*.
Van Nas, A., Guhathakurta, D., Wang, S. S., Yehya, N., Horvath, S., Zhang, B., et al. (2009). Elucidating the role of gonadal hormones in sexually dimorphic gene coexpression networks. *Endocrinology, 150*, 1235–1249.

Villa, A., Gelosa, P., Castiglioni, L., Cimino, M., Rizzi, N., Pepe, G., et al. (2018). Sex-specific features of microglia from adult mice. *Cell Reports*, *23*, 3501.

von Bartheld, C. S., Bahney, J., & Herculano-Houzel, S. (2016). The search for true numbers of neurons and glial cells in the human brain: A review of 150 years of cell counting. *The Journal of Comparative Neurology*, *524*, 3865.

von Vaupel-Klein, A. M., & Walsh, R. J. (2021). Considerations in genetic counseling of transgender patients: Cultural competencies and altered disease risk profiles. *Journal of Genetic Counseling*, *30*, 98–109.

Wang, M. (2011). Neurosteroids and GABA-A receptor function. *Frontiers in Endocrinology*, *2*, 44. Frontiers.

Wang, J., Sareddy, G. R., Lu, Y., Pratap, U. P., Tang, F., Greene, K. M., et al. (2020). Astrocyte-derived estrogen regulates reactive astrogliosis and is neuroprotective following ischemic brain injury. *Journal of Neuroscience*, *40*, 9751–9771.

Weber, C. M., & Clyne, A. M. (2021). Sex differences in the blood-brain barrier and neurodegenerative diseases. *APL Bioengineering*, *5*, 11509. American Institute of Physics Inc.

Wickens, M. M., Bangasser, D. A., & Briand, L. A. (2018). Sex differences in psychiatric disease: A focus on the glutamate system. *Frontiers in Molecular Neuroscience*, *11*, 197.

Wierman, M. E. (2007). Sex steroid effects at target tissues: mechanisms of action. *Advances in Physiology Education*, *31*, 26–33.

Wise, A. L., Gyi, L., & Manolio, T. A. (2013). eXclusion: Toward integrating the X chromosome in genome-wide association analyses. *American Journal of Human Genetics*, *92*, 643.

World Health Organisation. (2022). *Gender and health*. Retrieved from https://www.who.int/health-topics/gender.

Wu, C. T., & Morris, J. R. (2001). Genes, genetics, and epigenetics: A correspondence. *Science*, *293*, 1103–1105.

Yagi, S., & Galea, L. A. M. (2019). Sex differences in hippocampal cognition and neurogenesis. *Neuropsychopharmacology*, *44*, 200–213.

Yan, J., Read, S. J., Henderson, R. D., Hull, R., O'Sullivan, J. D., McCombe, P. A., et al. (2012). Frequency and function of regulatory T cells after ischaemic stroke in humans. *Journal of Neuroimmunology*, *243*, 89–94.

Yang, W., Warrington, N. M., Taylor, S. J., Whitmire, P., Carrasco, E., Singleton, K. W., et al. (2019). Sex differences in GBM revealed by analysis of patient imaging, transcriptome, and survival data. *Science Translational Medicine*, *11*.

Yanguas-Casás, N., Crespo-Castrillo, A., de Ceballos, M. L., Chowen, J. A., Azcoitia, I., Arevalo, M. A., et al. (2018). Sex differences in the phagocytic and migratory activity of microglia and their impairment by palmitic acid. *Glia*, *66*, 522–537.

Yasuda, K., Maki, T., Kinoshita, H., Kaji, S., Toyokawa, M., Nishigori, R., et al. (2020). Sex-specific differences in transcriptomic profiles and cellular characteristics of oligodendrocyte precursor cells. *Stem Cell Research*, *46*, 101866.

Young, A. M. H., Kumasaka, N., Calvert, F., Hammond, T. R., Knights, A., Panousis, N., et al. (2021). A map of transcriptional heterogeneity and regulatory variation in human microglia. *Nature Genetics*, *53*, 861–868.

Yurkovetskiy, L., Burrows, M., Khan, A. A., Graham, L., Volchkov, P., Becker, L., et al. (2013). Gender bias in autoimmunity is influenced by microbiota. *Immunity*, *39*, 400–412.

Yusipov, I., Bacalini, M. G., Kalyakulina, A., Krivonosov, M., Pirazzini, C., Gensous, N., et al. (2020). Age-related DNA methylation changes are sex-specific: a comprehensive assessment. *Aging (Albany NY)*, *12*, 24057.

Zajitschek, S. R. K., Zajitschek, F., Bonduriansky, R., Brooks, R. C., Cornwell, W., Falster, D. S., et al. (2020). Sexual dimorphism in trait variability and its eco-evolutionary and statistical implications. *ELife*, *9*, 1–17.
Zhang, X., Zhong, H., Li, Y., Shi, Z., Ren, H., Zhang, Z., et al. (2021). Sex- and age-related trajectories of the adult human gut microbiota shared across populations of different ethnicities. *Nature Aging*, *1*(1), 87–100.
Zhao, L., Mao, Z., Woody, S. K., & Brinton, R. D. (2016). Sex differences in metabolic aging of the brain: Insights into female susceptibility to Alzheimer's disease. *Neurobiology of Aging*, *42*, 69–79.

CHAPTER THREE

Sex differences in neurovascular disorders

Cheryl Carcel[a,b], Valeria Caso[c], Diana Aguiar de Sousa[d,e], and Else Charlotte Sandset[f,g,*]

[a]The George Institute for Global Health, Faculty of Medicine, University of New South Wales, Sydney, NSW, Australia
[b]Sydney School of Public Health, Sydney Medical School, The University of Sydney, Sydney, NSW, Australia
[c]Stroke Unit, Santa Maria della Misericordia Hospital University of Perugia, Perugia, Italy
[d]Stroke Center, Lisbon Central University Hospital, Lisbon, Portugal
[e]CEEM and Institute of Anatomy, Faculty of Medicine, University of Lisbon, Lisbon, Portugal
[f]Department of Neurology, Oslo University Hospital, Oslo, Norway
[g]The Norwegian Air Ambulance Foundation, Oslo, Norway
*Corresponding author: e-mail address: else@sandset.net

Contents

Abstract

Neurovascular disorders is a heterogenous group of diseases, including one of the most time critical disorders in emergency medicine; stroke. Sex differences are extensively described in neurovascular disorders, ranging from differences in symptom presentation, risk factors, treatment and outcomes. For example, women with stroke, more

International Review of Neurobiology, Volume 164
ISSN 0074-7742
https://doi.org/10.1016/bs.irn.2022.07.001

often present with generalized weakness, reduced consciousness and headache than men. Furthermore, there are differences in risk factors, outcomes and in the effect of secondary prevention. Women have a higher risk of cerebral venous thrombosis and developing cerebral aneurysms. In general, women have been underrepresented in trials on neurovascular disorders. This chapter provides an extensive overview of sex differences in stroke in general and in the differences specially seen in ischemic stroke, intracerebral hemorrhage, subarachnoid hemorrhage and in cerebral venous thrombosis.

1. Introduction

Stroke is defined as rapidly developing clinical signs of focal (or global) disturbance of cerebral function, lasting more than 24 h or leading to death, with no apparent cause other than of vascular origin. Stroke can be caused either by a blood clot (ischemic stroke) or a bleed in the brain (hemorrhagic stroke). Ischemic stroke accounts for around 85% of the stroke worldwide. Hemorrhagic stroke can be further subdivided into intracerebral hemorrhage, which is a sudden bleeding into brain parenchyma, its ventricles or both, or subarachnoid hemorrhage, defined as sudden bleeding in the space surrounding the brain.

Stroke continues to be a leading cause of death and disability worldwide, with women being more adversely affected by the global burden of stroke. According to the Global Burden of Disease study (Feigin et al., 2018), the highest lifetime risk of stroke among women was in both Eastern Europe and East Asia (Feigin et al., 2018). Age-specific stroke rates are higher in men, but women experience more frequent stroke events because of their long life expectancy and high stroke incidence at older ages. Data from other studies have suggested that the incidence of ischemic stroke is more significant in men than women until the age of 65 when the incidence of stroke equalizes between the sexes. Moreover, population-based data have shown that more than one-third of strokes in women occur after age 75, during which time there are three females for every two male patients with strokes (Vike, 1994). Strokes occurring in later lifetime lead to a less favorable outcome. Increasingly strokes occur in younger women (Ekker et al., 2019; Leppert et al., 2020), which is devastating for young stroke survivors and their families.

These worse stroke outcomes in women may be due to disease pathophysiology, access to care, provider bias, and socioeconomic status, in addition to biological differences that may explain the different manifestations of similar diseases in men and women, and the role of estrogen and other sex hormones likely plays a significant role in this regard.

2. Risk factors

The most significant non-specific modifiable stroke risk factors include hypertension, cardiac diseases, including atrial fibrillation (AF), smoking habit, diabetes mellitus and lipid metabolism disorders, alcohol abuse, being overweight, limited physical activity and emotional stress. However, the incidence and impact of these stroke risk factors in women are different compared to males, often due to women's reproductive status. Moreover, hypertension, AF, diabetes mellitus, abdominal obesity, migraine with aura, emotional stress and depression are more common risk factors in women. The stroke risk factors unique to women include age at menarche, pregnancy, gestational diabetes mellitus, hypertensive disorders of pregnancy (HDP) including preeclampsia, hormonal status disturbances, the current use of combined oral contraceptive pills and hormone replacement therapy.

2.1 General risk factors

Hypertension is more common in females older than 65 years of age (Howard et al., 2019) The case control study INTERSTROKE reported that overall, hypertension had a higher population attributable risk (PAR) in women (PAR 52.3% in women vs 45.2% in men) (O'Donnell et al., 2016). From the REGARDS study, the association between increasing hypertension severity and ischemic stroke incident was almost twice as significant in women compared with men (Madsen et al., 2019). Researchers from the REGARDS study found that for every 10 mm Hg increase in blood pressure (BP), the risk of stroke diverged between white women and men, and the risk of stroke across increasing levels of BP was about twice as high in women than men. Different biologic and/or hormonal mechanisms between sexes linking hypertension to vascular dysfunction, sex-specific synergistic effects between hypertension and other risk factors such as diabetes that contribute to vascular dysfunction and thus increase stroke risk, a difference in hypertension treatment and adherence were hypothesized as explanations (Madsen et al., 2019).

During pregnancy, whether or not secondary to pre-existing disease, having preeclampsia or eclampsia can be translated into a six- to ninefold increased long-term risk of stroke compared with women normotensive during pregnancy (Keteepe-Arachi & Sharma, 2017). In a prospective cohort study (California Teachers Study), enrolling 83,749 women, 4.9% had hypertensive disorders of pregnancy (HDP). Prior HDP was associated with a long-term risk of stroke, which was reduced by aspirin use (Miller et al., 2019). Current recommendations on antihypertensive treatment do not differ between women and men with regards to thresholds for initiating treatment, targets nor the choice of antihypertensive medication. A systematic review with meta-analysis of systematically investigated sex differences in cardiovascular medication prescription among patients at high risk or with established cardiovascular disease in primary care revealed that aspirin, statins, and angiotensin-converting enzyme inhibitors were less often prescribed in women then in men (Zhao et al., 2020). Fewer treated women achieve BP control compared to men, so further research should uncover reasons for why this is the case, e.g., less intense treatment in women, inappropriate drug choices, lack of compliance, treatment resistance, biological factors that account for hormonal status, and perhaps others (Bushnell et al., 2014).

Atrial fibrillation: The risk of stroke from AF increases with age. More women over the age of 75 years have AF compared to men. While men with AF have larger burden of coronary artery disease, women with AF tend to have a larger prevalence of heart failure and valvular heart disease (Kavousi, 2020). A meta-analysis involving 30 studies ($n = 4,371,714$) found that AF was associated with a higher risk of all-cause mortality in women and a significantly stronger risk of stroke, cardiovascular mortality, and cardiac events. The higher stroke risk persisted even after adjusting for the higher burden of hypertension and prior stroke in women (Emdin et al., 2016). Compared with men, women with AF also have greater risks of dependency and recurrence 36 months after a stroke after adjustment for stroke subtype, severity, and risk factors (Hong et al., 2017). Moreover, not only do women with AF have a greater risk of stroke than men, the AF-related stroke is also more severe (Wagstaff, Overvad, Lip, & Lane, 2014).

Migraine with aura: Migraine overwhelmingly affects women. Migraine with aura-associated ischemic stroke is more common in young (<45 years old) women with high frequency of migraine attacks, hormonal contraception use, and with pregnancy and preeclampsia. There is increasing evidence that cardioembolism, often in conjunction with thrombophilia, plays a prominent role in migraine with aura-associated cerebral infarction (Tietjen & Maly, 2020). In light of the higher risk of stroke in people with migraine with aura, it is important to identify and modify any vascular risk factor. There is currently no direct evidence to support that a migraine prophylactic treatment can reduce future stroke in people with migraine (Øie, Kurth, Gulati, & Dodick, 2020).

Obesity, abdominal obesity, and metabolic syndrome: The relative risk of ischemic stroke associated with metabolic syndrome (MetS) is higher in women than men, as shown a meta-analysis of 16 prospective cohort studies with 116,496 participants who were initially free of cardiovascular diseases (Li et al., 2017). Comparing the persons without MetS, those with MetS have a significantly higher risk of incident stroke, and the pooled relative risk (RR) was 1.70 (95% confidence interval (CI): 1.49–1.95). Subgroup analyses suggested that women were more sensitive to this effect (with an RR of 1.83, 95% CI: 1.31–2.56) than men (RR=1.47 (95% CI: 1.22–1.78). And those with MetS have a significantly higher risk of ischemic stroke (RR=2.12, 95% CI: 1.46–3.08) than hemorrhagic stroke (RR=1.48, 95% CI: 0.98–2.24) (Li et al., 2017). The Women's Health Initiative study randomized 48,835 postmenopausal women (who generally have an increased risk of stroke) to a low-fat dietary pattern intervention (targeting a maximum of 20% of energy from fat) or to a usual diet for 8.3 years. During the cumulative and post-intervention periods, there was no difference between groups in the rate of stroke among all 48,835 women (HR 1.00, 95% CI 0.91–1.10), the 1656 (3%) women with previous cardiovascular disease (HR 1.05, 95% CI 0.76–1.15), or the 19,984 (41%) women with hypertension but without previous cardiovascular disease (HR 0.92, 95% CI 0.81–1.04) (Howard et al., 2018).

Diabetes mellitus: In a systematic review and meta-analysis from 64 cohort studies, women with diabetes mellitus had a 27% higher risk of stroke than men with diabetes mellitus (Madsen et al., 2019). Female sex may increase the risk of fatal stroke among those with diabetes mellitus. In the UK Prospective Diabetes Study, women with diabetes mellitus had a twofold higher risk of fatal stroke than men with diabetes mellitus, and female sex remained a strong predictor of fatality even after adjustment for age, glucose control, and BP (Stevens et al., 2004). The risk of fatal stroke in women with diabetes mellitus is equivalent to that of women with prior stroke (Ho, Paultre, & Mosca, 2003). Despite evidence that diabetes mellitus carries more stroke risk for women than men, some data indicate that women are less likely than men to have their glycated hemoglobin controlled (defined by levels <7%) (Franzini et al., 2013), while other data conflict, showing no sex difference in glycated hemoglobin levels (Wong et al., 2016).

Dyslipidemia: In terms of treatment disparities, some data have demonstrated that women are at increased risk of not receiving statins for hyperlipidemia or not reaching recommended cholesterol goals (Spinler et al., 2015). Analysis of the national registry data on 5693 people eligible for statin treatment based on cholesterol management guidelines from the American Heart Association and American College of Cardiology revealed that women were considerably less likely than men—67.0% compared to 78.4%—to be treated with any statin therapy. Among the reasons cited, women were not offered the medication as often as men, were more likely to decline statin therapy when it was recommended and were more inclined to discontinue the treatment after starting it, usually citing side effects. Otherwise, women and men had different beliefs about the role of statins in relation to cardiovascular disease, the research found. Women were more likely to worry about having a heart attack or stroke—yet less likely to believe that having high cholesterol could contribute to one. Sex differences remained even after researchers adjusted for age, race and other demographics, as well as socioeconomic factors and health history. Sex differences were consistent across primary and secondary prevention indications for statin treatment (Nanna et al., 2019).

Carotid Stenosis: In recent analyses of another database comprising nearly 3.5 million patients in the United States who underwent voluntary screening for vascular disease, 3.9% were found to have carotid stenosis (Berger et al., 2013; Razzouk et al., 2015; Rockman et al., 2013).

Independent risk factors for ICA stenosis included advanced age, smoking, peripheral arterial disease, hypertension, coronary artery disease, diabetes, hypercholesterolemia, and abdominal aortic aneurysm. Women had a 3.4% prevalence of carotid stenosis of 50% or more, whereas males had a 4.2% prevalence (Rockman et al., 2013). Prevalence increased dramatically with age in both sexes (Sidawy & Perler, 2019).

A report from Berger et al. analyzed modifiable risk factors and their relationship with carotid artery stenosis (Berger et al., 2013). Risk factors included hypertension, hypercholesterolemia, smoking, diabetes, and sedentary lifestyle. The multivariate adjusted prevalence of carotid artery stenosis with 0 to 5 risk factors was, respectively, 1.41%, 2.36%, 3.72%, 5.73%, 8.48%, and 11.58%. Results were similar in men and women. For every additional modifiable risk factor that was present, the multivariate-adjusted odds of having carotid disease increased significantly (odds ratio, 1.57) (Berger et al., 2013).

2.2 Women specific stroke risk factors

Women-specific stroke risk factors are pregnancy, hypertensive disorders of pregnancy including preeclampsia, gestational diabetes mellitus (GDM), OCs use and postmenopausal hormonal replacement therapy. A meta-analysis of 3,488,160 women, including 198,252 with pre-eclampsia reported that after 10–15 years, women with pre-eclampsia had a 3.7-fold risk of hypertension, 2.2-fold risk of ischemic heart disease, 1.8-fold risk of stroke, and 1.5-fold risk of overall mortality (Bellamy, Casas, Hingorani, & Williams, 2007; de Havenon et al., 2021).

Analysis from the cohort study of 1435 women (169 had a history of preeclampsia), who participated in the Framingham Heart Study has shown that at baseline, women with preeclampsia were more likely to be younger, to be receiving cholesterol-lowering medications, to have lower cholesterol and higher diastolic BP, and to currently smoke. After adjustment for all vascular risk factors over the life course statistical analysis revealed that women with a history of preeclampsia had a higher risk of stroke in later life compared with women without a history of preeclampsia (relative risk, 3.79; 95% CI, 1.24–11.60) (Kramer, Campbell, & Retnakaran, 2019).

It is well known that women, who develop GDM have an elevated risk of type 2 diabetes in the years thereafter. Moreover, the recent studies show that women with GDM also have an increased future risk of cardiovascular disease. The meta-analysis of nine studies, involving more than 5 million women has revealed that women with GDM have a twofold higher risk of future cardiovascular events (RR 1.98 [95% CI 1.57, 2.50]). GDM conferred a 2.3-fold increased risk of cardiovascular events in the first decade postpartum (RR 2.31 [95% CI 1.57, 3.39]) (Kramer et al., 2019).

A Cochrane review showed that systemic hormonal contraceptive users had a 1.6 times higher relative risk for ischemic stroke and myocardial infarction, and more so with higher doses of estrogen (Roach et al., 2015). The risk of ischemic stroke in patients using combined oral contraceptives (OC) is increased in patients with additional stroke risk factors, including smoking, hypertension, and migraine with aura. Analysis of 179 female patients of <50 years has shown that contraceptive users with ischemic strokes are younger and have less tobacco use and migraine with aura and more hyperlipidemia. Their stroke mechanism is more often undetermined using a standardized work-up, and their adjusted long-term outcome is more favorable with less stroke recurrence (Correia et al., 2021).

3. Symptom presentation

Timely diagnosis of ischemic and hemorrhagic stroke saves time and leads to life-saving treatment. Early recognition of stroke symptoms is the critical first step in instituting rapid diagnostics and subsequent treatment. Cerebrovascular diseases have a wide range of symptoms including motor, sensory and language deficits but it is often the non-focal symptoms such as headache, pain and confusion that are more challenging to diagnose. In a hospital-based registry of patients transient ischemic attack (TIA) and non-disabling ischemic minor stroke in The Netherlands, 19% ($n=1265$) of patients had one or more non-focal symptoms (Plas et al., 2016). Patients with these underrecognized symptoms are at high risk for vascular events (Koudstaal et al., 1992).

Sex differences in presenting symptoms of women and men have been reported but evidence is lacking whether this influences a delay or missed diagnosis. A meta-analysis of 60 studies ($n=582{,}844$; 50% women) found sex differences in presentation of stroke in patients suspected of ischemic or hemorrhagic stroke or transient ischemic attack. In women, headache (pooled odds ratio [OR], 1.24 [95% CI, 1.11–1.39), changes in consciousness (OR, 1.38 [95% CI, 1.19–1.61]) and coma/stupor (OR, 1.39 [95% CI, 1.25–1.55]) occurred more often in men (Ali et al., 2022). Another recent systematic review and meta-analysis examined presenting symptoms of adult women and men with diagnosed stroke or TIA (Shajahan et al., 2022). Analysis from pooled results from 21 eligible studies showed that women and men presented with similar mean percentage of motor deficit (56% in women vs 56% in men) and speech deficit (41% in women vs 40% in men). Women had higher risk of presenting with a few non-focal symptoms: generalized weakness (relative risk [RR] 1.49, confidence interval [CI] 1.09–2.03); mental status change (RR 1.44 [1.22–1.71]); fatigue (RR 1.42 [1.05–1.92]) and loss of consciousness (RR 1.30 [1.12–1.51]). In contrast, women had lower risk of presenting with dysarthria (RR 0.89 [0.82–0.95]); dizziness (RR 0.87 [0.80–0.95]); gait disturbance (RR 0.79 [0.65–0.97]); and imbalance (RR 0.68 [0.57–0.81]). In this systematic review, there was only one study that linked sex differences in symptom presentation to having a definite stroke or TIA diagnosis. Women presenting with pain and unilateral sensory loss were less likely to receive a stroke/TIA diagnosis. This study highlights an important evidence gap on the link of sex differences in stroke symptoms to delayed or missed diagnosis.

There are a few potential explanations for these findings. Differences in patterns of stroke such as location, size and presence of white matter changes or other small vessel disease are seen in women and men. Using neuroimaging data, a study performed probabilistic lesion-symptom mapping technique with empirical lesion data from two large independent cohorts (555 for derivation and 503 for validation) of AIS patients (Bonkhoff, Karch, Weber, Wellmann, & Berger, 2021; Bonkhoff, Schirmir, et al., 2021). Eloquent lesion patterns were more widespread in female patients, implying that more regions contributed to stroke severity in women. These sex differences were more pronounced for lesions affecting some left-hemispheric regions such as: thalamus, hippocampus, and occipital cortical brain regions. The lesions in these regions were associated with a higher stroke severity in women only, indicating sex-specific distributions of functional deficits.

Another possible reason could be that age, cardiovascular risk factors and etiology of stroke may contribute to sex differences in symptom presentation. Older women have higher prevalence of atrial fibrillation, leading to a higher incidence of cardioembolic strokes which can result in more severe neurological deficits and cortical findings (Wang et al., 2020).

As we have seen in the most recent meta-analysis (Shajahan et al., 2022), the rank order of the symptoms of stroke were similar between the sexes. This suggests that symptoms that most often lead to a diagnosis of stroke are not different between the sexes. This is important for the applicability of public awareness campaigns to increase recognition of stroke in the community, irrespective of sex. In addition, while most studies in the meta-analysis evaluated individual symptoms, in clinical practice, patients often present with more than one symptom and these are evaluated as a whole.

4. Ischemic stroke

4.1 Biologic mechanisms of sex differences

The role of estrogen had been investigated as one of the key variables in sex differences in stroke. In pre-menopausal women, 17-β estradiol (E2) have been found to attenuate stroke effects (Girijala, Sohrabji, & Bush, 2016). Similar findings were seen in experimental models of cerebral ischemia where female rodents have smaller stroke volumes than males. The advantage is eliminated in ovariectomized rodents and restored by supplemental estrogen (Oertelt-Prigione & Regitz-Zagrosek, 2012). The protective effect of estrogen includes binding to vascular smooth muscle and endothelial receptors to facilitate vasodilatation and blood flow improvement; and induction of anti-inflammatory factors (Oertelt-Prigione & Regitz-Zagrosek, 2012).

In models of cerebral ischemia, sex differences in neuronal dysfunction and cell death are seen after primary ischemic damage (Roy-O'Reilly & McCullough, 2018). The neural cell mitochondria, crucial in maintaining cellular energy and regulating oxidative process, is disrupted during this process. When estrogen and progestin are present, they enhance mitochondrial respiration and augment antioxidative processes in neurons making female animals more efficient in oxidative respiration even after ischemic stroke. Ovariectomy in females impairs mitochondrial respiration and antioxidant capabilities, which can be reversed by the addition of exogenous 17β-estradiol. Upon energy failure, the activation of cell death pathways occurs. In ischemic stroke, there are important sex differences: cell death in females is dominated by activation of caspases making caspase-inhibiting therapies preferentially protective in female animals after experimental stroke. On the other hand, cell death in males depends on the formation of nitric oxide and activation of poly-ADP ribose polymerase 1 (PARP-1), causing mitochondrial depolarization and cell death. Inhibition of this pathway in animal models is protective only in males (Roy-O'Reilly & McCullough, 2018).

Another theory for biological differences includes chromosomal influences. In a study of aged mice with depleted or depleting estrogen, those with two X chromosome had larger infarcts, higher neurologic deficits and greater immune-cell infiltration and activation compared to those with an X chromosome and a Y chromosome. The investigators suggested that the second X chromosome has an unfavorable effect to the genetic makeup, which only became evident with senescence (Bushnell et al., 2018; Oertelt-Prigione & Regitz-Zagrosek, 2012).

4.2 Acute treatment of ischemic stroke

Intravenous thrombolysis (IVT) and endovascular thrombectomy (EVT) has been shown to improve 90-day functional outcomes and reduce disability among carefully selected patients with ischemic stroke. The utilization of IVT and EVT has drastically increased following incorporation of these treatments into guidelines. Sex differences have been reported in access and efficacy.

An updated meta-analysis reported sex-specific IVT with recombinant tissue plasminogen activator treatment rates for acute ischemic stroke studies published between 2008 and 2018. Studies were eligible if they were representative populations of patients with ischemic stroke such as those from hospital-based, registry-based, or administrative data (Strong, Lisabeth, & Reeves, 2020). Women had 13% lower odds of receiving IV rtPA treatment compared to men (summary unadjusted OR 0.87 [95% CI, 0.82–0.93]). Examination of time trends across 33 studies showed that the summary ORs for the earlier time periods were significantly different (Z test $P < 0.001$) wherein the 2000–2004 estimate (OR, 0.73 [95% CI, 0.67–0.80]) had lower odds compared to the 2014–2018 period (OR, 0.90 [95% CI, 0.84–0.97]). This narrowing of sex differences in IVT access throughout the years can be corroborated by the first meta-analysis (studies published between 2000 and 2008) that found women were 30% less likely to receive IVT (Reeves, Bhatt, Jajou, Brown, & Lisabeth, 2009). Despite this narrowing, treatment differences in IVT use are likely to translate into a large number of untreated women due to the high incidence of ischemic stroke in women.

Secondary analysis of large, international, randomized controlled trials comparing IV rtPA to placebo in ischemic stroke have not shown any sex differences in clinical outcome (Bushnell et al., 2018). Being post-hoc analysis, these studies were not specifically designed to test sex differences in efficacy; however, the findings are reassuring for both women and men.

Since the publication of the pivotal trials in 2015, a few studies have reported sex differences in the use of and efficacy of EVT. A recent retrospective, serial, cross-sectional study reported 10-year trends in the use of thrombolysis and thrombectomy in a random sample of all discharges from United States community hospitals (Otite et al., 2021). This data from the National Inpatient Sample included data prior to EVT becoming part of guideline recommendations and 2 years immediately after (2008–2017). Compared to men, women had noted greater odds of utilization of EVT among all patients in the later periods of the sample (2016–2017). In another study of over 1.11 million patients who were hospitalized for first or recurrent ischemic stroke in Germany from 2013 to 2017, sex differences were explored in in a nationwide administrative database (Bonkhoff, Karch, et al., 2021; Bonkhoff, Schirmer, et al., 2021). Women were found to have had an overall 26% (95% CI, 22%–30%) higher chance of receiving EVT compared to men. The reasons for this sex disparity are unclear and needs more research.

When Multicenter Randomized Clinical Trial of Endovascular Treatment for Acute Ischemic Stroke in the Netherlands (MR CLEAN trial) was first analyzed for sex differences, it was found not to be effective in women. Of the 500 patients (42% women), (de Ridder et al., 2016) the treatment effect (functional outcome at 90 days) was null in women (OR 0.99 [95% CI 0.60–1.66]) compared to protective in men (OR 2.39 [95% CI 1.55–3.68]). Women in the intervention group were also more likely to die, as well as have serious adverse events, than men in MR CLEAN. When this dataset was pooled together with five other randomized trials on large vessel ischemic stroke endovascular thrombectomy in an individual patient data meta-analysis,(Goyal et al., 2016) women and men equally benefitted from the treatment and there were no differences in on serious adverse events.

4.3 Secondary prevention after ischemic stroke

4.3.1 Antiplatelets and statins

Antiplatelet agents are considered to be equally effective for both sexes in primary and secondary stroke prevention (Acciarresi et al., 2011; Caso, Santalucia, Acciarresi, Pezzella, & Paciaroni, 2012). Whether statins have sex-related difference in terms of stroke protection is still unknown; however, SPARCL trial of high-dose statins for secondary stroke prevention did not find any sex by treatment interactions (Goldstein et al., 2008).

4.3.2 Anticoagulation in AF

Compared with men, women with AF also have greater risks of dependency and recurrence 36 months after a stroke after adjustment for stroke subtype, severity, and risk factors (Hong et al., 2017). Moreover, not only do women with AF have a greater risk of stroke than men, the AF-related stroke is also more severe (Wagstaff et al., 2014). Regarding treatment, several sex-related disparities have been identified. Women have been found to less frequently undergo cardiac ablation procedures compared with men (Ko et al., 2016). Catheter ablation, in some studies, has also been associated with more complications in women (Piccini et al., 2012). In 2010, the CHA2DS2-VASc risk scoring system added a female sex as a risk factor, was validated at better refining stroke risk, especially for people at intermediate risk based on older risk scoring systems. The PINNACLE National Cardiovascular Data Registry from 2008 to 2014 showed women were less likely to receive oral anticoagulation at all CHA2DS2-VASc scores, by a factor of up to 33% (Thompson et al., 2017). The representation of women in the major trials studying the novel oral anticoagulants, namely RE-LY, ROCKET-AF, ARISTOTLE and ENGAGE AF was 37% when all populations were pooled. There were no significant differences in the outcome between male and female sex in a subgroup analysis in any of these three trials (Connolly et al., 2009; Giugliano et al., 2013; Granger et al., 2011; Patel et al., 2011) (The PINNACLE National Cardiovascular Data Registry from 2008 to 2014 showed women were less likely to receive oral anticoagulation at all CHA2DS2-VASc scores, by a factor of up to 33% (Thompson et al., 2017). Results from the RAF-study of the 1029 patients (54.5% women) revealed that in patients with known AF, women were less likely to receive oral anticoagulants before index stroke ($P=0.026$) and were less likely to receive anticoagulants after stroke (71.3% vs 78.4%, $P=0.01$). There was no observed sex difference regarding the time of starting anticoagulant therapy between the two groups (6.4 ± 11.7 days for men vs 6.5 ± 12.4 days for women, $P=0.902$) (Antonenko et al., 2017). Date from the international registry program—Global Registry on Long-Term Oral Antithrombotic Treatment in Patients with Atrial Fibrillation (GLORIA-AF) with 15,092 consecutive patients (45.5% women), enrolled between 2011 and 2014), shown that globally, the prevalence of anticoagulant use is similar in women and men. Between-sex differences in anticoagulant use (lower use in women compared with men by decreasing order of magnitude of the difference) were found for CHA2DS2-VASc score = 1; CHADS2 score = 0; previous bleeding; age <65 years; no history of hypertension; myocardial infarction; coronary artery disease; North America region; and specialist office setting (Mazurek et al., 2018). Results from the RAF-study of the 1029 patients (54.5% women) revealed that in patients with known AF, women were less likely to receive oral anticoagulants before index stroke ($P=0.026$) and were less likely to receive anticoagulants after stroke (71.3% vs 78.4%, $P=0.01$). Moreover, data from the international registry program—Global Registry on Long-Term Oral Antithrombotic Treatment in Patients with Atrial Fibrillation (GLORIA-AF) with 15,092 consecutive patients (45.5% women), enrolled between 2011 and 2014), shown that globally, the prevalence of anticoagulant use is similar in women and men.

4.3.3 Hormone replacement treatment

A randomized, placebo-controlled study in women who experienced stroke and received hormone replacement therapy after menopause showed an increase in stroke mortality and a poorer prognosis in nonfatal strokes (Viscoli et al., 2001). Hormone replacement after menopause is not effective in the secondary stroke prevention in women and may increase mortality after stroke (Diener & Hankey, 2020).

4.3.4 Carotid artery stenting(CAS) and endarterectomy (CEA)

Sex-based differences in the treatment effects of carotid revascularization either in asymptomatic carotid stenosis, or symptomatic carotid stenosis is still an issue. In asymptomatic carotid stenosis, a pooled analysis from two surgical clinical trials: Asymptomatic Carotid Atherosclerosis Study (ACAS); and Asymptomatic Carotid Surgery Trial (ACST) found that endarterectomy had greater benefits in men on the 5-year risk of any stroke and perioperative death (OR 0.49; 95% CI 0.36–0.66), but uncertain benefits were seen in women (OR 0.96 95% CI 0.63–1.45) (Rothwell & Goldstein, 2004). The ACST-2 trial randomized 3625 patients between 2008 and Dec 31, 2020, to CAS and CEA, with a mean 5 years of follow-up. Overall, 1% had procedural related disabling stroke or death (15 in the CAS group vs 18 in the CEA group), while 2% had a non-disabling procedural stroke (48 in the CAS group vs 29 in the CEA group). Kaplan-Meier estimates of 5-year non-procedural stroke were 2.5% in each group for fatal or disabling stroke, and 5.3% with CAS vs 4.5% with CEA for any stroke (rate ratio [RR] 1.16, 95% CI 0.86–1·57; $P = 0{\cdot}33$) (Halliday et al., 2021). In the subgroup analysis according the sex, there were no differences between men and women. With ACST-2 there is more evidence regarding the appropriate treatment of among asymptomatic stenosis. A combined analysis of the European Carotid Surgery Trial (ECST) and the North American Symptomatic Carotid Endarterectomy Trial (NASCET) reported, for patients with 50% or higher levels of carotid stenosis, that the number of procedures needed to prevent one ipsilateral stroke over the following 5 years was nine for men and 36 for women (Rothwell & Goldstein, 2004). However, there are currently no sex-specific recommendations for the carotid revascularization of asymptomatic and symptomatic carotid stenosis. Women were underrepresented in all the CAS versus CEA trials (30% ACST-2, 34% ACAS, 34% ACST, 28% ECST, 30% NASCET,) (De Rango et al., 2013; Halliday et al., 2021).

In addition to trial data, there are multiple sources of data in the literature regarding the topic of sex as a determinant of outcome following CEA in both symptomatic and asymptomatic cases, particularly with regard to perioperative results. Many of these institutional and database studies have included significantly more female patients than in randomized trials. Studies have reported a wide range of outcomes, from a finding of no significant differences based on sex, to significantly higher perioperative stroke rates in female as compared to male patients (Rockman et al., 2011). Database from US comprising 54,658 carotid revascularization cases, of which 94.2% were CEA included 42.3% of women. Overall, male and female patients had nearly identical rates of postoperative stroke (1.1%), but male patients were more likely to die in the same hospitalization (0.7% vs 0.6%, $P=0.03$) (Rockman et al., 2011). Analysis was stratified by the presence of preoperative symptoms. Among asymptomatic patients undergoing CEA, there were no significant differences in the rate of postoperative stroke or in-hospital mortality between men and women. Among symptomatic patients who underwent CEA, women had a higher rate of postoperative stroke than men (3.4% vs 2.1%, $P=0.06$). No difference was noted based on sex in in-hospital mortality (Rockman et al., 2011). A subsequent larger analysis including over 200,000 patients who underwent CEA found no difference between men and women with regard to rates of perioperative stroke, cardiac complications, myocardial infarction, or death in symptomatic patients. Among asymptomatic patients, women had a trend toward increased perioperative stroke (0.6% vs 0.5%; $P=0.06$), but a lower rate of cardiac complications (1.5% vs 1.7%; $P=0.01$) and in-hospital mortality (0.26% vs 0.31%; $P=0.05$) (Kuy et al., 2014). A meta-analysis including a systematic review of all available publications ($n=58$) data on the association between sex and the procedural risk of stroke and/or death following CEA in both symptomatic and asymptomatic cases from 1980 to 2015 was reported (Luebke & Brunkwall, 2015). In the unselected overall meta-analysis, the incidence of stroke and death in the male and female groups differed significantly (OR 1.16, $P=0.001$) with a worse outcome for women. However, the meta-analysis of databases and case series with sex-related outcomes as a primary aim demonstrated no increased in the operative risk of stroke and death in women as compared to men.

Several additional studies have been published in recent years adding to the literature on this topic. A single institution review of 690 carotid procedures performed for both symptomatic and asymptomatic disease (the great majority, 92% consisting of CEA), including 32% women, noted that female sex was an independent predictor of periprocedural stroke with an OR of 8.5 (Dulai, Tawfick, Hynes, & Sultan, 2019). Mortality and cardiac morbidity did not differ between the sexes. Another large study from the NIS database including 1,083,912 CEA procedures as well as 158,776 CAS procedures included 41.5% female patients (Mayor et al., 2019). The report notes that the outcome of stroke/death/MI was more prevalent in men as compared to women ($P < 0.01$). Symptomatic women as compared to symptomatic men had higher rates of perioperative stroke after CEA (7.7% vs 6.2%). However, asymptomatic women had similar rates of periprocedural stroke as compared to asymptomatic men with either CEA (0.3% vs 0.3%) or CAS (0.4% vs 0.5%) (Mayor et al., 2019) Finally, a study from the VQI registry published in 2020 and including more than 71,000 CEA cases reports that there were no differences between males and females in major adverse events following CEA for asymptomatic disease. Following CEA for symptomatic disease, there was no difference in stroke/death rate or in-hospital stroke (Dansey et al., 2020).

5. Intracerebral hemorrhage, intracranial aneurysms and subarachnoid hemorrhage

5.1 ICH

Intracerebral hemorrhage is a severe subtype of stroke with up to 50% mortality with the first month, and up to 70% of patients have long-term disabling deficits (McGurgan, Ziai, Werring, Al-Shahi Salman, & Parry-Jones, 2021). There are currently no evidence-based medical treatments improving outcome. There is an increasing amount of research on sex differences in ICH, and the results are heterogenous. Differences across the populations such as age, ethnicity and location may contribute to the inconsistent results. Independent of age and stroke severity, women with ICH are treated less aggressively than men. Do-not-resuscitate orders are more common (Grundtvig et al., 2022; Nakagawa et al., 2013) and women are less likely to be admitted to intensive care units (Roquer et al., 2016). In secondary analyses of Intensive Blood Pressure Reduction in Acute Intracerebral Hemorrhage trial 1 and 2, there no differences in medical management, however, women were less likely to be intubated during the hospitalization (Sandset et al., 2020).

ICH outcome in women vs men is influenced by age (Umeano et al., 2013), pre-stroke functional status and stroke severity (Zhou et al., 2014).

Although individual studies have found increased risk of death or dependency (Roquer et al., 2016) and of dependency alone (Zhou et al., 2014) in women, a meta-analysis from 2010 including 4658 patients recruited pre-2000 found no difference in mortality (van Asch et al., 2010).

Hematoma characteristics may differ across sexes. In a prospective hospital registry of 515 ICH patients, lobar hematoma location was more common in women than in men (age adjusted OR 1.75, 95%CI 1.18–2.58) (Roquer et al., 2016). Female participants in the INTERACT trials smaller hematoma volumes at baseline, however, there were no difference in hematoma expansion nor perihematomal oedma (Sandset et al., 2020).

To date, there is no efficient acute treatment for ICH (Steiner et al., 2014). Overall, results of ICH trials have been neutral, with a majority of male participants and very few trials have reported sex-disaggregated results. The most promising treatment has been acute and intensive blood pressure lowering (Anderson et al., 2013). The INTERACT trials randomized patients to intensive vs guideline lowering blood pressure management within 6 h of symptom onset. Women in the INTERACT trials were older with more severe strokes. No differences were seen in between the intensive and guideline groups, and their difference in results in the sex-disaggregated analysis (Sandset et al., 2020). Up to 20% of patients with ICH are treated with oral anticoagulants at time of the event (Rexrode et al., 2022). Timely reversal of the anticoagulant effect is important for outcome (Steiner et al., 2014). In Danish cohort of 401 patients women received less reversal agents, but no differences were seen in surgical interventions (Grundtvig et al., 2022).

Large scale cohort studies are required to identify specificities in risk factors for ICH in women. Data are needed to confirm whether women have worse outcome following ICH than men, independent of age and pre-morbid functional status, and whether this is related to difference in care. Furthermore, results from ICH need to disaggregated according to sex.

5.2 Intracranial aneurysms

Women have more incidental findings of intracranial aneurysms on imaging, but also new aneurysm formations, even when adjusting for age (Fuentes, McGuire, & Amin-Hanjani, 2022). Both the number of aneurysms and the location of aneurysms appears to be influenced by sex. In an analysis of 1277 patients with ruptured aneurysms women were more likely to present with 2 aneurysms (relative risk ration 1.80, 95% CI 1.31–2.48) or with 3 or more aneurysms (relative risk ratio, 3.10, 95% CI 1.82–5.27) than men (McDowell et al., 2018). Whereas aneurysms in women are located along the internal carotid artery, aneurysms in men are more often located along the anterior cerebral artery. Interestingly, although 20% of patients with aneurysms or subarachnoid hemorrhage report a family history of these conditions, the role of sex in familial aneurysms remains uncertain (Fuentes et al., 2022).

Possible explanations for these differences include hormonal influences on vascular remodeling and potential increased frequency of vessel wall weakness. Decreasing estrogen levels following menopause is likely one of the main factors contributing to the sex discrepancy in incidence of unruptured aneurysms in individuals over 50 years of age (Fuentes et al., 2022). Associations between early age at menopause and presences of intracranial aneurysm has been seen in one case control study, suggestive of decreased levels of estrogen is a contributing factor to the increased incidence of aneurysms in older women (Ding, Toll, Ouyang, & Chen, 2013).

5.3 Subarachnoid hemorrhage

Subarachnoid hemorrhage accounts for approximately 3% of all stroke, and is more common in women than in men. There is inconsistent evidence regarding the risk of aneurysm rupture. In an individual patient data metaanalysis including 9940 patients from 9 cohorts women had higher risk of aneurysm rupture, independent of patient- and aneurysm-related factors (Zuurbier et al., 2022), whereas other have not shown sex to be a significant risk factor once an intracranial aneurysm has been developed (Turan et al., 2016). Data on sex differences in outcome after SAH is scarce with inconsistent results, however, preliminary data from a cohort of 8015 suggest an increased risk of poor functional outcome and delayed cerebral ischemia in women (Anadani, Kumar, Havenon, & MacDonald, 2021; Fuentes et al., 2022). No differences have been seen in the risk of rebleeding. With regards to treatment, no differences between men and women have been seen in outcomes following aneurysm coiling or clipping, nor in endovascular procedures alone (Fuentes et al., 2022). Again, the data is scarce, mostly registry based from single center registries.

6. Cerebral vein thrombosis

Cerebral venous and sinus thrombosis (CVT) is a relatively rare subtype of stroke that mainly affects young women at fertile age. This heavily skewed sex ratio (approximately 3 to 1) and decreased incidence with age is a very specific feature of CVT, both when compared with other forms of cerebrovascular disease or other locations of venous thrombosis (Ferro, Canhão, Stam, Bousser, & Barinagarrementeria, 2004).

This striking epidemiology is attributed to the strong association of CVT with sex-specific risk factors, particularly oral contraception and pregnancy or puerperium, and, less frequently, hormonal therapies, as replacement treatment, infertility interventions or hormonal medications used in oncology, as tamoxifen. In fact, the most common risk factors in adult CVT are female-specific, independently of the world region (Ferro et al., 2004; Khealani et al., 2008; Ruiz-Sandoval et al., 2012). A systematic review showed a shift in sex ratio over time toward women that parallels an increase in use of oral contraceptives (Zuurbier, Middeldorp, Stam, & Coutinho, 2016). However, in age groups that lack these sex-specific risk factors such as children and elderly patients, there is no sex predilection (deVeber et al., 2001; Ferro, Canhão, Bousser, Stam, & Barinagarrementeria, 2005).

In the ISCVT, the largest prospective study of CVST, a sex-specific risk factor was present in about two-thirds of women (Coutinho et al., 2009). However, women with CVT without a sex-specific risk factor were similar to men in clinical presentation, risk factor profile, and outcome. Importantly, absence of sex-specific risk factors was an independent predictor of poor outcome in women with CVST.

The overall risk of CVT associated with use of oral contraceptives was estimated to be seven times that of non-users (Amoozegar, Ronksley, Sauve, & Menon, 2015; Dentali, Crowther, & Ageno, 2006). However, the risk will be higher among those who carry other risk factors, as obesity or hereditary thrombophilia (Coutinho, de Bruijn, deVeber, & Stam, 2012; Dentali et al., 2006; Martinelli et al., 1998; Zuurbier et al., 2016).

Pregnancy is the second most common sex-specific risk factor for CVST (Coutinho et al., 2009). The risk of venous thrombosis was estimated to be increased by 4- to 5-fold during pregnancy and by 22-fold during the first 6 weeks after delivery, compared with women who are not pregnant (Heit et al., 2005; Pomp, Lenselink, Rosendaal, & Doggen, 2008; Sultan et al., 2012). While mechanical factors promoting venous stasis are also implicated in the increased risk of deep venous thrombosis of the lower extremities associated with pregnancy, in what concerns CVT the increase in risk should be particularly associated to the major physiological changes in the coagulation and fibrinolytic systems that occur during pregnancy (James, 2009; Patil, Ghosh, Satoskar, & Shetty, 2013; Sattar et al., 1999). Additionally, systemic conditions associated with pregnancy can exacerbate this increased risk, as dehydration, iron deficiency anemia, or pre-eclampsia (Dentali et al., 2006). The risk of pregnancy-related CVT is higher in third trimester and puerperium, especially in the first 6 weeks postpartum, as confirmed by a case-control study conducted in high-income countries (Silvis et al., 2019).

Treatment of pregnant and puerperal women with CVT raises some issues, particularly concerning safety of mother and fetus. Data is scarce, but treatment with low molecular weight heparin is commonly preferred due to its favorable safety and efficacy profile, although most evidence comes from studies in other conditions occurring in pregnancy (Bates et al., 2012; Ferro et al., 2017). Prior systematic reviews suggest that the incidence of bleeding in pregnant women receiving LMWH is low, including antepartum hemorrhage (0.43%, 95% CI 0.22–0.75%), postpartum hemorrhage (0.94%, 95%CI 0.36–0.98%) and wound hematoma (0.61%, 95% CI 0.36 0.98%) (Greer & Nelson-Piercy, 2005). With respect to fetal safety (teratogenicity, congenital malformations, fetal bleeding) there is ample experience with UFH and LMWH in pregnant women. These agents do not cross the placenta and are considered safe to use in pregnancy. Thus, mostly considering indirect evidence from other conditions occurring in pregnancy, LMWH is commonly preferred for treatment of pregnant women with established CVT due to its favorable safety and efficacy profile (Bates et al., 2012). Management of delivery options and possible discontinuation of anticoagulants prior to induction of labor or caesarean section (or expected time of neuraxial anesthesia) should be considered by a multidisciplinary team and follow the available obstetric guidelines (Douketis et al., 2012; Horlocker et al., 2010; McLintock et al., 2012).

Oral anticoagulants are contraindicated in pregnancy (Ferro et al., 2017). Vitamin K antagonists cross the placenta, are teratogenic and may lead to pregnancy loss (Chan, Anand, & Ginsberg, 2000). Pregnant women were excluded from participating in clinical trials evaluating the oral direct thrombin and factor Xa inhibitors. These agents are likely to cross the placenta and their human reproductive risks are unknown.

There are no studies assessing optimal duration of anticoagulant therapy for treatment of pregnancy-related CVT (Ferro et al., 2017). However, given the increased risk of CVT after delivery, the European guidelines follow the general recommendation for other venous thromboembolic conditions associated with pregnancy and suggest that anticoagulants are continued throughout the postpartum period, usually at least for 6 weeks (Coutinho et al., 2009; Demir, Inci, Özkan, Yıldız, & Özdemir, 2013). During the postpartum period, including for breast-feeding women, LMWH, UFH and warfarin are all acceptable (Clark, Porter, & West, 2000; Richter et al., 2001). Since there are no clinical data on the effect of non-vitamin K oral anticoagulants on breastfed infants and there is some evidence that these agents might be secreted into breast milk, use of new oral direct thrombin and factor Xa inhibitors should be avoided in breast-feeding women until further studies are available (Cohen, Arachchillage, Middeldorp, Beyer-Westendorf, & Abdul-Kadir, 2016).

7. Women in clinical stroke trials

Adequate participation of women in clinical trials is important to investigate for possible sex differences in treatment response. When participants of clinical trials do not have similar characteristics to the population targeted for treatment then issues of generalizability arise (Carcel & Reeves, 2021; Scott et al., 2018). Evidence for under-representation of women in clinical trials have been seen in most cardiovascular specialties. In a study of pivotal cardiovascular disease trials submitted to the U.S. Food and Drug Administration supporting marketing applications from 2005 to 2015, women were well represented in atrial fibrillation, hypertension, and pulmonary arterial hypertension trials but were under-represented in trials in heart failure, coronary artery disease, and acute coronary syndrome (Scott et al., 2018).

When stroke randomized controlled trials from 1990 to 2018 from ClinicalTrials.gov were analyzed, it was found that overall there were 40% females enrolled which has barely improved in the past 10 years. North American randomized controlled trials recruited the most women at 43% followed by Australia, Europe and Asia (Carcel et al., 2019). The under representation of women remained consistent even after correcting for sex-specific prevalence of stroke in the underlying population. In a study that quantified sex disparities in enrolment, the participation to prevalence ratio (PPR) was used and this is defined as the percentage of women participating in a trial against the prevalence of women in the disease population. A PPR of 1 denoted under-representation of women. For 281 published stroke trials between 1990 and 2020, women were represented at a lower proportion relative to their prevalence in the underlying population (mean PPR 0.84; 95% confidence interval [CI] (0.81 to 0.87)). The greatest disparity were seen in trials of intracerebral hemorrhage, trials with a mean age of participants <70 years, interventions started more than 24 h after symptoms onset and rehabilitation trials (Carcel et al., 2021). Similar findings were found in a meta-analysis of 115 phase 2 or 3 randomized clinical trials in stroke that were published in 9 major journals between 2010 and 2020 (Strong et al., 2021).

A few possible mechanisms have been identified in the under-representation of women in cardiovascular disease in general and these can be divided into trial-specific or patient-specific factors. Trial specific factors that may be biased toward women include eligibility criteria such as age cut-off, burden of long-term trials, unconscious or implicit bias toward recruitment of women and lack of women leaders in clinical trials. Several studies have shown that the more women ion chief investigator roles in a clinical trial, the more likely a trial will have a diverse trial population and results to be disaggregated by sex. Barriers that may limit the participation of women are familial responsibilities, cultural barriers, socioeconomic barriers, and concerns about risk (Carcel & Reeves, 2021; Scott et al., 2018). More research needs to be undertaken to identify barriers and facilitators of women participating in clinical trials specifically in stroke.

8. Conclusion

Sex differences in neurovascular disorders exist ranging from symptom presentation, risk factors, outcomes and treatment. Increased awareness and more research is needed to fully understand the extent of the differences.

Keypoints

- Increased awareness of distinctive symptom presentation in women is needed to detect all strokes
- Sex-specific risk factors such as hypertensive disorders of pregnancy should be included in risk assessments of stroke
- Access to care remains a barrier to acute treatment for women in some regions
- Equal representation of men and women stroke trials is warranted

References

Acciarresi, M., De Rango, P., Pezzella, F. R., Santalucia, P., Amici, S., Paciaroni, M., et al. (2011). Secondary stroke prevention in women. *Women's Health*, 7(3), 391–397. https://doi.org/10.2217/whe.11.20.

Ali, M., van Os, H. J. A., van der Weerd, N., Schoones, J. W., Heymans, M. W., Kruyt, N. D., et al. (2022). Sex differences in presentation of stroke: A systematic review and meta-analysis. *Stroke*, *53*(2), 345–354. https://doi.org/10.1161/STROKEAHA.120.034040.

Amoozegar, F., Ronksley, P. E., Sauve, R., & Menon, B. K. (2015). Hormonal contraceptives and cerebral venous thrombosis risk: A systematic review and meta-analysis [review]. *Frontiers in Neurology*, *6*. https://doi.org/10.3389/fneur.2015.00007.

Anadani, M., Kumar, A., Phuah, C.-I., Havenon, A. H. D., & MacDonald, R. (2021). Abstract P434: Sex differences in outcomes after subarachnoid hemorrhage: Insight from the subarachnoid hemorrhage international trialists repository. *Stroke*, *52*(Suppl_1). https://doi.org/10.1161/str.52.suppl_1.P434.

Anderson, C. S., Heeley, E., Huang, Y., Wang, J., Stapf, C., Delcourt, C., et al. (2013). Rapid blood-pressure lowering in patients with acute intracerebral hemorrhage. *The New England Journal of Medicine*, *368*(25), 2355–2365. https://doi.org/10.1056/NEJMoa1214609 (doi] (Not in File).

Antonenko, K., Paciaroni, M., Agnelli, G., Falocci, N., Becattini, C., Marcheselli, S., et al. (2017). Sex-related differences in risk factors, type of treatment received and outcomes in patients with atrial fibrillation and acute stroke: Results from the RAF-study (early recurrence and cerebral bleeding in patients with acute ischemic stroke and atrial fibrillation). *European Stroke Journal*, *2*(1), 46–53. https://doi.org/10.1177/2396987316679577.

Bates, S. M., Greer, I. A., Middeldorp, S., Veenstra, D. L., Prabulos, A. M., & Vandvik, P. O. (2012). VTE, thrombophilia, antithrombotic therapy, and pregnancy: Antithrombotic therapy and prevention of thrombosis, 9th ed: American College of Chest Physicians Evidence-Based Clinical Practice Guidelines. *Chest*, *141*(2 Suppl), e691S–e736S. https://doi.org/10.1378/chest.11-2300.

Bellamy, L., Casas, J. P., Hingorani, A. D., & Williams, D. J. (2007). Pre-eclampsia and risk of cardiovascular disease and cancer in later life: Systematic review and meta-analysis. *BMJ*, *335*(7627), 974. https://doi.org/10.1136/bmj.39335.385301.BE.

Berger, J. S., Hochman, J., Lobach, I., Adelman, M. A., Riles, T. S., & Rockman, C. B. (2013). Modifiable risk factor burden and the prevalence of peripheral artery disease in different vascular territories. *Journal of Vascular Surgery*, *58*(3), 673–681. https://doi.org/10.1016/j.jvs.2013.01.053.

Bonkhoff, A. K., Karch, A., Weber, R., Wellmann, J., & Berger, K. (2021). Female stroke: Sex differences in acute treatment and early outcomes of acute ischemic stroke. *Stroke*, *52*(2), 406–415. https://doi.org/10.1161/strokeaha.120.032850.

Bonkhoff, A. K., Schirmer, M. D., Bretzner, M., Hong, S., Regenhardt, R. W., Brudfors, M., et al. (2021). Outcome after acute ischemic stroke is linked to sex-specific lesion patterns. *Nature Communications*, *12*(1), 3289. https://doi.org/10.1038/s41467-021-23492-3.

Bushnell, C. D., Chaturvedi, S., Gage, K. R., Herson, P. S., Hurn, P. D., Jiménez, M. C., et al. (2018). Sex differences in stroke: Challenges and opportunities. *Journal of Cerebral Blood Flow & Metabolism*, *38*(12), 2179–2191. https://doi.org/10.1177/0271678X18793324.

Bushnell, C., Howard, V. J., Lisabeth, L., Caso, V., Gall, S., Kleindorfer, D., et al. (2018). Sex differences in the evaluation and treatment of acute ischaemic stroke. *The Lancet Neurology*, *17*(7), 641–650. https://doi.org/10.1016/S1474-4422(18)30201-1.

Bushnell, C., McCullough, L. D., Awad, I. A., Chireau, M. V., Fedder, W. N., Furie, K. L., et al. (2014). Guidelines for the prevention of stroke in women: A statement for healthcare professionals from the American Heart Association/American Stroke Association. *Stroke, 45*(5), 1545–1588. https://doi.org/10.1161/01.str.0000442009.06663.48.

Carcel, C., Harris, K., Peters, S. A. E., Sandset, E. C., Balicki, G., Bushnell, C. D., et al. (2021). Representation of women in stroke clinical trials: A review of 281 trials involving more than 500,000 participants. *Neurology, 97*(18), e1768–e1774. https://doi.org/10.1212/wnl.0000000000012767.

Carcel, C., & Reeves, M. (2021). Under-enrollment of women in stroke clinical trials. *Stroke, 52*(2), 452–457. https://doi.org/10.1161/STROKEAHA.120.033227.

Carcel, C., Woodward, M., Balicki, G., Koroneos, G. L., de Sousa, D. A., Cordonnier, C., et al. (2019). Trends in recruitment of women and reporting of sex differences in large-scale published randomized controlled trials in stroke. *International Journal of Stroke, 14*(9), 931–938. https://doi.org/10.1177/1747493019851292.

Caso, V., Santalucia, P., Acciarresi, M., Pezzella, F. R., & Paciaroni, M. (2012). Antiplatelet treatment in primary and secondary stroke prevention in women. *European Journal of Internal Medicine, 23*(7), 580–585. https://doi.org/10.1016/j.ejim.2012.04.010.

Chan, W. S., Anand, S., & Ginsberg, J. S. (2000). Anticoagulation of pregnant women with mechanical heart valves: A systematic review of the literature. *Archives of Internal Medicine, 160*(2), 191–196. https://doi.org/10.1001/archinte.160.2.191.

Clark, S. L., Porter, T. F., & West, F. G. (2000). Coumarin derivatives and breast-feeding. *Obstetrics and Gynecology, 95*(6 Pt 1), 938–940. https://doi.org/10.1016/s0029-7844(00)00809-7.

Cohen, H., Arachchillage, D. R., Middeldorp, S., Beyer-Westendorf, J., & Abdul-Kadir, R. (2016). Management of direct oral anticoagulants in women of childbearing potential: Guidance from the SSC of the ISTH. *Journal of Thrombosis and Haemostasis, 14*(8), 1673–1676. https://doi.org/10.1111/jth.13366.

Connolly, S. J., Ezekowitz, M. D., Yusuf, S., Eikelboom, J., Oldgren, J., Parekh, A., et al. (2009). Dabigatran versus warfarin in patients with atrial fibrillation. *The New England Journal of Medicine, 361*(12), 1139–1151. https://doi.org/10.1056/NEJMoa0905561.

Correia, P., Machado, S., Meyer, I., Amiguet, M., Eskandari, A., & Michel, P. (2021). Ischemic stroke on hormonal contraceptives: Characteristics, mechanisms and outcome. *European Stroke Journal, 6*(2), 205–212. https://doi.org/10.1177/23969873211019586.

Coutinho, J. M., de Bruijn, S. F. T. M., deVeber, G., & Stam, J. (2012). Anticoagulation for cerebral venous sinus thrombosis. *Stroke, 43*(4), e41–e42. https://doi.org/10.1161/STROKEAHA.111.648162.

Coutinho, J. M., Ferro, J. M., Canhão, P., Barinagarrementeria, F., Cantú, C., Bousser, M. G., et al. (2009). Cerebral venous and sinus thrombosis in women. *Stroke, 40*(7), 2356–2361. https://doi.org/10.1161/strokeaha.108.543884.

Dansey, K. D., Pothof, A. B., Zettervall, S. L., Swerdlow, N. J., Liang, P., Schneider, J. R., et al. (2020). Clinical impact of sex on carotid revascularization. *Journal of Vascular Surgery, 71*(5), 1587–1594. https://doi.org/10.1016/j.jvs.2019.07.088.

de Havenon, A., Delic, A., Stulberg, E., Sheibani, N., Stoddard, G., Hanson, H., et al. (2021). Association of Preeclampsia with Incident Stroke in later life among women in the Framingham heart study. *JAMA Network Open, 4*(4), e215077. https://doi.org/10.1001/jamanetworkopen.2021.5077.

De Rango, P., Brown, M. M., Leys, D., Didier, L., Howard, V. J., Moore, W. S., et al. (2013). Management of carotid stenosis in women: Consensus document. *Neurology, 80*(24), 2258–2268. https://doi.org/10.1212/WNL.0b013e318296e952.

de Ridder, I. R., Fransen, P. S., Beumer, D., Berkhemer, O. A., van den Berg, L. A., Wermer, M. J., et al. (2016). Is intra-arterial treatment for acute ischemic stroke less effective in women than in men? *Interventional Neurology*, *5*(3–4), 174–178. https://doi.org/10.1159/000447331.

Demir, C. F., Inci, M. F., Özkan, F., Yıldız, M., & Özdemir, H. (2013). Clinical and radiological management and outcome of pregnancies complicated by cerebral venous thrombosis: A review of 19 cases. *Journal of Stroke and Cerebrovascular Diseases*, *22*(8), 1252–1257. https://doi.org/10.1016/j.jstrokecerebrovasdis.2012.07.004.

Dentali, F., Crowther, M., & Ageno, W. (2006). Thrombophilic abnormalities, oral contraceptives, and risk of cerebral vein thrombosis: A meta-analysis. *Blood*, *107*(7), 2766–2773. https://doi.org/10.1182/blood-2005-09-3578.

deVeber, G., Andrew, M., Adams, C., Bjornson, B., Booth, F., Buckley, D. J., et al. (2001). Cerebral Sinovenous thrombosis in children. *New England Journal of Medicine*, *345*(6), 417–423. https://doi.org/10.1056/NEJM200108093450604.

Diener, H. C., & Hankey, G. J. (2020). Primary and secondary prevention of ischemic stroke and cerebral hemorrhage: JACC focus seminar. *Journal of the American College of Cardiology*, *75*(15), 1804–1818. https://doi.org/10.1016/j.jacc.2019.12.072.

Ding, C., Toll, V., Ouyang, B., & Chen, M. (2013). Younger age of menopause in women with cerebral aneurysms. *Journal of NeuroInterventional Surgery*, *5*(4), 327. https://doi.org/10.1136/neurintsurg-2012-010364.

Douketis, J. D., Spyropoulos, A. C., Spencer, F. A., Mayr, M., Jaffer, A. K., Eckman, M. H., et al. (2012). Perioperative management of antithrombotic therapy: Antithrombotic therapy and prevention of thrombosis, 9th ed: American College of Chest Physicians Evidence-Based Clinical Practice Guidelines. *Chest*, *141*(2 Suppl), e326S–e350S. https://doi.org/10.1378/chest.11-2298.

Dulai, M., Tawfick, W., Hynes, N., & Sultan, S. (2019). Female gender as a risk factor for adverse outcomes after carotid revascularization. *Annals of Vascular Surgery*, *60*, 254–263. https://doi.org/10.1016/j.avsg.2019.03.007.

Ekker, M. S., Verhoeven, J. I., Vaartjes, I., van Nieuwenhuizen, K. M., Klijn, C. J. M., & de Leeuw, F. E. (2019). Stroke incidence in young adults according to age, subtype, sex, and time trends. *Neurology*, *92*(21), e2444–e2454. https://doi.org/10.1212/wnl.0000000000007533.

Emdin, C. A., Wong, C. X., Hsiao, A. J., Altman, D. G., Peters, S. A., Woodward, M., et al. (2016). Atrial fibrillation as risk factor for cardiovascular disease and death in women compared with men: Systematic review and meta-analysis of cohort studies. *BMJ*, *532*, h7013. https://doi.org/10.1136/bmj.h7013.

Feigin, V. L., Nguyen, G., Cercy, K., Johnson, C. O., Alam, T., Parmar, P. G., et al. (2018). Global, regional, and country-specific lifetime risks of stroke, 1990 and 2016. *The New England Journal of Medicine*, *379*(25), 2429–2437. https://doi.org/10.1056/NEJMoa1804492.

Ferro, J. M., Bousser, M. G., Canhão, P., Coutinho, J. M., Crassard, I., Dentali, F., et al. (2017). European stroke organization guideline for the diagnosis and treatment of cerebral venous thrombosis - endorsed by the European academy of neurology. *European Journal of Neurology*, *24*(10), 1203–1213. https://doi.org/10.1111/ene.13381.

Ferro, J. M., Canhão, P., Bousser, M. G., Stam, J., & Barinagarrementeria, F. (2005). Cerebral vein and dural sinus thrombosis in elderly patients. *Stroke*, *36*(9), 1927–1932. https://doi.org/10.1161/01.Str.0000177894.05495.54.

Ferro, J. M., Canhão, P., Stam, J., Bousser, M. G., & Barinagarrementeria, F. (2004). Prognosis of cerebral vein and dural sinus thrombosis: Results of the international study on cerebral vein and Dural sinus thrombosis (ISCVT). *Stroke*, *35*(3), 664–670. https://doi.org/10.1161/01.Str.0000117571.76197.26.

Franzini, L., Ardigò, D., Cavalot, F., Miccoli, R., Rivellese, A. A., Trovati, M., et al. (2013). Women show worse control of type 2 diabetes and cardiovascular disease risk factors than men: Results from the MIND.IT study Group of the Italian Society of diabetology. *Nutrition, Metabolism, and Cardiovascular Diseases, 23*(3), 235–241. https://doi.org/10.1016/j.numecd.2011.12.003.

Fuentes, A. M., McGuire, L. S., & Amin-Hanjani, S. (2022). Sex differences in cerebral aneurysms and subarachnoid hemorrhage. *Stroke, 53*(2), 624–633. https://doi.org/doi: https://doi.org/10.1161/STROKEAHA.121.037147.

Girijala, R. L., Sohrabji, F., & Bush, R. L. (2016). Sex differences in stroke: Review of current knowledge and evidence. *Vascular Medicine, 22*(2), 135–145. https://doi.org/10.1177/1358863X16668263.

Giugliano, R. P., Ruff, C. T., Braunwald, E., Murphy, S. A., Wiviott, S. D., Halperin, J. L., et al. (2013). Edoxaban versus warfarin in patients with atrial fibrillation. *The New England Journal of Medicine, 369*(22), 2093–2104. https://doi.org/10.1056/NEJMoa1310907.

Goldstein, L. B., Amarenco, P., Lamonte, M., Gilbert, S., Messig, M., Callahan, A., et al. (2008). Relative effects of statin therapy on stroke and cardiovascular events in men and women: Secondary analysis of the stroke prevention by aggressive reduction in cholesterol levels (SPARCL) study. *Stroke, 39*(9), 2444–2448. https://doi.org/10.1161/strokeaha.107.513747.

Goyal, M., Menon, B. K., van Zwam, W. H., Dippel, D. W., Mitchell, P. J., Demchuk, A. M., et al. (2016). Endovascular thrombectomy after large-vessel ischaemic stroke: A meta-analysis of individual patient data from five randomised trials. *Lancet, 387*(10029), 1723–1731. https://doi.org/10.1016/s0140-6736(16)00163-x.

Granger, C. B., Alexander, J. H., McMurray, J. J. V., Lopes, R. D., Hylek, E. M., Hanna, M., et al. (2011). Apixaban versus warfarin in patients with atrial fibrillation. *New England Journal of Medicine, 365*(11), 981–992. https://doi.org/10.1056/NEJMoa1107039.

Greer, I. A., & Nelson-Piercy, C. (2005). Low-molecular-weight heparins for thromboprophylaxis and treatment of venous thromboembolism in pregnancy: A systematic review of safety and efficacy. *Blood, 106*(2), 401–407. https://doi.org/10.1182/blood-2005-02-0626.

Grundtvig, J., Ovesen, C., Steiner, T., Carcel, C., Gaist, D., Christensen, L., et al. (2022). Sex-differences in Oral anticoagulant-related intracerebral hemorrhage [original research]. *Frontiers in Neurology, 13*. https://doi.org/10.3389/fneur.2022.832903.

Halliday, A., Bulbulia, R., Bonati, L. H., Chester, J., Cradduck-Bamford, A., Peto, R., et al. (2021). Second asymptomatic carotid surgery trial (ACST-2): A randomised comparison of carotid artery stenting versus carotid endarterectomy. *Lancet, 398*(10305), 1065–1073. https://doi.org/10.1016/s0140-6736(21)01910-3.

Heit, J. A., Kobbervig, C. E., James, A. H., Petterson, T. M., Bailey, K. R., & Melton, L. J., 3rd. (2005). Trends in the incidence of venous thromboembolism during pregnancy or postpartum: A 30-year population-based study. *Annals of Internal Medicine, 143*(10), 697–706. https://doi.org/10.7326/0003-4819-143-10-200511150-00006.

Ho, J. E., Paultre, F., & Mosca, L. (2003). Is diabetes mellitus a cardiovascular disease risk equivalent for fatal stroke in women? *Stroke, 34*(12), 2812–2816. https://doi.org/doi:https://doi.org/10.1161/01.STR.0000102901.41780.5C.

Hong, Y., Yang, X., Zhao, W., Zhang, X., Zhao, J., Yang, Y., et al. (2017). Sex differences in outcomes among stroke survivors with non-valvular atrial fibrillation in China [original research]. *Frontiers in Neurology, 8*. https://doi.org/10.3389/fneur.2017.00166.

Horlocker, T. T., Wedel, D. J., Rowlingson, J. C., Enneking, F. K., Kopp, S. L., Benzon, H. T., et al. (2010). Regional anesthesia in the patient receiving antithrombotic or thrombolytic therapy: American Society of Regional Anesthesia and Pain Medicine evidence-based guidelines (third edition). *Regional Anesthesia and Pain Medicine, 35*(1), 64–101. https://doi.org/10.1097/aap.0b013e3181c15c70.

Howard, B. V., Aragaki, A. K., Tinker, L. F., Allison, M., Hingle, M. D., Johnson, K. C., et al. (2018). A low-fat dietary pattern and diabetes: A secondary analysis from the Women's Health Initiative dietary modification trial. *Diabetes Care*, *41*(4), 680–687. https://doi.org/10.2337/dc17-0534.

Howard, V. J., Madsen, T. E., Kleindorfer, D. O., Judd, S. E., Rhodes, J. D., Soliman, E. Z., et al. (2019). Sex and race differences in the Association of Incident Ischemic Stroke with Risk Factors. *JAMA Neurology*, *76*(2), 179–186. https://doi.org/10.1001/jamaneurol.2018.3862.

James, A. H. (2009). Venous thromboembolism in pregnancy. *Arteriosclerosis, Thrombosis, and Vascular Biology*, *29*(3), 326–331. https://doi.org/10.1161/atvbaha.109.184127.

Kavousi, M. (2020). Differences in epidemiology and risk factors for atrial fibrillation between women and men [mini review]. *Frontiers in Cardiovascular Medicine*, 7. https://doi.org/10.3389/fcvm.2020.00003.

Keteepe-Arachi, T., & Sharma, S. (2017). Preventing stroke and assessing risk in women. *Practitioner*, *261*(1802), 13–17.

Khealani, B. A., Wasay, M., Saadah, M., Sultana, E., Mustafa, S., Khan, F. S., et al. (2008). Cerebral venous thrombosis: A descriptive multicenter study of patients in Pakistan and Middle East. *Stroke*, *39*(10), 2707–2711. https://doi.org/10.1161/strokeaha.107.512814.

Ko, D., Rahman, F., Schnabel, R. B., Yin, X., Benjamin, E. J., & Christophersen, I. E. (2016). Atrial fibrillation in women: Epidemiology, pathophysiology, presentation, and prognosis. *Nature Reviews. Cardiology*, *13*(6), 321–332. https://doi.org/10.1038/nrcardio.2016.45.

Koudstaal, P. J., Algra, A., Pop, G. A., Kappelle, L. J., van Latum, J. C., & van Gijn, J. (1992). Risk of cardiac events in atypical transient ischaemic attack or minor stroke. The Dutch TIA Study Group. *Lancet*, *340*(8820), 630–633. https://doi.org/10.1016/0140-6736(92)92170-k.

Kramer, C. K., Campbell, S., & Retnakaran, R. (2019). Gestational diabetes and the risk of cardiovascular disease in women: A systematic review and meta-analysis. *Diabetologia*, *62*(6), 905–914. https://doi.org/10.1007/s00125-019-4840-2.

Kuy, S., Dua, A., Desai, S. S., Rossi, P. J., Seabrook, G. R., Lewis, B. D., et al. (2014). Carotid endarterectomy national trends over a decade: Does sex matter? *Annals of Vascular Surgery*, *28*(4), 887–892. https://doi.org/10.1016/j.avsg.2013.08.016.

Leppert, M. H., Ho, P. M., Burke, J., Madsen, T. E., Kleindorfer, D., Sillau, S., et al. (2020). Young women had more strokes than Young men in a large, United States claims sample. *Stroke*, *51*(11), 3352–3355. https://doi.org/doi:10.1161/STROKEAHA.120.030803.

Li, X., Li, X., Lin, H., Fu, X., Lin, W., Li, M., et al. (2017). Metabolic syndrome and stroke: A meta-analysis of prospective cohort studies. *Journal of Clinical Neuroscience*, *40*, 34–38. https://doi.org/10.1016/j.jocn.2017.01.018.

Luebke, T., & Brunkwall, J. (2015). Meta- analysis and meta-regression analysis of the associations between sex and the operative outcomes of carotid endarterectomy. *BMC Cardiovascular Disorders*, *15*(1), 32. https://doi.org/10.1186/s12872-015-0029-x.

Madsen, T. E., Howard, G., Kleindorfer, D. O., Furie, K. L., Oparil, S., Manson, J. E., et al. (2019). Sex differences in hypertension and stroke risk in the REGARDS study: A longitudinal cohort study. *Hypertension*, *74*(4), 749–755. https://doi.org/10.1161/hypertensionaha.119.12729.

Martinelli, I., Sacchi, E., Landi, G., Taioli, E., Duca, F., & Mannucci, P. M. (1998). High risk of cerebral-vein thrombosis in carriers of a prothrombin-gene mutation and in users of Oral contraceptives. *New England Journal of Medicine*, *338*(25), 1793–1797. https://doi.org/10.1056/NEJM199806183382502.

Mayor, J. M., Salemi, J. L., Dongarwar, D., Salihu, H. M., Montero-Baker, M., Mills, J. L., et al. (2019). Sex-based differences in ten-year Nationwide outcomes of carotid revascularization. *Journal of the American College of Surgeons*, *229*(1), 38–46. https://doi.org/10.1016/j.jamcollsurg.2019.02.054.

Mazurek, M., Huisman, M. V., Rothman, K. J., Paquette, M., Teutsch, C., Diener, H. C., et al. (2018). Gender differences in antithrombotic treatment for newly diagnosed atrial fibrillation: The GLORIA-AF registry program. *The American Journal of Medicine, 131*(8), 945–955. https://doi.org/10.1016/j.amjmed.2018.03.024.

McDowell, M. M., Zhao, Y., Kellner, C. P., Barton, S. M., Sussman, E., Claassen, J., et al. (2018). Demographic and clinical predictors of multiple intracranial aneurysms in patients with subarachnoid hemorrhage. *Journal of Neurosurgery, 128*(4), 961–968. https://doi.org/10.3171/2017.1.Jns162785.

McGurgan, I. J., Ziai, W. C., Werring, D. J., Al-Shahi Salman, R., & Parry-Jones, A. R. (2021). Acute intracerebral haemorrhage: Diagnosis and management. *Practical Neurology, 21*(2), 128. https://doi.org/10.1136/practneurol-2020-002763.

McLintock, C., Brighton, T., Chunilal, S., Dekker, G., McDonnell, N., McRae, S., et al. (2012). Recommendations for the prevention of pregnancy-associated venous thromboembolism. *The Australian & New Zealand Journal of Obstetrics & Gynaecology, 52*(1), 3–13. https://doi.org/10.1111/j.1479-828X.2011.01357.x.

Miller, E. C., Boehme, A. K., Chung, N. T., Wang, S. S., Lacey, J. V., Jr., Lakshminarayan, K., et al. (2019). Aspirin reduces long-term stroke risk in women with prior hypertensive disorders of pregnancy. *Neurology, 92*(4), e305–e316. https://doi.org/10.1212/wnl.0000000000006815.

Nakagawa, K., Vento, M. A., Seto, T. B., Koenig, M. A., Asai, S. M., Chang, C. W., et al. (2013). Sex differences in the use of early do-not-resuscitate orders after intracerebral hemorrhage. *Stroke, 44*(11), 3229–3231. https://doi.org/10.1161/STROKEAHA.113.002814.

Nanna, M. G., Wang, T. Y., Xiang, Q., Goldberg, A. C., Robinson, J. G., Roger, V. L., et al. (2019). Sex differences in the use of statins in community practice. *Circulation: Cardiovascular Quality and Outcomes, 12*(8), e005562. https://doi.org/10.1161/CIRCOUTCOMES.118.005562.

O'Donnell, M. J., Chin, S. L., Rangarajan, S., Xavier, D., Liu, L., Zhang, H., et al. (2016). Global and regional effects of potentially modifiable risk factors associated with acute stroke in 32 countries (INTERSTROKE): A case-control study. *Lancet, 388*(10046), 761–775. https://doi.org/10.1016/S0140-6736(16)30506-2.

Oertelt-Prigione, S., & Regitz-Zagrosek, V. (2012). *Sex and gender aspects in clinical medicine*. Springer. https://doi.org/10.1007/978-0-85729-832-4.

Øie, L. R., Kurth, T., Gulati, S., & Dodick, D. W. (2020). Migraine and risk of stroke. *Journal of Neurology, Neurosurgery, and Psychiatry, 91*(6), 593–604. https://doi.org/10.1136/jnnp-2018-318254.

Otite, F. O., Saini, V., Sur, N. B., Patel, S., Sharma, R., Akano, E. O., et al. (2021). Ten-year trend in age, sex, and racial disparity in tPA (alteplase) and thrombectomy use following stroke in the United States. *Stroke, 52*(8), 2562–2570. https://doi.org/10.1161/STROKEAHA.120.032132.

Patel, M. R., Mahaffey, K. W., Garg, J., Pan, G., Singer, D. E., Hacke, W., et al. (2011). Rivaroxaban versus warfarin in nonvalvular atrial fibrillation. *The New England Journal of Medicine, 365*(10), 883–891. https://doi.org/10.1056/NEJMoa1009638.

Patil, R., Ghosh, K., Satoskar, P., & Shetty, S. (2013). Elevated procoagulant endothelial and tissue factor expressing microparticles in women with recurrent pregnancy loss. *PLoS One, 8*(11), e81407. https://doi.org/10.1371/journal.pone.0081407.

Piccini, J. P., Sinner, M. F., Greiner, M. A., Hammill, B. G., Fontes, J. D., Daubert, J. P., et al. (2012). Outcomes of Medicare beneficiaries undergoing catheter ablation for atrial fibrillation. *Circulation, 126*(18), 2200–2207. https://doi.org/10.1161/circulationaha.112.109330.

Plas, G. J., Booij, H. A., Brouwers, P. J., Brusse-Keizer, M., Koudstaal, P. J., Dippel, D. W., et al. (2016). Nonfocal symptoms in patients with transient ischemic attack or ischemic stroke: Occurrence, clinical determinants, and association with cardiac history. *Cerebrovascular Diseases*, *42*(5–6), 439–445. https://doi.org/10.1159/000448148.

Pomp, E. R., Lenselink, A. M., Rosendaal, F. R., & Doggen, C. J. (2008). Pregnancy, the postpartum period and prothrombotic defects: Risk of venous thrombosis in the MEGA study. *Journal of Thrombosis and Haemostasis*, *6*(4), 632–637. https://doi.org/10.1111/j.1538-7836.2008.02921.x.

Razzouk, L., Rockman, C. B., Patel, M. R., Guo, Y., Adelman, M. A., Riles, T. S., et al. (2015). Co-existence of vascular disease in different arterial beds: Peripheral artery disease and carotid artery stenosis- -data from life line screening(®). *Atherosclerosis*, *241*(2), 687–691. https://doi.org/10.1016/j.atherosclerosis.2015.06.029.

Reeves, M., Bhatt, A., Jajou, P., Brown, M., & Lisabeth, L. (2009). Sex differences in the use of intravenous rt-PA thrombolysis treatment for acute ischemic stroke: A meta-analysis. *Stroke*, *40*(5), 1743–1749. https://doi.org/10.1161/strokeaha.108.543181.

Rexrode, K. M., Madsen, T. E., Yu, A. Y. X., Carcel, C., Lichtman, J. H., & Miller, E. C. (2022). The impact of sex and gender on stroke. *Circulation Research*, *130*(4), 512–528. https://doi.org/doi:https://doi.org/10.1161/CIRCRESAHA.121.319915.

Richter, C., Sitzmann, J., Lang, P., Weitzel, H., Huch, A., & Huch, R. (2001). Excretion of low molecular weight heparin in human milk. *British Journal of Clinical Pharmacology*, *52*(6), 708–710. https://doi.org/10.1046/j.0306-5251.2001.01517.x.

Roach, R. E., Helmerhorst, F. M., Lijfering, W. M., Stijnen, T., Algra, A., & Dekkers, O. M. (2015). Combined oral contraceptives: The risk of myocardial infarction and ischemic stroke. *Cochrane Database of Systematic Reviews*, *2015*(8), Cd011054. https://doi.org/10.1002/14651858.CD011054.pub2.

Rockman, C. B., Garg, K., Jacobowitz, G. R., Berger, J. S., Mussa, F. F., Cayne, N. S., et al. (2011). Outcome of carotid artery interventions among female patients, 2004 to 2005. *Journal of Vascular Surgery*, *53*(6), 1457–1464. https://doi.org/10.1016/j.jvs.2011.02.029.

Rockman, C. B., Hoang, H., Guo, Y., Maldonado, T. S., Jacobowitz, G. R., Talishinskiy, T., et al. (2013). The prevalence of carotid artery stenosis varies significantly by race. *Journal of Vascular Surgery*, *57*(2), 327–337. https://doi.org/10.1016/j.jvs.2012.08.118.

Roquer, J., Rodriguez-Campello, A., Jimenez-Conde, J., Cuadrado-Godia, E., Giralt-Steinhauer, E., Vivanco Hidalgo, R. M., et al. (2016). Sex-related differences in primary intracerebral hemorrhage. *Neurology*, *87*(3), 257–262. https://doi.org/10.1212/WNL.0000000000002792.

Rothwell, P. M., & Goldstein, L. B. (2004). Carotid endarterectomy for asymptomatic carotid stenosis: Asymptomatic carotid surgery trial. *Stroke*, *35*(10), 2425–2427. https://doi.org/10.1161/01.STR.0000141706.50170.a7.

Roy-O'Reilly, M., & McCullough, L. D. (2018). Age and sex are critical factors in ischemic stroke pathology. *Endocrinology*, *159*(8), 3120–3131. https://doi.org/10.1210/en.2018-00465.

Ruiz-Sandoval, J. L., Chiquete, E., Bañuelos-Becerra, L. J., Torres-Anguiano, C., González-Padilla, C., Arauz, A., et al. (2012). Cerebral venous thrombosis in a Mexican multicenter registry of acute cerebrovascular disease: The RENAMEVASC study. *Journal of Stroke and Cerebrovascular Diseases*, *21*(5), 395–400. https://doi.org/10.1016/j.jstrokecerebrovasdis.2011.01.001.

Sandset, E. C., Wang, X., Carcel, C., Sato, S., Delcourt, C., Arima, H., et al. (2020). Sex differences in treatment, radiological features and outcome after intracerebral haemorrhage: Pooled analysis of intensive blood pressure reduction in acute cerebral Haemorrhage trials 1 and 2. *European Stroke Journal, 5*(4), 345–350. https://doi.org/10.1177/2396987320957513.

Sattar, N., Greer, I. A., Rumley, A., Stewart, G., Shepherd, J., Packard, C. J., et al. (1999). A longitudinal study of the relationships between Haemostatic, lipid, and Oestradiol changes during Normal human pregnancy. *Thrombosis and Haemostasis, 81*(01), 71–75.

Scott, P. E., Unger, E. F., Jenkins, M. R., Southworth, M. R., McDowell, T.-Y., Geller, R. J., et al. (2018). Participation of women in clinical trials supporting FDA approval of cardiovascular drugs. *Journal of the American College of Cardiology, 71*(18), 1960–1969. https://doi.org/https://doi.org/10.1016/j.jacc.2018.02.070.

Shajahan, S., Sun, L., Harris, K., Wang, X., Sandset, E. C., Yu, A. Y. X., et al. (2022). Sex differences in the symptom presentation of stroke: A systematic review and meta-analysis. *International Journal of Stroke*. https://doi.org/10.1177/17474930221090133.

Sidawy, A. P., & Perler, B. A. (2019). *Rutherford's vascular surgery and endovascular therapy*. Elsevier.

Silvis, S. M., Lindgren, E., Hiltunen, S., Devasagayam, S., Scheres, L. J., Jood, K., et al. (2019). Postpartum period is a risk factor for cerebral venous thrombosis. *Stroke, 50*(2), 501–503. https://doi.org/10.1161/STROKEAHA.118.023017.

Spinler, S. A., Cziraky, M. J., Willey, V. J., Tang, F., Maddox, T. M., Thomas, T., et al. (2015). Frequency of attainment of low-density lipoprotein cholesterol and non-high-density lipoprotein cholesterol goals in cardiovascular clinical practice (from the National Cardiovascular Data Registry PINNACLE registry). *The American Journal of Cardiology, 116*(4), 547–553. https://doi.org/10.1016/j.amjcard.2015.05.011.

Steiner, T., Al-Shahi, S. R., Beer, R., Christensen, H., Cordonnier, C., Csiba, L., et al. (2014). European stroke organisation (ESO) guidelines for the management of spontaneous intracerebral hemorrhage. *International Journal of Stroke*. https://doi.org/10.1111/ijs.12309.

Steiner, T., Al-Shahi Salman, R., Beer, R., Christensen, H., Cordonnier, C., Csiba, L., et al. (2014). European stroke organisation (ESO) guidelines for the management of spontaneous intracerebral hemorrhage. *International Journal of Stroke*. https://doi.org/10.1111/ijs.12309.

Stevens, R. J., Coleman, R. L., Adler, A. I., Stratton, I. M., Matthews, D. R., & Holman, R. R. (2004). Risk factors for myocardial infarction case fatality and stroke case fatality in type 2 diabetes: UKPDS 66. *Diabetes Care, 27*(1), 201–207. https://doi.org/10.2337/diacare.27.1.201.

Strong, B., Lisabeth, L. D., & Reeves, M. (2020). Sex differences in IV thrombolysis treatment for acute ischemic stroke: A systematic review and meta-analysis. *Neurology, 95*(1), e11–e22. https://doi.org/10.1212/wnl.0000000000009733.

Strong, B., Pudar, J., Thrift, A. G., Howard, V. J., Hussain, M., Carcel, C., et al. (2021). Sex disparities in enrollment in recent randomized clinical trials of acute stroke: A meta-analysis. *JAMA Neurology, 78*(6), 666–677. https://doi.org/10.1001/jamaneurol.2021.0873.

Sultan, A. A., West, J., Tata, L. J., Fleming, K. M., Nelson-Piercy, C., & Grainge, M. J. (2012). Risk of first venous thromboembolism in and around pregnancy: A population-based cohort study. *British Journal of Haematology, 156*(3), 366–373. https://doi.org/10.1111/j.1365-2141.2011.08956.x.

Thompson, L. E., Maddox, T. M., Lei, L., Grunwald, G. K., Bradley, S. M., Peterson, P. N., et al. (2017). Sex differences in the use of Oral anticoagulants for atrial fibrillation: A report from the National Cardiovascular Data Registry (NCDR(®)) PINNACLE registry. *Journal of the American Heart Association, 6*(7). https://doi.org/10.1161/jaha.117.005801.

Tietjen, G. E., & Maly, E. F. (2020). Migraine and ischemic stroke in women. A narrative review. *Headache, 60*(5), 843–863. https://doi.org/10.1111/head.13796.

Turan, N., Heider, R. A.-J., Zaharieva, D., Ahmad, F. U., Barrow, D. L., & Pradilla, G. (2016). Sex differences in the formation of intracranial aneurysms and incidence and outcome of subarachnoid hemorrhage: Review of experimental and human studies. *Translational Stroke Research*, 7(1), 12–19. https://doi.org/10.1007/s12975-015-0434-6.

Umeano, O., Phillips-Bute, B., Hailey, C. E., Sun, W., Gray, M. C., Roulhac-Wilson, B., et al. (2013). Gender and age interact to affect early outcome after intracerebral hemorrhage. *PLoS One, 8*(11), e81664. https://doi.org/10.1371/journal.pone.0081664.

van Asch, C. J., Luitse, M. J., Rinkel, G. J., van der Tweel, I., Algra, A., & Klijn, C. J. (2010). Incidence, case fatality, and functional outcome of intracerebral haemorrhage over time, according to age, sex, and ethnic origin: A systematic review and meta-analysis. *Lancet Neurology, 9*(2), 167–176. https://doi.org/10.1016/S1474-4422(09)70340-0.

Vike, J. (1994). Stroke and its effect in women. *Journal of the American Medical Women's Association (1972), 49*(6), 198–201 (206).

Viscoli, C. M., Brass, L. M., Kernan, W. N., Sarrel, P. M., Suissa, S., & Horwitz, R. I. (2001). A clinical trial of estrogen-replacement therapy after ischemic stroke. *The New England Journal of Medicine, 345*(17), 1243–1249. https://doi.org/10.1056/NEJMoa010534.

Wagstaff, A. J., Overvad, T. F., Lip, G. Y., & Lane, D. A. (2014). Is female sex a risk factor for stroke and thromboembolism in patients with atrial fibrillation? A systematic review and meta-analysis. *QJM, 107*(12), 955–967. https://doi.org/10.1093/qjmed/hcu054.

Wang, X., Phan, H. T., Li, J., Reeves, M. J., Thrift, A. G., Cadilhac, D. A., et al. (2020). Sex differences in disease profiles, management, and outcomes among people with atrial fibrillation after ischemic stroke: Aggregated and individual participant data meta-analyses. *Women's Health Reports, 1*(1), 190–202. https://doi.org/10.1089/whr.2020.0029.

Wong, N. D., Zhao, Y., Patel, R., Patao, C., Malik, S., Bertoni, A. G., et al. (2016). Cardiovascular risk factor targets and cardiovascular disease event risk in diabetes: A pooling project of the atherosclerosis risk in communities study, multi-ethnic study of atherosclerosis, and Jackson heart study. *Diabetes Care, 39*(5), 668–676. https://doi.org/10.2337/dc15-2439.

Zhao, M., Woodward, M., Vaartjes, I., Millett, E. R. C., Klipstein-Grobusch, K., Hyun, K., et al. (2020). Sex differences in cardiovascular medication prescription in primary care: A systematic review and meta-analysis. *Journal of the American Heart Association, 9*(11), e014742. https://doi.org/10.1161/jaha.119.014742.

Zhou, J., Zhang, Y., Arima, H., Zhao, Y., Zhao, H., Zheng, D., et al. (2014). Sex differences in clinical characteristics and outcomes after intracerebral haemorrhage: Results from a 12-month prospective stroke registry in Nanjing, China. *BMC Neurology, 14*, 172. https://doi.org/10.1186/s12883-014-0172-5.

Zuurbier, S. M., Arnold, M., Middeldorp, S., Broeg-Morvay, A., Silvis, S. M., Heldner, M. R., et al. (2016). Risk of cerebral venous thrombosis in obese women. *JAMA Neurology, 73*(5), 579–584. https://doi.org/10.1001/jamaneurol.2016.0001.

Zuurbier, S. M., Middeldorp, S., Stam, J., & Coutinho, J. M. (2016). Sex differences in cerebral venous thrombosis: A systematic analysis of a shift over time. *International Journal of Stroke, 11*(2), 164–170. https://doi.org/10.1177/1747493015620708.

Zuurbier, C. C. M., Molenberg, R., Mensing, L. A., Wermer, M. J. H., Juvela, S., Lindgren, A. E., et al. (2022). Sex difference and rupture rate of intracranial aneurysms: An individual patient data meta-analysis. *Stroke, 53*(2), 362–369. https://doi.org/10.1161/strokeaha.121.035187.

CHAPTER FOUR

Sex and gender differences in movement disorders: Parkinson's disease, essential tremor, dystonia and chorea

Gennarina Arabia[a,*], Antonio De Martino[a], and Elena Moro[b]
[a]Magna Graecia University, Movement Disorders Center, Neurology Unit, Catanzaro, Italy
[b]Grenoble Alpes University, CHU of Grenoble, Division of Neurology, Grenoble Institute of Neurosciences, Grenoble, France
*Corresponding author: e-mail address: g.arabia@unicz.it

Contents

Abstract

Sex and gender-based differences in epidemiology, clinical features and therapeutical responses are emerging in several movement disorders, even though they are still not widely recognized. In this chapter, we summarize the most relevant evidence

International Review of Neurobiology, Volume 164
ISSN 0074-7742
https://doi.org/10.1016/bs.irn.2022.06.010

concerning these differences in Parkinson's disease, essential tremor, dystonia and chorea. Indeed, both sex-related biological (hormonal levels fluctuations) and gender-related variables (socio-cultural and environmental factors) may differently impact symptoms manifestation and severity, phenotype and disease progression of movement disorders on men and women. Moreover, sex differences in treatment responses should be taken into account in any therapeutical planning. Physicians need to be aware of these major differences between men and women that will eventually have a major impact on better tailoring prevention, treatment, or even delaying progression of the most common movement disorders.

1. Sex and gender-specific differences in Parkinson's disease

Parkinson's disease (PD) is the second most frequent age-related neurodegenerative disorder, and affects more than 8.5 million of people worldwide with an age-standardized prevalence rate of 109 (95% UI 90–131) per 100,000 (Deuschl et al., 2020), and an annual incidence of 15 per 100,000 people in the United States (Ascherio & Schwarzschild, 2016). It is characterized by motor symptoms (bradykinesia, rigidity, tremor at rest) as well as nonmotor symptoms (cognitive impairment, mood disorders, olfaction and autonomic dysfunctions, gastrointestinal, urinary and sleep disorders). Neuropathological hallmark is the presence of intraneuronal Lewy's bodies (aggregates mainly consisting of α-synuclein protein) in the substantia nigra pars compacta (Spillantini, Crowther, Jakes, Hasegawa, & Goedert, 1998), and in other areas of the central and enteric nervous system. This has suggested a double possibility (olfactory system-brain and gut-to-brain) of prion-like spreading of pathologic aggregated α-synuclein, possibly linked to complex interaction of environmental toxins, infections, and genetic predisposition (Braak et al., 2003; Holmqvist et al., 2014). The classical triad of motor symptoms (bradykinesia, rigidity and resting tremor) is related to the selective degeneration of dopaminergic neurons located in the substantia nigra and innervating the striatum, forming the nigrostriatal dopaminergic system.

Sex differences have been reported in many aspects of PD, including the effect size of PD risk factors, the disease onset, the type and prevalence of motor and non-motor symptoms, medication use, levodopa bioavailability, neuropsychiatric and cognitive changes, caregiver reliance, access to health care and quality of life (Cabezas et al., 2014; Colombo et al., 2015; Dahodwala, Pei, & Schmidt, 2016; Meoni, Macerollo, & Moro, 2020).

1.1 Epidemiology

Sex differences in both prevalence and incidence of PD are largely reported. Together with aging, male sex is the strongest risk factor for PD. Indeed, the risk for PD is about 1.5-fold greater in men than in women at all ages (Haaxma et al., 2007; Picillo et al., 2017). However, the age-standardized male:female (M:F) incidence ratios varies among different populations. A systematic analysis of the global burden of neurological diseases (GBD 2016 Neurology Collaborators, 2019) spanning from 1990 to 2016 has reported a M:F age-standardized prevalence ratio of 1:40, consistent with previous studies (Abbas, Xu, & Tan, 2017; Hirsch, Jette, Frolkis, Steeves, & Pringsheim, 2016; Moisan et al., 2016; Pringsheim, Jette, Frolkis, & Steeves, 2014). Higher M:F incidence ratios are reported in Western and South America, ranging from 1.3 to 2.0 (de Lau, Verbaan, Marinus, & van Hilten, 2014; Hirsch et al., 2016; Moisan et al., 2016; Taylor, Cook, & Counsell, 2007), in comparison to Asian populations (0.95–1.20) (Abbas et al., 2017; Maeda et al., 2017; Xu, Gong, Man, & Fan, 2014). Both M:F incidence and prevalence ratios for PD increase with age but in men, the increased incidence of PD does not parallel with a similar increase of prevalence. There are some possible explanations for this, including that men tend to show earlier PD onset than women and that mortality increases with disease duration. Indeed, recent studies have reported a higher mortality risk only for men with PD than in the general population (Diem-Zangerl et al., 2009).

Some studies showed an increased risk of PD in men probably due to toxic exposure, head trauma and habit behaviors, such as smoking (Wooten, Currie, Bovbjerg, Lee, & Patrie, 2004). For instance, pesticide exposure as well as head trauma have been associated with the development of PD only in men and not in women (Frigerio et al., 2006; Savica, Grossardt, Bower, Ahlskog, & Rocca, 2013; Wooten et al., 2004). Other habits, such as alcohol consumption, seem to have a PD risk-lowering effect in women more than in men (Kim, Yoo, Jung, Han, & Lee, 2020) whereas caffeine consumption is linked to a lower risk of PD among men but not women (Ascherio et al., 2003). Interestingly, a recent population-based cohort study has shown lower PD risk for current male smokers compared to current female smokers, with the same level of exposure (Kim et al., 2020). However, all this data may reflect differences in methodological approaches or sex differences in genetics, ethnicity or behavioral factors influencing a differential risk for PD development.

1.2 Sex hormones and neuroprotection for PD

Sex steroid hormones induce specific and differential responses in the nigrostriatal dopaminergic system, possibly because of the distinctive organization and regulatory system in the two sexes (Gillies & McArthur, 2010). Experimental studies have shown a neuroprotective effect of estrogen on the nigrostriatal dopaminergic system (Brann, Dhandapani, Wakade, Mahesh, & Khan, 2007). Indeed, PD was consistently observed to occur at a lesser frequency in women than in men and with a more benign phenotype, suggesting the possible beneficial effect of estrogen (Picillo et al., 2017). Although estrogen may serve as a neuroprotective factor for neurodegenerative diseases such as PD, the mechanisms and the cellular targets involved in this potential estrogen's protective influence are under investigation. Studies performed on developing midbrain dopaminergic cells support the role of estrogen as an important regulator of plasticity and function of these neurons (Küppers, Ivanova, Karolczak, & Beyer, 2000). More precisely, dopaminergic neurons are direct target for estrogen. Estrogen stimulates extension and branching of the neurites and the expression of tyrosine hydroxylase, key enzyme in dopamine synthesis. All these evidences, along with others coming from in vivo studies, suggest that estrogen is required for the plasticity and activity of the developing and adult nigrostriatal system. Estrogen effects on nigrostriatal cell function are not only mediated by classical nuclear estrogen receptors but also depend on other actions through putative membrane receptors coupled to diverse intracellular signaling cascades (Küppers et al., 2000).

Estrogens increase antioxidant responses and potentiating respiratory chain activity in the female brain, by inducing the expression of mitochondrial proteins and enzymes. On the other hand, the role of androgens in the nigrostriatal dopaminergic system in PD is still unclear (Holmes, Abbassi, Su, Singh, & Cunningham, 2013; Wang et al., 2015).

Following brain iron overload in females, estrogens were shown to reduce autophagy and severity of neuronal injury, while dopaminergic neurons from male mice are more prone to autophagy activation and more vulnerable to toxicity (Wang et al., 2015).

Although neuroprotective effects of estrogens in the brain seems to involve inflammatory cells, little information is available on the effects of sex hormones on inflammation in PD models (Arevalo, Azcoitia, & Garcia-Segura, 2015; Villa, Rizzi, Vegeto, Ciana, & Maggi, 2015). A recent study suggests that estrogens modify the reactivity of microglia to 6-OHDA in females, inducing a protective anti-inflammatory phenotype (Siani et al., 2017). Exiguous are the studies on progesterone and PD, and no data are available concerning risk of PD and progesterone levels.

Therefore, whether estrogens prevent neurodegeneration or rather potentiate neurotransmission of resilient neurons remains unclear. Future studies will help to understand the role of estrogens in the sexual dimorphism of neurodegeneration in PD and in addressing the role of pathogenic mechanisms, such as protein aggregation, neuroinflammation or extra-nigral pathology (Rodríguez-Navarro et al., 2008).

Clinical studies have showed that higher estrogens levels inversely correlate with the severity of PD symptoms in women, while women who experienced early natural or surgical menopause show a higher risk of PD (Benedetti et al., 2001; Rocca et al., 2008). Some investigations showed that there is an association between postmenopausal estrogen replacement therapy (ERT) and a lower risk of PD (Currie, Harrison, Trugman, Bennett, & Wooten, 2004; Ragonese, D'Amelio, & Savettieri, 2006), while others did not observe such association (Nicoletti et al., 2011; Rugbjerg, Christensen, Tjønneland, & Olsen, 2013). It appears that estrogen may play a favorable role against development and severity of motor symptoms, delaying the onset and mitigating the symptoms of PD, although confounding factors such as timing, dosage, composition, and route of administration of hormonal replacement therapy in women may limit the appreciation of hormonal effects (Dye, Miller, Singer, & Levine, 2012). Similarly, studies evaluating the relationship between the risk of PD and the ERT in women have yielded contrasting results. A recent meta-analysis on the effect of ERT on PD in postmenopausal women has showed that ERT significantly decreased the risk of onset and/or development of PD (OR: 0.470; 95% CI: 0.368–0.600; $P < 0.001$) compared with the control group (Song et al., 2020). By contrast, other meta-analyses do not support this protective role of ERT in female PD development, even though a positive association (RR: 1.16, 95% CI: 1.02–1.31) was found for the high-quality studies group included in the analysis, at the contrary of low-quality studies group (RR: 0.58, 95% CI: 0.40–0.82) (Wang et al., 2015). Indeed, a "critical period hypothesis" for neuroprotection by ERT has been proposed (Henderson, 2007; Sherwin, 2007). This hypothesis emphasized the importance of timing ERT at the onset of menopause, and subsequently opened new avenues for exploring the multifaceted and complex role of estrogen treatment in postmenopausal women of different ages. If menopausal women experience a period of long-term estrogen deprivation (e.g., 10 years post menopause) and they are then administered a delayed hormone therapy, estrogen may become detrimental to them, with increased risk of venous thromboembolism, ischemic stroke, and dementia (Rocca, Grossardt, & Shuster, 2011). At the contrary, ERT may exert a profound neuroprotective function in early menopausal women in whom the E2 receptors and signaling mechanisms at the neuronal membrane are not yet diminished. E2 mediate regulation of neurotrophic and antiapoptotic pathways in the brain, of dendritic spines necessary for synaptic contact, and potentially interact with some integral proteins such as caudlin-5 and occludin from the tight junctions of endothelial cells necessary for maintaining integrity of the blood brain barrier (Guo et al., 2020). These findings have supported the relevance of timing ERT for neuroprotection.

ERT has been also suggested to have an impact on parkinsonian disability. As an adjunct therapy to existing antiparkinsonian treatment, ERT may reduce motor disability in postmenopausal women with PD associated with motor fluctuations (Tsang, Ho, & Lo, 2000). Moreover, some studies have showed that ERT administration significantly improved peak-dose dyskinesia without worsening motor disability, thus suggesting a possible benefit on PD dyskinetic women (Nicoletti et al., 2007).

ERT potential beneficial effects have been reported not only on PD motor functions but also on cognition. In particular, in a large observational study on 10,145 elderly women with PD, nursing home residents in five US states, estrogen users were found, independently of age, less cognitively impaired and more independent in their activities of daily living (Fernandez & Lapane, 2000).

1.3 Sex and gender-related differences in clinical features of PD

Many studies have suggested that the phenotypic characteristics of PD are sexually dimorphic. It is well established that PD symptomatology goes beyond the merely motor aspects, encompassing several non-motor symptoms which involve different domains, such as psychiatric and behavioral problems, cognitive dysfunction, sleep disturbance, gastrointestinal problems, sexual dysfunction, and cardiovascular symptoms. Sex differences have been reported both in motor and in non-motor symptoms frequencies and severity.

Men tend to present with more severe motor symptoms than women. For instance, several studies have reported significantly worse rigidity in men with PD as compared to women with PD (Baba, Putzke, Whaley,

Wszolek, & Uitti, 2005), and more frequent symptoms such as writing difficulties, speech and gait problems, and sialorrhea among males than females (Scott, Borgman, Engler, Johnels, & Aquilonius, 2000). In addition, men tend to show an earlier age at symptom onset in comparison to women, with a difference of approximately two years (Alves et al., 2009; Haaxma et al., 2007). Indeed, a milder phenotype, more often a tremor-dominant, and a slower rate of decline of motor impairment are typically more frequent in women with PD (Haaxma et al., 2007). On the other hand, women have a higher prevalence of motor fluctuations and dyskinesia (Chandran et al., 2014).

Studies investigating the potential gender differences in non-motor symptoms of PD have not always reported consistent findings (Barone et al., 2009; Guo et al., 2013; Khedr, El Fetoh, Khalifa, Ahmed, & El Beh, 2013; Liu et al., 2015; Martinez-Martin et al., 2012; Solla et al., 2012; Szewczyk-Krolikowski et al., 2014). Some studies, in fact, found higher prevalence or severity of fatigue, anxiety (Barone et al., 2009; Martinez-Martin et al., 2012; Solla et al., 2012) and pain (Guo et al., 2013; Martinez-Martin et al., 2012; Picillo et al., 2013; Solla et al., 2012; Szewczyk-Krolikowski et al., 2014) in female patients, whereas sexual dysfunction (Barone et al., 2009; Martinez-Martin et al., 2012; Picillo et al., 2013; Szewczyk-Krolikowski et al., 2014) was more prevalent in male patients. However, some other studies reported not significant differences in non-motor symptoms between men and women (Khedr et al., 2013; Krishnan, Sarma, Sarma, & Kishore, 2011).

The most consistent data is related to the presence of depression (mood/apathy domain), that in almost all studies was more frequent among parkinsonian women. Along with depression, nervousness and constipation are more prevalent in women whereas men suffer more from rapid eye movement behavior disorder, daytime sleepiness, dribbling and sex-related symptoms, in particular decreased libido (Martinez-Martin et al., 2012; Nicoletti et al., 2017). Recent studies have reported an emerging sex-specific pattern for PD-associated cognitive changes, with deficits in verbal fluency and recognition of facial emotions being more prevalent in men, whilst a reduction in visuospatial cognition occur more frequently in women (Miller & Cronin-Golomb, 2010).

1.4 Sex and gender-related differences in management and treatment

For both medical and surgical treatment for PD, differential responses can be found for men and women. Numerous studies have highlighted the differences in levodopa pharmacokinetics and pharmacodynamics between the sexes (Arabia et al., 2002; Kumagai et al., 2014; Sharma, Ross, Rascol, & Brooks, 2008). Compared with men, women show higher levodopa plasma concentrations, likely related to lower body weight (Arabia et al., 2002), and reduced rates of levodopa clearance (Martinelli et al., 2003) resulting in greater levodopa bioavailability (Kompoliti et al., 2002; Kumagai et al., 2014). These findings could be related to sex discrepancy in levodopa-related complications (Warren Olanow et al., 2013), such as the higher rate of dyskinesia and the greater severity of motor and non-motor fluctuations in women than in men. Indeed, other factors other than body weight need to be taken into account to explain differences in levodopa complications observed between sexes, including abnormal plastic responses to levodopa, differences in energy metabolism (Montaurier et al., 2007) and even dimorphism in genes implicated in dopamine metabolism (Sampaio et al., 2018) or dopamine receptor profiles (Zappia et al., 2005). To date, however, no recommendations have been formulated about sex and gender-specific management of medical treatment in PD.

Only a few studies on sex differences in the response to deep brain stimulation (DBS) have been conducted. Evidences showed reduced utilization and later access to surgery in women than in men (Hariz et al., 2011; Willis et al., 2014). These data has been related to more severe anxiety related to surgery and a lower DBS referral rate among women in countries with low socioeconomic status. The benefits from DBS are reported to be similar between sexes, although women tend to show better health-related quality of life outcomes (Hariz et al., 2013).

More in general, management of people living with PD need a more personalized approach considering sex, gender and cultural context to accurately select drugs and device and organize the most appropriate medical, psychological and social support for each affected individual (Subramanian et al., 2022).

2. Sex and gender-specific differences in essential tremor

Essential tremor (ET) is considered the most common movement disorder although it is difficult to establish its prevalence with precision. It was estimated that ET affects about 1–5% of the population above the age of 60 years (Louis et al., 2020). ET is phenotypically heterogeneous and it is defined as isolated postural or kinetic bilateral tremor with variable frequency range (4–12 Hz) in the absence of other neurological signs, such as dystonia, ataxia, or parkinsonism (Bhatia et al., 2018). Typically tremor affects upper-limbs but it can also involve other location such as head, larynx (voice tremor), or lower limbs. The age of onset has a bimodal distribution with two peaks in the second and sixth decade of life (Louis & Dogu, 2007)

A review on epidemiological studies found that 11 out of 18 studies reported no sex or gender difference in ET prevalence (M:F ratio ranging from 1.19:1 to 0.78:1; median 0.95:1) (Louis & Ferreira, 2010). Six of these studies showed a statistically significant gender difference with higher prevalence of ET among men probably due to clinical, epidemiological and pathological associations between ET and PD. Instead, one of these studies showed higher prevalence among women (Louis & Ferreira, 2010). However, ET prevalence in pediatric population seems to be three-fold higher in boys compared to girls (Louis, Fernandez-Alvarez, Dure 4th, Frucht, & Ford, 2005).

Only few studies analyzed the clinical differences in ET between men and women. Female gender was strongly associated with head tremor; moreover, the association of female gender and severe hand tremor could increase the risk to develop head and/or voice tremor in ET. Instead, mid-line tremors (e.g., facial or palatal tremor) were stronger associate with hand tremors severity in males than woman (Chen et al., 2017). Male patients seem to present with more severe postural hands tremor compared to women (Hubble, Busenbark, Pahwa, Lyons, & Koller, 1997).

2.1 Sex differences in treatment response in ET

No sex-related differences have been reported concerning ET treatment. However, different therapy strategies should be considered for women with ET who would like to get pregnant. Indeed, commonly used medication for ET such as propranolol, botulinum toxin injections, primidone and topiramate have potential teratogenic side effects (Sarah M. Kranick, Mowry, Colcher, Horn, & Golbe, 2010). Drug diffusion to breast milk and placenta have been reported (Ohman, Vitols, Luef, Söderfeldt, & Tomson, 2002). Moreover, low level of topiramate has been found in breastfeeding infants (Ba & Miyasaki, 2020). However, most women with ET prefer stop medication during pregnancy instead of taking the risk of potential birth teratogenic effects from drug exposure (Zesiewicz et al., 2011).

If tremor is severe and disabling, a possible alternative can be thalamic DBS surgery or MRI-guided focal ultrasound thalamotomy before pregnancy (Ba & Miyasaki, 2020).

3. Sex and gender-specific differences in dystonia

3.1 Epidemiology

Conflicting data reports on prevalence of different dystonia subtypes, probably due to methodological differences in epidemiological studies (Steeves, Day, Dykeman, Jette, & Pringsheim, 2012). Different prevalence frequencies were reported in clinical based studies of isolated and generalized dystonia in European and Japan population, but this heterogeneity could be also attributed to non-uniform studies methodologies (Steeves et al., 2012). One study found a prevalence of focal dystonia in adult age of about 16 per 100,000 (Steeves et al., 2012). Another study estimated a higher prevalence of 732 per 100,000 individuals: according to authors this greater value is probably due to undiagnosed cases (Müller et al., 2002).

Dystonia is characterized by abnormal movements and/or postures that are caused by sustained or intermittent muscle contractions (Bhatia et al., 2018). In people with isolated idiopathic (IID) and genetic dystonia (GD), dystonia is the unique or the most disabling symptom (Das et al., 2007). Focal limb dystonia has usually the onset in the fourth decade of life, and presents no sex differences in prevalence, except for musician dystonia which shows an earlier onset and higher prevalence in men (Williams et al., 2017). Other forms of dystonia, such as blepharospasm, oromandibular dystonia, Meige syndrome, cervical and laryngeal dystonia, are more common in women than men (Asgeirsson, Jakobsson, Hjaltason, Jonsdottir, & Sveinbjornsdottir, 2006; Soland, Bhatia, & Marsden, 1996; Williams et al., 2017). Conversely tardive dystonia seems to be more prevalent in young men (Savitt & Jankovic, 2018). For the most common forms, such as cervical dystonia, focal hand dystonia and musician dystonia, men had an earlier mean age at onset than women (Williams et al., 2017). Lifestyle and environments factor plays an important role in focal dystonia in susceptible individuals: car accidents requiring hospitalization and surgical procedures were found to be risk factor for development of cervical dystonia (Molloy et al., 2015). Moreover, individuals living in areas with elevated sun exposure levels had higher prevalence of the blepharospasm (Molloy et al., 2016). In Ireland population, indeed, low prevalence of individuals with blepharospasm was linked to lower sunlight exposure.

However, these findings do not explain why females have three times higher risk to develop blepharospasm than males (Molloy et al., 2016) and eighth times than Puglia region population in Southern Italy (Defazio et al., 2001).

3.2 Sex-specific differences in genetics of dystonia

Genetics studies were initially hindered by scarcity of families with hereditary form of dystonia and by wide phenotypic variability of presentation (Balint et al., 2018). The first dystonia's gene TOR1A, causative of genetic DYT-1 dystonia, was discovered in 1997 in Ashkenazi Jewish population, and it is considered the most frequent genetic cause of childhood-onset generalized dystonia (Ozelius et al., 1997). Subsequently, THAP1, the second gene associated with primary dystonia, was identified in three Amish-Mennonite families (Fuchs et al., 2009). THAP1 dystonia seems to be slightly more prevalent in females (LeDoux et al., 2012). The advent of next-generation sequencing has had a remarkable impact on finding new genes associated to dystonia such as GNAL (Fuchs et al., 2013) and ANO3 (Miltgen et al., 2016). This new technology has revolutionized the approach to genetic pathologies and also of isolated dystonia. Because of its wide clinical heterogeneity, exome sequencing could be more useful compared to "gene by gene" analysis (Balint et al., 2018). However, this new technology has intrinsic limit such as the ability to interpret the wide sporadic variant found in studies. Indeed, some missenses variants were identified in both ANO3 and TOR1A healthy and affected individuals but their pathogenetic role requires future investigations (Zech et al., 2014; Zech et al., 2016).

New evidence suggests that the sex chromosome genes (i.e., X- and Y-linked genes) also play an important role in sex-related differences in dystonia incidence. In a recent study, it has been postulated that "sex-determining region on the Y" gene (SRY) at Yp11.3 could influence the modulation of dopamine biosynthesis because its expression in dopamine-abundant brain regions (i.e., thalamus, cortical areas, and substantia nigra pars compacta in males), where it regulates dopamine biosynthesis and functions. Given the localization and function of SRY in male dopamine neurons, abnormal regulation of this gene could be linked to the increased risk of movement disorders in males (Loke, Harley, & Lee, 2015; Rafee, O'Riordan, Reilly, & Hutchinson, 2021).

GTP cyclohydrolase 1 (GCH1) deficiency induces childhood-onset dystonia, with a typical excellent response to low doses of levodopa, and it is less frequent in males compared to females (Ichinose et al., 1994).

Dystonia associated to DYT-1, DYT-6 and DYT-11 mutation present with no sex prevalence, though DYT-11 is associated with reduced penetrance on maternal transmission (Zimprich et al., 2001). DYT-3 (Lubag syndrome) is a X-linked dystonia that has been reported as more prevalent in men than women (Lee et al., 2011).

Another possible correlation is related to gamma aminobutyric acid (GABA). This amino acid is the main inhibitory neurotransmitter in the brain and its deficiency in animals models is related to dystonia (Beales, Lorden, Walz, & Oltmans, 1990; Levy & Hallett, 2002). Young women presented with higher GABA level but a more dramatic age-related decline compared to men (Loke et al., 2015).

3.3 Pregnancy in women with dystonia

Women can be affected by dystonia during their reproductive age. It has been described a case of abrupt cervical dystonia onset during the first trimester that improved in the third one (Buccoliero, Palmeri, Malandrini, Dotti, & Federico, 2007). This could add to other reports suggesting that estrogen fluctuations could play a role in movement disorders onset or worsening (Cersósimo, Bertoti, Roca, & Micheli, 2004; Smith & Evatt, 2004) both during pregnancy and contraceptive therapy, likely related to their effects on basal ganglia circuit (Cogen & Zimmerman, 1979; Hruska & Silbergeld, 1980). In a study including 62 women with dystonia, among 27 who had pregnancy four of them had worsening of dystonia (Gwinn-Hardy, Adler, Weaver, Fish, & Newman, 2000). In a similar study with ten woman affected by dystonia during pregnancy, two cases reported on symptoms worsening while three had an improvment of symptoms (Rogers & Fahn, 1994).

Levodopa-responsive dystonia, Wilson's disease and other young-onset genetic dystonia (i.e., DYT-1) are the most common types affecting women in pregnancy (Bordelon & Smith, 2007; Kranick et al., 2010). Treatment of dystonia during pregnancy needs to be carefully evaluated. It should be reminded that neuroleptics have been responsible of drug-induced dystonia in patients taking medications for hyperemesis gravidarum (Ba & Miyasaki, 2020). Levodopa use was widely accepted during pregnancy (Nageshwaran, Smith, & Bordelon, 2016). Studies on levodopa-responsive dystonia demonstrated that a low dose of levodopa (up to 300 mg/day) can be administered in pregnancy without adverse events (Dostal, Weber-Schoendorfer, Sobesky, & Schaefer, 2013; Nomoto, Kaseda, Iwata, Osame, & Fukuda, 1997), but its FDA class risk remains C. It has been showed that L-Dopa crosses the placenta but not the peripheral decarboxylase inhibitor (Merchant et al., 1995). A study reported a case of a young woman who got pregnant while using L-Dopa/carbidopa for levodopa-responsive dystonia. During this combination therapy she had two abortions, she was shifted to monotherapy only with L-Dopa and she had a good pregnancy outcome (Nomoto et al., 1997).

Literature review reported on 114 patients treated with levodopa during pregnancies, including PD and levodopa-responsive dystonia, without association with an increase rate of spontaneous abortion, adverse event or teratogenicity, thus supporting its use in pregnant women with severe motor symptoms (Seier & Hiller, 2017).

Benzodiazepine exposure during pregnancy do not seems to have teratogenic effect, but they may be used careful if associated with antidepressant because of congenital malformations reported in this combination therapy (Grigoriadis et al., 2019). Anticholinergic drugs have limited data on risk during pregnancy to the fetus and for this reason caution is recommended (Seier & Hiller, 2017).

Botulinum neuro-toxin (BoNT) injections might be used in focal dystonia during pregnancy. BoNT is an high weight purified toxin (Botox® product monograph). In a correct procedure, it is unlikely that it can enter in systemic circulation. Moreover, it has been demonstrated that those molecules larger than 500 Daltons can't cross the placenta (Pacifici & Nottoli, 1995). Both botulin toxin A and B are not linked to complications in pregnant women and fetuses, but its FDA pregnancy risk category remains C (Ba & Miyasaki, 2020; Li & Tang, 2020).

Limited evidence supports safety of globus pallidus internus DBS in patients who became pregnant after insertion of the device (Scelzo et al., 2015).

3.4 Functional dystonia

Functional dystonia (FD) is associated with some specific psychiatric features, including childhood trauma, anxiety depression, and dissociative disorders (Ibrahim et al., 2009; Kranick et al., 2011). Women are affected by functional disorders more often than men. In large cohort studies, the percentage of females was estimated between 60% and 90% (Ani, Reading, Lynn, Forlee, & Garralda, 2013; Perez et al., 2021; Stephen, Fung, Lungu, & Espay, 2021; Tinazzi et al., 2020). However, some movement disorders, for example, functional myoclonus and functional parkinsonism, seem to present similar frequency in men and women (Cubo et al., 2005).

In women affected by FD, mood disorders usually anticipate the motor symptoms onset and are associated with an early onset of motor symptoms in comparison with women with no mood disorders (Ndukwe, O'Riordan, Walsh, & Hutchinson, 2020).

To note, the diagnosis of FD is still difficult, but some clinical characteristic can help physician to evaluate the etiology of dystonia. It is reasonable to suspect FD when it is fixed at onset whereas organic dystonia is often fluctuating at early stage. Often the appearance of functional motor symptoms is linked to a minor physical injury. Atypical localization and age of onset (i.e. adult onset foot dystonia) or spreading to other body parts can help physician to recognize FD (Galli, Béreau, Magnin, Moulin, & Aybek, 2020).

4. Sex and gender-specific differences in chorea: Huntington's disease

4.1 Epidemiology

Huntington's disease (HD) is an autosomal-dominant, progressive neurodegenerative disorder characterized by chorea, dystonia, cognitive decline, and psychiatric symptoms. It is caused by triplet expanded CAG repeat on the huntingtin (HTT) gene on chromosome 4; when these repeats exceed the 41 one the disease is fully penetrant (Ho et al., 2001; Rubinsztein et al., 1996; Walker, 2007). HD presents with equal penetrance and prevalence in both sexes even if a large American cohort study showed a slightly prevalence among women (Bruzelius et al., 2019). Earlier HD onset in males is associated to genotype apolipoprotein E ε2ε3 genotype (Kehoe, Krawczak, Harper, Owen, & Jones, 1999). Paternal transmission is associated with large expansions of CAG repeats and it is likely due to major instability in course of spermatogenesis than oogenesis (Ranen et al., 1995). These findings could explain the occurrence of anticipation and the likelihood of paternal inheritance in children with early onset symptoms whereas maternal inheritance is more frequently associated with later onset of the disease (Farrer, Cupples, Kiely, Conneally, & Myers, 1992; Walker, 2007).

4.2 Sex-related differences in HD clinical course

The sex of the parent transmitting the mutation plays a crucial role in determining the clinical features of HD in children. Indeed, an early onset, more characterized by a parkinsonian rather than a choreiform phenotype, is linked to paternal inheritance (Ba & Miyasaki, 2020). Paternal age at the time of conception correlates with an earlier onset of the disease in the offspring. The higher the father's age, the earlier the disease manifests in the child. The same is not observed for maternal inheritance. This is probably due to a poor capacity of silencing of the abnormal sequences that are normally silenced and to an expansion of them in women (Wizemann, 2001).

Women have a more severe disease course and faster progression, with particular impairment in motor function. On the other hand, men tend to present a longer disease duration, and no significant differences in age of onset (Zielonka et al., 2013). In one study (Zielonka et al., 2013) collecting data from 1267 patients with HD (636 women) from the registry project of European Huntington's Disease Network (EHDN), gender differences were found in disease course: women showed lower motor and functional Unified Huntington's Disease Rating Scale (UHDRS) scores at the age of onset, lower disease burden, shorter duration and less alcohol and tobacco use. However, women presented with faster rate of progression compared to men with HD onset in middle-age in the same domains. Furthermore, they confirmed the negative correlation between the number of CAG repeats in the HTT gene with the age of onset of the disease, but no gender difference was observed. In this study, authors also demonstrated that a higher percentage of female than male presented with symptoms and signs of HD if corrected for CAG repeat triplets (Zielonka et al., 2013).

The role of depression is predominant in patient with HD and is correlated with worsening of cognitive motor and functional features (Aziz, Anguelova, Marinus, Lammers, & Roos, 2010; Wittchen et al., 2011). Data from the European Huntington's disease cohort registry did not find independent effect of depression in HD patients. Same results were obtained analyzing anxiety and depressive symptoms in European HD patients, and no sex differences have been found (Van Duijn et al., 2014). Although suicide is the cause of death in about 9% of cases (Frank, 2014), the homogeneity of depressive symptoms between males and females in HD patients compared to the general population has been found, and it could be due to the higher level of disability of HD patients. Suicidal ideation and suicide attempts and psychotic symptoms similarly did not show any sex or gender-related differences in prevalence or incidence rates (Dale et al., 2016; Zielonka et al., 2013).

4.3 Sex hormones and HD

As in other neurological disorders, estrogens might play a protective role on neurodegeneration related to HD (Bode et al., 2008; Padovan-Neto et al., 2019).

Lifelong exposure and substitutive estrogen therapy has been limitedly evaluated because of early onset of the disease. Plasmatic levels of testosterone and dehydroepiandrosterone have been correlated to depression in women patients affected with HD (Markianos, Panas, Kalfakis, & Vassilopoulos, 2007). In men, low plasma testosterone levels have been associated with high disease severity and dementia but not with depression or psychotic features in men with HD (Markianos, Panas, Kalfakis, & Vassilopoulos, 2005).

Interestingly, sex differences have also been reported in metabolism functions of patients affected by HD. Female showed lower bone mineral density compared to healthy controls and males with HD. Men affected by HD presented with a significant reduction in lean body mass; on the other hand, women seem to suffer less than men regarding body mass alteration. this can be possible due to alteration of metabolism and energy consumption between sexes (Costa de Miranda et al., 2019).

4.4 Pregnancy in women with HD

Managing the desire for pregnancy in pre-symptomatic or non-symptomatic high-risk women is an important issue that needs to be addressed through genetic counseling. Carrying out a genetic test in the absence of symptoms constitutes a great ethical dilemma due to the fatality of the disease and may involve several family members who may not have chosen to perform the test. HD does not adversely affect the course of pregnancy or childbirth. In a study involving 868 couples, 9% conducted intrauterine tests and most of them chose to continue pregnancy (Bordelon & Smith, 2007).

HD treatment consists of dopamine antagonists, neuroleptics and amantadine. The latter is associated with cardiac malformations in the unborn child (Bordelon & Smith, 2007) whereas neuroleptics are designated FDA pregnancy class C (Robottom & Weiner, 2011). Some reports described the use of the aloperidolo, clorpromazina and benzodiazepine in the most severe cases of chorea (Quinn & Schrag, 1998; Robottom & Weiner, 2011).

Tricyclic antidepressants, nortriptyline and desipramine are the most used antidepressants during pregnancy (Quinn, 2001). Fluoxamine should be avoided due to its possible adverse effects on the fetus (Huntington & Zantop, 2004; Wisner, Gelenberg, Leonard, Zarin, & Frank, 1999). To date, there are insufficient safety data to recommend other antidepressants (i.e., bupropione, paroxetina, sertralina, venlafaxina) or MAO inhibitors in pregnant women. As for breastfeeding, it is known that neuroleptics pass into breast milk as well as tetrabenazine but to date there are no guidelines on this topic yet (Drugs and Lactation Database: LactMed, 2006).

4.5 Chorea gravidarum

Chorea gravidarum (CG) is a syndrome characterized by choreiform movement during pregnancy described for the first time in 1661 by Horstius (Willson & Preece, 1932). Rheumatic cardiopathy was found in 86% of CG cases (Willson & Preece, 1932). It was linked to 33% of maternal death risk and 50% of fetal death risk. This percentage was dramatically decreased thanks to the availability of penicillin and the incidence has been estimated around one to 139,000 (Zegart & Schwarz, 1968).

CG was supposed to be induced by the effect of pregnancy hormones on basal ganglia damaged from rheumatic fever (Maia et al., 2012), and it was observed also in women who used oral contraceptive hormones also in absence of history of previous rheumatic fever (Nausieda, Koller, Weiner, & Klawans, 1979). The movements are similar to choreiform movement in HD, characterized by flitting, dance-like and random movements mostly involving the limbs and face and disappearing during sleep (Parsons, 2006). Most women who develop CG presented with symptoms within the first trimester of pregnancy (Cardoso, 2002; Willson & Preece, 1932), and in most cases there was a spontaneously remission before delivery (Palanivelu, 2007). Psychiatric symptoms could also be associated to CG like personality change, depression, hallucination and chronic cognitive deficits (Brockington, 2006).

Dopamine blockers like tetrabenazine are first choice treatment is but they are FDA class C and should be avoided in the first trimester of pregnancy. Haloperidol and clorpromazine are also class C drugs but there is some evidence that they could be used at low dose during pregnancy (Robottom & Weiner, 2011). The American Academy of Pediatrics suggests haloperidol instead of dopamine blockers because of its less teratogenic effects in the severe manifestation of CG (Committee on Drugs. American Academy of Pediatrics, 2000).

5. Conclusions

Sex and gender-sensitive approach to diagnosis, treatment, and management of movement disorders is an innovative and essential component towards a more efficient and comprehensive care of people suffering from a heterogeneous group of diseases.

Sex and gender-related variables need to be better considered in medical sciences, starting from research development design to analysis and interpretation of data.

This "call" towards a personalized medicine should provide practical guidance on prevention, care and management of the different issues that can dramatically impact the quality of life of men and women with movement disorders.

6. Key points

- Distribution of movement disorders is heterogeneous between sexes (higher prevalence in men for PD, ET and focal limb dystonia; in women for functional disorders and HD).
- Both biological and socio-cultural factors differently impact clinical care and quality of life of men and women suffering from movement disorders.
- Treatment of movement disorders during pregnancy needs to be carefully evaluated and planned.
- Sex and gender-related differences in epidemiology, clinical features and therapeutical responses of the most common movement disorders should result in optimization of clinical practice.

References

Abbas, M. M., Xu, Z., & Tan, L. (2017). Epidemiology of Parkinson's disease—East versus West. *Movement Disorders Clinical Practice*, *5*(1), 14–28.

Alves, G., Müller, B., Herlofson, K., HogenEsch, I., Telstad, W., Aarsland, D., et al. (2009). Incidence of Parkinson's disease in Norway: The Norwegian ParkWest study. *Journal of Neurology, Neurosurgery, and Psychiatry*, *80*(8), 851–857.

Ani, C., Reading, R., Lynn, R., Forlee, S., & Garralda, E. (2013). Incidence and 12-month outcome of non-transient childhood conversion disorder in the U.K. and Ireland. *The British Journal of Psychiatry: the Journal of Mental Science*, *202*, 413–418.

Arabia, G., Zappia, M., Bosco, D., Crescibene, L., Bagalà, A., Bastone, L., et al. (2002). Body weight, levodopa pharmacokinetics and dyskinesia in Parkinson's disease. *Neurological Sciences: Official Journal of the Italian Neurological Society and of the Italian Society of Clinical Neurophysiology*, *23*(Suppl. 2), S53–S54.

Arevalo, M. A., Azcoitia, I., & Garcia-Segura, L. M. (2015). The neuroprotective actions of oestradiol and oestrogen receptors. *Nature Reviews. Neuroscience*, *16*(1), 17–29.

Ascherio, A., Chen, H., Schwarzschild, M. A., Zhang, S. M., Colditz, G. A., & Speizer, F. E. (2003). Caffeine, postmenopausal estrogen, and risk of Parkinson's disease. *Neurology*, *60*(5), 790–795.

Ascherio, A., & Schwarzschild, M. A. (2016). The epidemiology of Parkinson's disease: Risk factors and prevention. *The Lancet. Neurology*, *15*(12), 1257–1272.

Asgeirsson, H., Jakobsson, F., Hjaltason, H., Jonsdottir, H., & Sveinbjornsdottir, S. (2006). Prevalence study of primary dystonia in Iceland. *Movement Disorders: Official Journal of the Movement Disorder Society*, *21*(3), 293–298.

Aziz, N. A., Anguelova, G. V., Marinus, J., Lammers, G. J., & Roos, R. A. (2010). Sleep and circadian rhythm alterations correlate with depression and cognitive impairment in Huntington's disease. *Parkinsonism & Related Disorders*, *16*(5), 345–350.

Ba, F., & Miyasaki, J. M. (2020). Movement disorders in pregnancy. *Handbook of Clinical Neurology*, *172*, 219–239.

Baba, Y., Putzke, J. D., Whaley, N. R., Wszolek, Z. K., & Uitti, R. J. (2005). Gender and the Parkinson's disease phenotype. *Journal of Neurology*, *252*(10), 1201–1205.

Balint, B., Mencacci, N. E., Valente, E. M., Pisani, A., Rothwell, J., Jankovic, J., et al. (2018). Dystonia. *Nature Reviews. Disease Primers*, *4*(1), 25.

Barone, P., Antonini, A., Colosimo, C., Marconi, R., Morgante, L., Avarello, T. P., et al. (2009). The PRIAMO study: A multicenter assessment of nonmotor symptoms and their impact on quality of life in Parkinson's disease. *Movement Disorders: Official Journal of the Movement Disorder Society*, *24*(11), 1641–1649.

Beales, M., Lorden, J. F., Walz, E., & Oltmans, G. A. (1990). Quantitative autoradiography reveals selective changes in cerebellar GABA receptors of the rat mutant dystonic. *The Journal of Neuroscience: The Official Journal of the Society for Neuroscience*, *10*(6), 1874–1885.

Benedetti, M. D., Maraganore, D. M., Bower, J. H., McDonnell, S. K., Peterson, B. J., Ahlskog, J. E., et al. (2001). Hysterectomy, menopause, and estrogen use preceding Parkinson's disease: An exploratory case-control study. *Movement Disorders: Official Journal of the Movement Disorder Society*, *16*(5), 830–837.

Bhatia, K. P., Bain, P., Bajaj, N., Elble, R. J., Hallett, M., Louis, E. D., et al. (2018). Consensus statement on the classification of tremors. From the task force on tremor of the International Parkinson and Movement Disorder Society. *Movement Disorders: Official Journal of the Movement Disorder Society*, *33*(1), 75–87.

Bode, F. J., Stephan, M., Suhling, H., Pabst, R., Straub, R. H., Raber, K. A., et al. (2008). Sex differences in a transgenic rat model of Huntington's disease: Decreased 17beta-estradiol levels correlate with reduced numbers of DARPP32+ neurons in males. *Human Molecular Genetics*, *17*(17), 2595–2609.

Bordelon, Y. M., & Smith, M. (2007). Movement disorders in pregnancy. *Seminars in Neurology*, *27*(5), 467–475. https://doi.org/10.1055/s-2007-991128. Botox cosmetic [product monograph]. Irvine, CA: Allergan Inc; 2011.

Braak, H., Del Tredici, K., Rüb, U., de Vos, R. A., Jansen Steur, E. N., & Braak, E. (2003). Staging of brain pathology related to sporadic Parkinson's disease. *Neurobiology of Aging*, *24*(2), 197–211.

Brann, D. W., Dhandapani, K., Wakade, C., Mahesh, V. B., & Khan, M. M. (2007). Neurotrophic and neuroprotective actions of estrogen: Basic mechanisms and clinical implications. *Steroids*, *72*(5), 381–405.
Brockington, I. (2006). Psychosis complicating Chorea gravidarum. *Archives of Women's Mental Health*, *9*(2), 113–114.
Bruzelius, E., Scarpa, J., Zhao, Y., Basu, S., Faghmous, J. H., & Baum, A. (2019). Huntington's disease in the United States: Variation by demographic and socioeconomic factors. *Movement Disorders: Official Journal of the Movement Disorder Society*, *34*(6), 858–865.
Buccoliero, R., Palmeri, S., Malandrini, A., Dotti, M. T., & Federico, A. (2007). A case of dystonia with onset during pregnancy. *Journal of the Neurological Sciences*, *260*(1-2), 265–266.
Cabezas, R., Avila, M., Gonzalez, J., El-Bachá, R. S., Báez, E., García-Segura, L. M., et al. (2014). Astrocytic modulation of blood brain barrier: Perspectives on Parkinson's disease. *Frontiers in Cellular Neuroscience*, *8*, 211.
Cardoso, F. (2002). Chorea gravidarum. *Archives of Neurology*, *59*(5), 868–870.
Cersósimo, M. G., Bertoti, A., Roca, C. U., & Micheli, F. (2004). Botulinum toxin in a case of hemimasticatory spasm with severe worsening during pregnancy. *Clinical Neuropharmacology*, *27*(1), 6–8.
Chandran, S., Krishnan, S., Rao, R. M., Sarma, S. G., Sarma, P. S., & Kishore, A. (2014). Gender influence on selection and outcome of deep brain stimulation for Parkinson's disease. *Annals of Indian Academy of Neurology*, *17*(1), 66–70.
Chen, W., Hopfner, F., Szymczak, S., Granert, O., Müller, S. H., Kuhlenbäumer, G., et al. (2017). Topography of essential tremor. *Parkinsonism & Related Disorders*, *40*, 58–63.
Cogen, P. H., & Zimmerman, E. A. (1979). Ovarian steroid hormones and cerebral function. *Advances in Neurology*, *26*, 123–133.
Colombo, D., Abbruzzese, G., Antonini, A., Barone, P., Bellia, G., Franconi, F., et al. (2015). The "gender factor" in wearing-off among patients with Parkinson's disease: A post hoc analysis of DEEP study. *TheScientificWorldJOURNAL*, *2015*, 787451.
Committee on Drugs. American Academy of Pediatrics. (2000). Use of psychoactive medication during pregnancy and possible effects on the fetus and newborn. *Pediatrics*, *105*(4 Pt 1), 880–887.
Costa de Miranda, R., Di Lorenzo, N., Andreoli, A., Romano, L., De Santis, G. L., Gualtieri, P., et al. (2019). Body composition and bone mineral density in Huntington's disease. *Nutrition (Burbank, Los Angeles County, Calif.)*, *59*, 145–149.
Cubo, E., Hinson, V. K., Goetz, C. G., Garcia Ruiz, P., Garcia de Yebenes, J., Marti, M. J., et al. (2005). Transcultural comparison of psychogenic movement disorders. *Movement Disorders: Official Journal of the Movement Disorder Society*, *20*(10), 1343–1345.
Currie, L. J., Harrison, M. B., Trugman, J. M., Bennett, J. P., & Wooten, G. F. (2004). Postmenopausal estrogen use affects risk for Parkinson disease. *Archives of Neurology*, *61*(6), 886–888.
Dahodwala, N., Pei, Q., & Schmidt, P. (2016). Sex differences in the clinical progression of Parkinson's disease. *Journal of Obstetric, Gynecologic, and Neonatal Nursing: JOGNN*, *45*(5), 749–756.
Dale, M., Maltby, J., Shimozaki, S., Cramp, R., Rickards, H., & REGISTRY Investigators of the European Huntington's Disease Network. (2016). Disease stage, but not sex, predicts depression and psychological distress in Huntington's disease: A European population study. *Journal of Psychosomatic Research*, *80*, 17–22.
Das, S. K., Banerjee, T. K., Biswas, A., Roy, T., Raut, D. K., Chaudhuri, A., et al. (2007). Community survey of primary dystonia in the city of Kolkata, India. *Movement Disorders: Official Journal of the Movement Disorder Society*, *22*(14), 2031–2036.

de Lau, L. M., Verbaan, D., Marinus, J., & van Hilten, J. J. (2014). Survival in Parkinson's disease. Relation with motor and non-motor features. *Parkinsonism & Related Disorders*, *20*(6), 613–616.

Defazio, G., Livrea, P., De Salvia, R., Manobianca, G., Coviello, V., Anaclerio, D., et al. (2001). Prevalence of primary blepharospasm in a community of Puglia region, Southern Italy. *Neurology*, *56*(11), 1579–1581.

Deuschl, G., Beghi, E., Fazekas, F., Varga, T., Christoforidi, K. A., Sipido, E., et al. (2020). The burden of neurological diseases in Europe: An analysis for the Global Burden of Disease Study 2017. *The Lancet. Public health*, *5*(10), e551–e567.

Diem-Zangerl, A., Seppi, K., Wenning, G. K., Trinka, E., Ransmayr, G., Oberaigner, W., et al. (2009). Mortality in Parkinson's disease: A 20-year follow-up study. *Movement Disorders: official journal of the Movement Disorder Society*, *24*(6), 819–825.

Dostal, M., Weber-Schoendorfer, C., Sobesky, J., & Schaefer, C. (2013). Pregnancy outcome following use of levodopa, pramipexole, ropinirole, and rotigotine for restless legs syndrome during pregnancy: A case series. *European Journal of Neurology*, *20*(9), 1241–1246.

Drugs and lactation database: LactMed. (2006). J Electron Resour Med Libr. In *Vol. 9. Drugs and lactation database (LactMed) [Internet]* (pp. 272–277). Bethesda (MD): National Library of Medicine (US). Available from: https://www.ncbi.nlm.nih.gov/books/NBK501922/.

Dye, R. V., Miller, K. J., Singer, E. J., & Levine, A. J. (2012). Hormone replacement therapy and risk for neurodegenerative diseases. *International Journal of Alzheimer's Disease*, *2012*, 258454.

Farrer, L. A., Cupples, L. A., Kiely, D. K., Conneally, P. M., & Myers, R. H. (1992). Inverse relationship between age at onset of Huntington disease and paternal age suggests involvement of genetic imprinting. *American Journal of Human Genetics*, *50*(3), 528–535.

Fernandez, H. H., & Lapane, K. L. (2000). Estrogen use among nursing home residents with a diagnosis of Parkinson's disease. *Movement Disorders: Official Journal of the Movement Disorder Society*, *15*(6), 1119–1124.

Frank, S. (2014). Treatment of Huntington's disease. *Neurotherapeutics: The Journal of the American Society for Experimental NeuroTherapeutics*, *11*(1), 153–160.

Frigerio, R., Sanft, K. R., Grossardt, B. R., Peterson, B. J., Elbaz, A., Bower, J. H., et al. (2006). Chemical exposures and Parkinson's disease: A population-based case-control study. *Movement Disorders: Official Journal of the Movement Disorder Society*, *21*(10), 1688–1692.

Fuchs, T., Gavarini, S., Saunders-Pullman, R., Raymond, D., Ehrlich, M. E., Bressman, S. B., et al. (2009). Mutations in the THAP1 gene are responsible for DYT6 primary torsion dystonia. *Nature Genetics*, *41*(3), 286–288.

Fuchs, T., Saunders-Pullman, R., Masuho, I., Luciano, M. S., Raymond, D., Factor, S., et al. (2013). Mutations in GNAL cause primary torsion dystonia. *Nature Genetics*, *45*(1), 88–92.

Galli, S., Béreau, M., Magnin, E., Moulin, T., & Aybek, S. (2020). Functional movement disorders. *Revue Neurologique*, *176*(4), 244–251.

GBD 2016 Neurology Collaborators. (2019). Global, regional, and national burden of neurological disorders, 1990-2016: A systematic analysis for the Global Burden of Disease Study 2016. *The Lancet. Neurology*, *18*(5), 459–480.

Gillies, G. E., & McArthur, S. (2010). Estrogen actions in the brain and the basis for differential action in men and women: A case for sex-specific medicines. *Pharmacological Reviews*, *62*(2), 155–198.

Grigoriadis, S., Graves, L., Peer, M., Mamisashvili, L., Dennis, C. L., Vigod, S. N., et al. (2019). Benzodiazepine use during pregnancy alone or in combination with an antidepressant and congenital malformations: Systematic review and meta-analysis. *The Journal of Clinical Psychiatry*, *80*(4), 18r12412.

Guo, H., Liu, M., Zhang, L., Wang, L., Hou, W., Ma, Y., et al. (2020). The critical period for neuroprotection by estrogen replacement therapy and the potential underlying mechanisms. *Current Neuropharmacology*, *18*(6), 485–500.

Guo, X., Song, W., Chen, K., Chen, X., Zheng, Z., Cao, B., et al. (2013). Gender and onset age-related features of non-motor symptoms of patients with Parkinson's disease- -a study from Southwest China. *Parkinsonism & Related Disorders*, *19*(11), 961–965.

Gwinn-Hardy, K. A., Adler, C. H., Weaver, A. L., Fish, N. M., & Newman, S. J. (2000). Effect of hormone variations and other factors on symptom severity in women with dystonia. *Mayo Clinic Proceedings*, *75*(3), 235–240.

Haaxma, C. A., Bloem, B. R., Borm, G. F., Oyen, W. J., Leenders, K. L., Eshuis, S., et al. (2007). Gender differences in Parkinson's disease. *Journal of Neurology, Neurosurgery, and Psychiatry*, *78*(8), 819–824.

Hariz, G. M., Limousin, P., Zrinzo, L., Tripoliti, E., Aviles-Olmos, I., Jahanshahi, M., et al. (2013). Gender differences in quality of life following subthalamic stimulation for Parkinson's disease. *Acta Neurologica Scandinavica*, *128*(4), 281–285.

Hariz, G. M., Nakajima, T., Limousin, P., Foltynie, T., Zrinzo, L., Jahanshahi, M., et al. (2011). Gender distribution of patients with Parkinson's disease treated with subthalamic deep brain stimulation; a review of the 2000-2009 literature. *Parkinsonism & Related Disorders*, *17*(3), 146–149.

Henderson, V. W. (2007). Alzheimer's disease and other neurological disorders. *Climacteric: The Journal of the International Menopause Society*, *10*(Suppl 2), 92–96.

Hirsch, L., Jette, N., Frolkis, A., Steeves, T., & Pringsheim, T. (2016). The incidence of Parkinson's disease: A systematic review and meta-analysis. *Neuroepidemiology*, *46*(4), 292–300.

Ho, L. W., Carmichael, J., Swartz, J., Wyttenbach, A., Rankin, J., & Rubinsztein, D. C. (2001). The molecular biology of Huntington's disease. *Psychological Medicine*, *31*(1), 3–14.

Holmes, S., Abbassi, B., Su, C., Singh, M., & Cunningham, R. L. (2013). Oxidative stress defines the neuroprotective or neurotoxic properties of androgens in immortalized female rat dopaminergic neuronal cells. *Endocrinology*, *154*(11), 4281–4292.

Holmqvist, S., Chutna, O., Bousset, L., Aldrin-Kirk, P., Li, W., Björklund, T., et al. (2014). Direct evidence of Parkinson pathology spread from the gastrointestinal tract to the brain in rats. *Acta Neuropathologica*, *128*(6), 805–820.

Hruska, R. E., & Silbergeld, E. K. (1980). Increased dopamine receptor sensitivity after estrogen treatment using the rat rotation model. *Science (New York, N.Y.)*, *208*(4451), 1466–1468.

Hubble, J. P., Busenbark, K. L., Pahwa, R., Lyons, K., & Koller, W. C. (1997). Clinical expression of essential tremor: Effects of gender and age. *Movement Disorders: Official Journal of the Movement Disorder Society*, *12*(6), 969–972.

Huntington, J., & Zantop, V. (2004). Antidepressant medications in pregnancy. *American Family Physician*, *70*(11), 2195–2196.

Ibrahim, N. M., Martino, D., van de Warrenburg, B. P., Quinn, N. P., Bhatia, K. P., Brown, R. J., et al. (2009). The prognosis of fixed dystonia: A follow-up study. *Parkinsonism & Related Disorders*, *15*(8), 592–597.

Ichinose, H., Ohye, T., Takahashi, E., Seki, N., Hori, T., Segawa, M., et al. (1994). Hereditary progressive dystonia with marked diurnal fluctuation caused by mutations in the GTP cyclohydrolase I gene. *Nature Genetics*, *8*(3), 236–242.

Kehoe, P., Krawczak, M., Harper, P. S., Owen, M. J., & Jones, A. L. (1999). Age of onset in Huntington disease: Sex specific influence of apolipoprotein E genotype and normal CAG repeat length. *Journal of Medical Genetics*, *36*(2), 108–111.

Khedr, E. M., El Fetoh, N. A., Khalifa, H., Ahmed, M. A., & El Beh, K. M. (2013). Prevalence of non motor features in a cohort of Parkinson's disease patients. *Clinical Neurology and Neurosurgery*, *115*(6), 673–677.

Kim, R., Yoo, D., Jung, Y. J., Han, K., & Lee, J. Y. (2020). Sex differences in smoking, alcohol consumption, and risk of Parkinson's disease: A nationwide cohort study. *Parkinsonism & Related Disorders*, *71*, 60–65.

Kompoliti, K., Adler, C. H., Raman, R., Pincus, J. H., Leibowitz, M. T., Ferry, J. J., et al. (2002). Gender and pramipexole effects on levodopa pharmacokinetics and pharmacodynamics. *Neurology*, *58*(9), 1418–1422.

Kranick, S., Ekanayake, V., Martinez, V., Ameli, R., Hallett, M., & Voon, V. (2011). Psychopathology and psychogenic movement disorders. *Movement Disorders: Official Journal of the Movement Disorder Society*, *26*(10), 1844–1850.

Kranick, S. M., Mowry, E. M., Colcher, A., Horn, S., & Golbe, L. I. (2010). Movement disorders and pregnancy: A review of the literature. *Movement Disorders: Official Journal of the Movement Disorder Society*, *25*(6), 665–671.

Krishnan, S., Sarma, G., Sarma, S., & Kishore, A. (2011). Do nonmotor symptoms in Parkinson's disease differ from normal aging? *Movement Disorders: Official Journal of the Movement Disorder Society*, *26*(11), 2110–2113.

Kumagai, T., Nagayama, H., Ota, T., Nishiyama, Y., Mishina, M., & Ueda, M. (2014). Sex differences in the pharmacokinetics of levodopa in elderly patients with Parkinson disease. *Clinical Neuropharmacology*, *37*(6), 173–176.

Küppers, E., Ivanova, T., Karolczak, M., & Beyer, C. (2000). Estrogen: A multifunctional messenger to nigrostriatal dopaminergic neurons. *Journal of Neurocytology*, *29*(5-6), 375–385.

LeDoux, M. S., Xiao, J., Rudzińska, M., Bastian, R. W., Wszolek, Z. K., Van Gerpen, J. A., et al. (2012). Genotype-phenotype correlations in THAP1 dystonia: Molecular foundations and description of new cases. *Parkinsonism & Related Disorders*, *18*(5), 414–425.

Lee, L. V., Rivera, C., Teleg, R. A., Dantes, M. B., Pasco, P. M., Jamora, R. D., et al. (2011). The unique phenomenology of sex-linked dystonia parkinsonism (XDP, DYT3, "Lubag"). *The International Journal of Neuroscience*, *121*(Suppl 1), 3–11.

Levy, L. M., & Hallett, M. (2002). Impaired brain GABA in focal dystonia. *Annals of Neurology*, *51*(1), 93–101.

Li, W., & Tang, M. (2020). Application of botulinum toxin in pregnancy and its impact on female reproductive health. *Expert Opinion on Drug Safety*, *19*(1), 83–91.

Liu, R., Umbach, D. M., Peddada, S. D., Xu, Z., Tröster, A. I., Huang, X., et al. (2015). Potential sex differences in nonmotor symptoms in early drug-naive Parkinson disease. *Neurology*, *84*(21), 2107–2115.

Loke, H., Harley, V., & Lee, J. (2015). Biological factors underlying sex differences in neurological disorders. *The International Journal of Biochemistry & Cell Biology*, *65*, 139–150.

Louis, E. D., Bares, M., Benito-Leon, J., Fahn, S., Frucht, S. J., Jankovic, J., et al. (2020). Essential tremor-plus: A controversial new concept. *The Lancet. Neurology*, *19*(3), 266–270.

Louis, E. D., & Dogu, O. (2007). Does age of onset in essential tremor have a bimodal distribution? Data from a tertiary referral setting and a population-based study. *Neuroepidemiology*, *29*(3-4), 208–212.

Louis, E. D., Fernandez-Alvarez, E., Dure, L. S., 4th, Frucht, S., & Ford, B. (2005). Association between male gender and pediatric essential tremor. *Movement Disorders: Official Journal of the Movement Disorder Society*, *20*(7), 904–906.

Louis, E. D., & Ferreira, J. J. (2010). How common is the most common adult movement disorder? Update on the worldwide prevalence of essential tremor. *Movement Disorders: Official Journal of the Movement Disorder Society*, *25*(5), 534–541.

Maeda, T., Shimo, Y., Chiu, S. W., Yamaguchi, T., Kashihara, K., Tsuboi, Y., et al. (2017). Clinical manifestations of nonmotor symptoms in 1021 Japanese Parkinson's disease patients from 35 medical centers. *Parkinsonism & Related Disorders*, *38*, 54–60.

Maia, D. P., Fonseca, P. G., Camargos, S. T., Pfannes, C., Cunningham, M. C., & Cardoso, F. (2012). Pregnancy in patients with Sydenham's Chorea. *Parkinsonism & Related Disorders*, *18*(5), 458–461.

Markianos, M., Panas, M., Kalfakis, N., & Vassilopoulos, D. (2005). Plasma testosterone in male patients with Huntington's disease: Relations to severity of illness and dementia. *Annals of Neurology*, *57*(4), 520–525.

Markianos, M., Panas, M., Kalfakis, N., & Vassilopoulos, D. (2007). Plasma testosterone, dehydroepiandrosterone sulfate, and cortisol in female patients with Huntington's disease. *Neuro Endocrinology Letters*, *28*(2), 199–203.

Martinelli, P., Contin, M., Scaglione, C., Riva, R., Albani, F., & Baruzzi, A. (2003). Levodopa pharmacokinetics and dyskinesias: Are there sex-related differences? *Neurological Sciences: Official Journal of the Italian Neurological Society and of the Italian Society of Clinical Neurophysiology*, *24*(3).

Martinez-Martin, P., Falup Pecurariu, C., Odin, P., van Hilten, J. J., Antonini, A., Rojo-Abuin, J. M., et al. (2012). Gender-related differences in the burden of non-motor symptoms in Parkinson's disease. *Journal of Neurology*, *259*(8), 1639–1647.

Meoni, S., Macerollo, A., & Moro, E. (2020). Sex differences in movement disorders. Nature reviews. *Neurology*, *16*(2), 84–96.

Merchant, C. A., Cohen, G., Mytilineou, C., DiRocco, A., Moros, D., Molinari, S., et al. (1995). Human transplacental transfer of carbidopa/levodopa. *Journal of Neural Transmission. Parkinson's Disease and Dementia Section*, *9*(2-3), 239–242.

Miller, I. N., & Cronin-Golomb, A. (2010). Gender differences in Parkinson's disease: Clinical characteristics and cognition. *Movement Disorders: Official Journal of the Movement Disorder Society*, *25*(16), 2695–2703.

Miltgen, M., Blanchard, A., Mathieu, H., Kreisler, A., Jean-Pierre-Desvignes, Salgado, D., et al. (2016). Novel heterozygous mutation in ANO3 responsible for craniocervical dystonia. *Movement Disorders: Official Journal of the Movement Disorder Society*, *31*(8), 1251–1252.

Moisan, F., Kab, S., Mohamed, F., Canonico, M., Le Guern, M., Quintin, C., et al. (2016). Parkinson disease male-to-female ratios increase with age: French nationwide study and meta-analysis. *Journal of Neurology, Neurosurgery, and Psychiatry*, *87*(9), 952–957.

Molloy, A., Kimmich, O., Williams, L., Butler, J. S., Byrne, N., Molloy, F., et al. (2015). An evaluation of the role of environmental factors in the disease penetrance of cervical dystonia. *Journal of Neurology, Neurosurgery, and Psychiatry*, *86*(3), 331–335.

Molloy, A., Williams, L., Kimmich, O., Butler, J. S., Beiser, I., McGovern, E., et al. (2016). Sun exposure is an environmental factor for the development of blepharospasm. *Journal of Neurology, Neurosurgery, and Psychiatry*, *87*(4), 420–424.

Montaurier, C., Morio, B., Bannier, S., Derost, P., Arnaud, P., Brandolini-Bunlon, M., et al. (2007). Mechanisms of body weight gain in patients with Parkinson's disease after subthalamic stimulation. *Brain: A Journal of Neurology*, *130*(Pt. 7), 1808–1818.

Müller, J., Kiechl, S., Wenning, G. K., Seppi, K., Willeit, J., Gasperi, A., et al. (2002). The prevalence of primary dystonia in the general community. *Neurology*, *59*(6), 941–943.

Nageshwaran, S., Smith, M., & Bordelon, Y. M. (2016). Movement disorders and pregnancy. In *Neurological illness in pregnancy: Principles and practice* (pp. 179–190). Hoboken: Wiley-Blackwell.

Nausieda, P. A., Koller, W. C., Weiner, W. J., & Klawans, H. L. (1979). Chorea induced by oral contraceptives. *Neurology*, *29*(12), 1605–1609.

Ndukwe, I., O'Riordan, S., Walsh, C. B., & Hutchinson, M. (2020). Mood disorder affects age at onset of adult-onset cervical dystonia. *Clinical Parkinsonism & Related Disorders*, *3*, 100049.

Nicoletti, A., Arabia, G., Pugliese, P., Nicoletti, G., Torchia, G., Condino, F., et al. (2007). Hormonal replacement therapy in women with Parkinson disease and levodopa-induced dyskinesia: A crossover trial. *Clinical Neuropharmacology*, *30*(5), 276–280. https://doi.org/10.1097/wnf.0b013e318050c9f9.

Nicoletti, A., Nicoletti, G., Arabia, G., Annesi, G., De Mari, M., Lamberti, P., et al. (2011). Reproductive factors and Parkinson's disease: A multicenter case-control study. *Movement Disorders: Official Journal of the Movement Disorder Society*, *26*(14), 2563–2566.

Nicoletti, A., Vasta, R., Mostile, G., Nicoletti, G., Arabia, G., Iliceto, G., et al. (2017). Gender effect on non-motor symptoms in Parkinson's disease: Are men more at risk? *Parkinsonism & Related Disorders*, *35*, 69–74.

Nomoto, M., Kaseda, S., Iwata, S., Osame, M., & Fukuda, T. (1997). Levodopa in pregnancy. *Movement Disorders: Official Journal of the Movement Disorder Society*, *12*(2), 261.

Ohman, I., Vitols, S., Luef, G., Söderfeldt, B., & Tomson, T. (2002). Topiramate kinetics during delivery, lactation, and in the neonate: Preliminary observations. *Epilepsia*, *43*(10), 1157–1160.

Ozelius, L. J., Hewett, J. W., Page, C. E., Bressman, S. B., Kramer, P. L., Shalish, C., et al. (1997). The early-onset torsion dystonia gene (DYT1) encodes an ATP-binding protein. *Nature Genetics*, *17*(1), 40–48.

Pacifici, G. M., & Nottoli, R. (1995). Placental transfer of drugs administered to the mother. *Clinical Pharmacokinetics*, *28*(3), 235–269.

Padovan-Neto, F. E., Jurkowski, L., Murray, C., Stutzmann, G. E., Kwan, M., Ghavami, A., et al. (2019). Age- and sex-related changes in cortical and striatal nitric oxide synthase in the Q175 mouse model of Huntington's disease. *Nitric Oxide: Biology and Chemistry*, *83*, 40–50.

Palanivelu, L. M. (2007). Chorea gravidarum. *Journal of Obstetrics and Gynaecology: The Journal of the Institute of Obstetrics and Gynaecology*, *27*(3), 310.

Parsons, L. M. (2006). Neurologic disease in women, second edition. *Journal of Neurology, Neurosurgery, and Psychiatry*, 77(4), 567–568.

Perez, D. L., Aybek, S., Popkirov, S., Kozlowska, K., Stephen, C. D., Anderson, J., et al. (2021). A review and expert opinion on the neuropsychiatric assessment of motor functional neurological disorders. *The Journal of Neuropsychiatry and Clinical Neurosciences*, *33*(1), 14–26.

Picillo, M., Amboni, M., Erro, R., Longo, K., Vitale, C., Moccia, M., et al. (2013). Gender differences in non-motor symptoms in early, drug naïve Parkinson's disease. *Journal of Neurology*, *260*(11), 2849–2855.

Picillo, M., Nicoletti, A., Fetoni, V., Garavaglia, B., Barone, P., & Pellecchia, M. T. (2017). The relevance of gender in Parkinson's disease: A review. *Journal of Neurology*, *264*(8), 1583–1607.

Pringsheim, T., Jette, N., Frolkis, A., & Steeves, T. D. (2014). The prevalence of Parkinson's disease: A systematic review and meta-analysis. *Movement Disorders: Official Journal of the Movement Disorder Society*, *29*(13), 1583–1590.

Quinn, D. (2001). Use of psychoactive medication during pregnancy and possible effects on the fetus and newborn. *Pediatrics*, *107*(6), 1498.

Quinn, N., & Schrag, A. (1998). Huntington's disease and other choreas. *Journal of Neurology*, *245*(11), 709–716.

Rafee, S., O'Riordan, S., Reilly, R., & Hutchinson, M. (2021). We must talk about sex and focal dystonia. *Movement Disorders: Official Journal of the Movement Disorder Society*, *36*(3), 604–608.

Ragonese, P., D'Amelio, M., & Savettieri, G. (2006). Implications for estrogens in Parkinson's disease: An epidemiological approach. *Annals of the New York Academy of Sciences*, *1089*, 373–382.

Ranen, N. G., Stine, O. C., Abbott, M. H., Sherr, M., Codori, A. M., Franz, M. L., et al. (1995). Anticipation and instability of IT-15 (CAG)n repeats in parent-offspring pairs with Huntington disease. *American Journal of Human Genetics*, *57*(3), 593–602.
Robottom, B. J., & Weiner, W. J. (2011). Chorea gravidarum. *Handbook of Clinical Neurology*, *100*, 231–235.
Rocca, W. A., Bower, J. H., Maraganore, D. M., Ahlskog, J. E., Grossardt, B. R., de Andrade, M., et al. (2008). Increased risk of parkinsonism in women who underwent oophorectomy before menopause. *Neurology*, *70*(3), 200–209.
Rocca, W. A., Grossardt, B. R., & Shuster, L. T. (2011). Oophorectomy, menopause, estrogen treatment, and cognitive aging: Clinical evidence for a window of opportunity. *Brain Research*, *1379*, 188–198.
Rodríguez-Navarro, J. A., Solano, R. M., Casarejos, M. J., Gomez, A., Perucho, J., de Yébenes, J. G., et al. (2008). Gender differences and estrogen effects in parkin null mice. *Journal of Neurochemistry*, *106*(5), 2143–2157.
Rogers, J. D., & Fahn, S. (1994). Movement disorders and pregnancy. *Advances in Neurology*, *64*, 163–178.
Rubinsztein, D. C., Leggo, J., Coles, R., Almqvist, E., Biancalana, V., Cassiman, J. J., et al. (1996). Phenotypic characterization of individuals with 30-40 CAG repeats in the Huntington disease (HD) gene reveals HD cases with 36 repeats and apparently normal elderly individuals with 36-39 repeats. *American Journal of Human Genetics*, *59*(1), 16–22.
Rugbjerg, K., Christensen, J., Tjønneland, A., & Olsen, J. H. (2013). Exposure to estrogen and women's risk for Parkinson's disease: A prospective cohort study in Denmark. *Parkinsonism & Related Disorders*, *19*(4), 457–460.
Sampaio, T. F., Dos Santos, E., de Lima, G., Dos Anjos, R., da Silva, R. C., Asano, A., et al. (2018). MAO-B and COMT genetic variations associated with levodopa treatment response in patients with Parkinson's disease. *Journal of Clinical Pharmacology*, *58*(7), 920–926.
Savica, R., Grossardt, B. R., Bower, J. H., Ahlskog, J. E., & Rocca, W. A. (2013). Risk factors for Parkinson's disease may differ in men and women: An exploratory study. *Hormones and Behavior*, *63*(2), 308–314.
Savitt, D., & Jankovic, J. (2018). Tardive syndromes. *Journal of the Neurological Sciences*, *389*, 35–42.
Scelzo, E., Mehrkens, J. H., Bötzel, K., Krack, P., Mendes, A., Chabardès, S., et al. (2015). Deep brain stimulation during pregnancy and delivery: Experience from a series of "DBS Babies". *Frontiers in Neurology*, *6*, 191.
Scott, B., Borgman, A., Engler, H., Johnels, B., & Aquilonius, S. M. (2000). Gender differences in Parkinson's disease symptom profile. *Acta Neurologica Scandinavica*, *102*(1), 37–43.
Seier, M., & Hiller, A. (2017). Parkinson's disease and pregnancy: An updated review. *Parkinsonism & Related Disorders*, *40*, 11–17.
Sharma, J. C., Ross, I. N., Rascol, O., & Brooks, D. (2008). Relationship between weight, levodopa and dyskinesia: The significance of levodopa dose per kilogram body weight. *European Journal of Neurology*, *15*(5), 493–496.
Sherwin, B. B. (2007). The critical period hypothesis: Can it explain discrepancies in the oestrogen-cognition literature? *Journal of Neuroendocrinology*, *19*(2), 77–81.
Siani, F., Greco, R., Levandis, G., Ghezzi, C., Daviddi, F., Demartini, C., et al. (2017). Influence of estrogen modulation on glia activation in a murine model of Parkinson's disease. *Frontiers in Neuroscience*, *11*, 306.
Smith, M. S., & Evatt, M. L. (2004). Movement disorders in pregnancy. *Neurologic Clinics*, *22*(4), 783–798.
Soland, V. L., Bhatia, K. P., & Marsden, C. D. (1996). Sex prevalence of focal dystonias. *Journal of Neurology, Neurosurgery, and Psychiatry*, *60*(2), 204–205.

Solla, P., Cannas, A., Ibba, F. C., Loi, F., Corona, M., Orofino, G., et al. (2012). Gender differences in motor and non-motor symptoms among Sardinian patients with Parkinson's disease. *Journal of the Neurological Sciences*, *323*(1-2), 33–39.

Song, Y. J., Li, S. R., Li, X. W., Chen, X., Wei, Z. X., Liu, Q. S., et al. (2020). The effect of estrogen replacement therapy on Alzheimer's disease and Parkinson's disease in postmenopausal women: A meta-analysis. *Frontiers in Neuroscience*, *14*, 157.

Spillantini, M. G., Crowther, R. A., Jakes, R., Hasegawa, M., & Goedert, M. (1998). alpha-Synuclein in filamentous inclusions of Lewy bodies from Parkinson's disease and dementia with Lewy bodies. *Proceedings of the National Academy of Sciences of the United States of America*, *95*(11), 6469–6473.

Steeves, T. D., Day, L., Dykeman, J., Jette, N., & Pringsheim, T. (2012). The prevalence of primary dystonia: A systematic review and meta-analysis. *Movement disorders: Official Journal of the Movement Disorder Society*, *27*(14), 1789–1796.

Stephen, C. D., Fung, V., Lungu, C. I., & Espay, A. J. (2021). Assessment of emergency department and inpatient use and costs in adult and pediatric functional neurological disorders. *JAMA Neurology*, *78*(1), 88–101.

Subramanian, I., Mathur, S., Oosterbaan, A., Flanagan, R., Keener, A. M., & Moro, E. (2022). Unmet needs of women living with Parkinson's disease: Gaps and controversies. *Movement Disorders: Official Journal of the Movement Disorder Society*, *37*(3), 444–455.

Szewczyk-Krolikowski, K., Tomlinson, P., Nithi, K., Wade-Martins, R., Talbot, K., Ben-Shlomo, Y., et al. (2014). The influence of age and gender on motor and non-motor features of early Parkinson's disease: Initial findings from the Oxford Parkinson Disease Center (OPDC) discovery cohort. *Parkinsonism & Related Disorders*, *20*(1), 99–105.

Taylor, K. S., Cook, J. A., & Counsell, C. E. (2007). Heterogeneity in male to female risk for Parkinson's disease. *Journal of Neurology, Neurosurgery, and Psychiatry*, *78*(8), 905–906.

Tinazzi, M., Morgante, F., Marcuzzo, E., Erro, R., Barone, P., Ceravolo, R., et al. (2020). Clinical correlates of functional motor disorders: An Italian Multicenter Study. *Movement Disorders Clinical Practice*, 7(8), 920–929.

Tsang, K. L., Ho, S. L., & Lo, S. K. (2000). Estrogen improves motor disability in parkinsonian postmenopausal women with motor fluctuations. *Neurology*, *54*(12), 2292–2298.

Van Duijn, E., Craufurd, D., Hubers, A. A., Giltay, E. J., Bonelli, R., Rickards, H., et al. (2014). Neuropsychiatric symptoms in a European Huntington's disease cohort (REGISTRY). *Journal of Neurology, Neurosurgery, and Psychiatry*, *85*(12), 1411–1418.

Villa, A., Rizzi, N., Vegeto, E., Ciana, P., & Maggi, A. (2015). Estrogen accelerates the resolution of inflammation in macrophagic cells. *Scientific Reports*, *5*, 15224.

Walker, F. O. (2007). Huntington's disease. *Lancet (London, England)*, *369*(9557), 218–228.

Wang, L. F., Yokoyama, K. K., Chen, T. Y., Hsiao, H. W., Chiang, P. C., Hsieh, Y. C., et al. (2015). Male-specific alleviation of iron-induced striatal injury by inhibition of autophagy. *PLoS One*, *10*(7), e0131224.

Warren Olanow, C., Kieburtz, K., Rascol, O., Poewe, W., Schapira, A. H., Emre, M., et al. (2013). Factors predictive of the development of Levodopa-induced dyskinesia and wearing-off in Parkinson's disease. *Movement Disorders: Official Journal of the Movement Disorder Society*, *28*(8), 1064–1071.

Williams, L., McGovern, E., Kimmich, O., Molloy, A., Beiser, I., Butler, J. S., et al. (2017). Epidemiological, clinical and genetic aspects of adult onset isolated focal dystonia in Ireland. *European Journal of Neurology*, *24*(1), 73–81.

Willis, A. W., Schootman, M., Kung, N., Wang, X. Y., Perlmutter, J. S., & Racette, B. A. (2014). Disparities in deep brain stimulation surgery among insured elders with Parkinson disease. *Neurology*, *82*(2), 163–171.

Willson, P., & Preece, A. A. (1932). Chorea gravidarum: A statistical study of 951 collected cases, 846 from the literature and 105 previously unreported. *Archives of Internal Medicine (Chicago, Ill.)*, *49*(3), 471–533.
Wisner, K. L., Gelenberg, A. J., Leonard, H., Zarin, D., & Frank, E. (1999). Pharmacologic treatment of depression during pregnancy. *JAMA*, *282*(13), 1264–1269.
Wittchen, H. U., Jacobi, F., Rehm, J., Gustavsson, A., Svensson, M., Jönsson, B., et al. (2011). The size and burden of mental disorders and other disorders of the brain in Europe 2010. *European Neuropsychopharmacology: The Journal of the European College of Neuropsychopharmacology*, *21*(9), 655–679.
Wizemann, T. M. (2001). Exploring the biological contributions to human health: Does sex matter? *Journal of Women's Health & Gender-Based Medicine*, *10*(5), 433–439.
Wooten, G. F., Currie, L. J., Bovbjerg, V. E., Lee, J. K., & Patrie, J. (2004). Are men at greater risk for Parkinson's disease than women? *Journal of Neurology, Neurosurgery, and Psychiatry*, *75*(4), 637–639.
Xu, J., Gong, D. D., Man, C. F., & Fan, Y. (2014). Parkinson's disease and risk of mortality: Meta-analysis and systematic review. *Acta Neurologica Scandinavica*, *129*(2), 71–79.
Zappia, M., Annesi, G., Nicoletti, G., Arabia, G., Annesi, F., Messina, D., et al. (2005). Sex differences in clinical and genetic determinants of levodopa peak-dose dyskinesias in Parkinson disease: An exploratory study. *Archives of Neurology*, *62*(4), 601–605.
Zech, M., Gross, N., Jochim, A., Castrop, F., Kaffe, M., Dresel, C., et al. (2014). Rare sequence variants in ANO3 and GNAL in a primary torsion dystonia series and controls. *Movement Disorders: Official Journal of the Movement Disorder Society*, *29*(1), 143–147.
Zech, M., Jochim, A., Boesch, S., Weber, S., Meindl, T., Peters, A., et al. (2016). Systematic TOR1A non-c.907_909delGAG variant analysis in isolated dystonia and controls. *Parkinsonism & Related Disorders*, *31*, 119–123.
Zegart, K. N., & Schwarz, R. H. (1968). Chorea gravidarum. *Obstetrics and Gynecology*, *32*(1), 24–27.
Zesiewicz, T. A., Elble, R. J., Louis, E. D., Gronseth, G. S., Ondo, W. G., Dewey, R. B., Jr., et al. (2011). Evidence-based guideline update: Treatment of essential tremor: Report of the Quality Standards subcommittee of the American Academy of Neurology. *Neurology*, 77(19), 1752–1755.
Zielonka, D., Marinus, J., Roos, R. A., De Michele, G., Di Donato, S., Putter, H., et al. (2013). The influence of gender on phenotype and disease progression in patients with Huntington's disease. *Parkinsonism & Related Disorders*, *19*(2), 192–197.
Zimprich, A., Grabowski, M., Asmus, F., Naumann, M., Berg, D., Bertram, M., et al. (2001). Mutations in the gene encoding epsilon-sarcoglycan cause myoclonus-dystonia syndrome. *Nature Genetics*, *29*(1), 66–69.

CHAPTER FIVE

Sex and gender differences in autoimmune demyelinating CNS disorders: Multiple sclerosis (MS), neuromyelitis optica spectrum disorder (NMOSD) and myelin-oligodendrocyte-glycoprotein antibody associated disorder (MOGAD)

Lara Diem[a], Helly Hammer[a], Robert Hoepner[a], Max Pistor[a], Jana Remlinger[a,b], and Anke Salmen[a,*]

[a]Department of Neurology, Inselspital, Bern University Hospital and University of Bern, Freiburgstrasse, Bern, Switzerland

[b]Department of Biomedical Research and Graduate School for Cellular and Biomedical Sciences, University of Bern, Bern, Switzerland

*Corresponding author: e-mail address: anke.salmen@insel.ch

Contents

International Review of Neurobiology, Volume 164
ISSN 0074-7742
https://doi.org/10.1016/bs.irn.2022.06.011

Abstract

Multiple sclerosis (MS), Neuromyelitis optica spectrum disorder (NMOSD) and Myelin-Oligodendrocyte-Glycoprotein antibody associated disorder (MOGAD) are demyelinating disorders of the central nervous system (CNS) of autoimmune origin. Here, we summarize general considerations on sex-specific differences in the immunopathogenesis and hormonal influences as well as key clinical and epidemiological elements.

Gender-specific issues are widely neglected starting with the lacking separation of sex as a biological variable and gender comprising the sociocultural components.

As for other autoimmune diseases, female preponderance is common in MS and NMOSD. However, sex distribution in MOGAD seems equal.

As in MS, immunotherapy in NMOSD and MOGAD is crucial to prevent further disease activity. Therefore, we assessed data on sex differences of the currently licensed disease-modifying treatments for efficacy and safety. This topic seems widely neglected with only fragmented information resulting from post-hoc analyses of clinical trials or real-world post-marketing studies afflicted with lacking power and/or inherent sources of bias.

In summary, biological hypotheses of sex differences including genetic factors, the constitution of the immune system and hormonal influences are based upon human and preclinical data, especially for the paradigmatic disease of MS whereas specific data for NMOSD and MOGAD are widely lacking. Epidemiological and clinical differences between men and women are well described for MS and to some extent for NMOSD, yet, with remaining contradictory findings. MOGAD needs further detailed investigation. Sex-specific analyses of safety and efficacy of long-term immunotherapies need to be addressed in future studies designed and powered to answer the pressing questions and to optimize and individualize treatment.

1. Background: Pathophysiology of CNS autoimmunity in the context of sex-specific differences

Sex differences play an important role in prevalence, immunopathogenesis, clinical phenotype and prognosis of autoimmune disorders of the central nervous system (CNS) like multiple sclerosis (MS), neuromyelitis optica spectrum disorder (NMOSD) and Myelin-Oligodendrocyte-Glycoprotein (MOG-) antibody associated disorder (MOGAD).

Epidemiologically, the life-time prevalence of many autoimmune disorders is higher in women, in some reports as high as 2.8-times (Jacobson, Gange, Rose, & Graham, 1997). The female preponderance in autoimmune disorders may reach 86% (Systemic Lupus Erythematodes (SLE)) to 90% (Hashimoto Thyroiditis) (Feldman et al., 2013; Ramagopalan, Goldacre, Skingsley, Conlon, & Goldacre, 2013).

In autoimmune CNS disorders, an increasing female-to-male ratio for MS, the most prevalent of these disorders, is observed. Whereas in 1980 a ratio of 2:1 (Confavreux, Aimard, & Devic, 1980) was reported, the ratio increased to 3:1 in more recent reports (Koch-Henriksen & Sorensen, 2010). The reason for these findings is not entirely clear.

Sex differences of these disorders are linked to inherent properties in the chromosomal and hormonal basis of the immune system affecting its functional constitution. Sex as a biological variable refers to the chromosomal definition (XX vs XY genotype) and reproductive organ structures and sex hormone levels (Klein, Pekosz, Passaretti, Anker, & Olukoya, 2010). Gender refers to sociocultural and behavioral components.

1.1 Sex-specific genetic and constitutional aspects of the immune system

The immune system consists of two major inter-connected components. The evolutionarily older component, the so-called innate immune system, is represented by complement system, phagocytes and cellular "pattern recognition receptors" (PRR), recognizing typical microbiological structures. The adaptive immune system consists of the humoral part (B-cells and plasma cells producing antibodies) and the cellular part (cytotoxic and helper T-cells).

The genetic bases of sex differences of the immune system and in autoimmune diseases are only partially understood. Many immune regulatory genes like the innate PRR Toll-like receptor 7 (TLR-7) and TLR-8 are coded on the X-chromosomes (Bianchi, Lleo, Gershwin, & Invernizzi, 2012). This implies a role of sex chromosomes and their organization in the function of the immune system and thus immunopathogenesis of disorders like MS (Libert, Dejager, & Pinheiro, 2010). Incomplete inactivation of an X-chromosome can cause differences in X-chromosomal gene dosages, which occurs in around 5% of X-chromosomal genes (Voskuhl, Sawalha, & Itoh, 2018).

An incomplete inactivation of genes like TLR-7 in B- and myeloid lineage cells of XX- and XXY-genotype (Klinefelter syndrome) results in a more pronounced B-cell response (Souyris et al., 2018) which may have implications especially in primary B-cell-driven autoimmune diseases like SLE or NMOSD.

In addition, a sex-specific expression of the Forkhead-Box-Protein P3 (FOXP-3), the main transcription factor necessary for the development of regulatory T-cells (T_{reg}), is described. Here, differences in the epigenetic regulation via less methylation of the maternal X-chromosome cause a higher FOXP3-expression in male vs female murine $CD4^+$ T-cells resulting in different regulatory T-cell counts and susceptibility for autoimmunity (Voskuhl et al., 2018).

The immune system in women displays a more pro-inflammatory immune response as compared to men: As part of the innate immune system, macrophages display a more activated phenotype with higher pro-inflammatory cytokine expression levels and phagocytic activity. Peptide presentation by antigen-presenting cells (APC) is more efficient (Klein & Flanagan, 2016). The adaptive cellular immune reactivity in response to unspecific *in vitro* stimulation is stronger (Klein & Flanagan, 2016). Numbers of Cluster of Differentiation 4 positive (CD4+) T-lymphocytes and CD4/CD8 ratios (Abdullah et al., 2012; Bouman, Schipper, Heineman, & Faas, 2004) are higher in women while numbers of T_{reg} are lower compared to men (Afshan, Afzal, & Qureshi, 2012). B-cell counts, the antibody production in response to vaccines and basal immunoglobulin levels, are significantly higher in women than men (Abdullah et al., 2012; Furman et al., 2014) which may help to explain sex differences in the prevalence of NMOSD as a B-cell and auto-antibody driven disease.

Women are prone to develop a more pronounced cellular and humoral response also to infectious and vaccinal antigens (Cook, 2008; Klein, Jedlicka, & Pekosz, 2010; vom Steeg & Klein, 2016). *In vitro* stimulated $CD4^+$ and $CD8^+$ T-cells of women show a stronger cytotoxic activity with upregulation of antiviral and pro-inflammatory genes (Hewagama, Patel, Yarlagadda, Strickland, & Richardson, 2009).

In response to parasitic infections, $CD4^+$ T-cells of female experimental mice produce higher levels of T-helper cell 1 (Th1) cytokines, which are generally considered as pro-inflammatory (Roberts, Walker, & Alexander, 2001) as opposed to Th2 patterns.

The differential constitution of the innate and adaptive immune system physiologically implies a more pronounced reactivity to stimuli, and in turn predisposes for a higher risk of autoimmune disorders in women.

1.2 MS as a model disease of CNS autoimmunity

As MS is the most frequent and more detailed studied disorder among the here discussed entities, both in the human and in experimental settings, we here focus on the data on the differential pathophysiology of MS between male and female patients or in experimental conditions. Data on NMOSD and MOGAD are still sparse and rather refer back to the general hypotheses as given above.

In vitro, peripheral extracted T-cells of female MS patients express a Th1-cytokine profile (e.g., Interferon γ (IFN-γ)) in response to proteolipid protein (PLP), myelin basic protein (MBP) and myelin oligodendrocyte glycoprotein (MOG), which are all parts of human CNS myelin, compared to men, who rather demonstrate a Th2-response (Moldovan, Cotleur, Zamor, Butler, & Pelfrey, 2008; Pelfrey, Cotleur, Lee, & Rudick, 2002). In a spontaneous experimental autoimmune encephalomyelitis (EAE) in SJL-mice, expression of testosterone-dependent interleukin- (IL-) 33 by innate immune cells like mast cells promotes a Th2-response and thereby nearly completely suppressed the development of clinical signs of EAE in male mice (Russi, Ebel, Yang, & Brown, 2018) as a potential explanation for the human *in vitro* findings.

On a genetic basis, the so-called Fore-Core Model helps to experimentally address questions of genotype, sex hormones and reproductive organs, separately. With the help of different combinations of deleted or non-deleted Y-encoded *Sex Determining Region Y* (*SRY*, the gene inducing the development of a male phenotype including hormonal status embryologically), mice with XY-genotype can be created, which are phenotypically female including ovaries and sex hormone profile (XY^{-}).

TLR-7 expression may cause neurodegeneration in neuronal cell cultures (Lehmann et al., 2012). Within the aforementioned model, XY^{-} mice have a higher neuronal TLR-7 expression. In an active immunization EAE model, the most commonly used MS mouse model, this results in a more severe disease course and histologically more neuroaxonal degeneration (Du et al., 2014). As discussed above, the expression of TLR-7 in B-cells is higher in women as compared to men. In neurons, this ratio is inversed, which points to a cell line specific expression pattern (linked to epigenetic regulation and X-inactivation). Taken together, these findings may elucidate the phenomenon of higher inflammatory responses in women, but higher neurodegenerative potential in men via the differential genetic regulation of TLR-7 expression.

These insights on a differential functioning of the immune system but also direct influences of sex-specific genetic factors on neuronal structures, supported by experimental data, to some extent might be translated back to the human situation in MS:

Whereas women more frequently develop a relapsing-remitting phenotype, characterized by focal inflammation indicating enhanced immune function, men more commonly present with a progressive MS phenotype with more pronounced neurodegeneration (Koch-Henriksen & Sorensen, 2010) (Fig. 1).

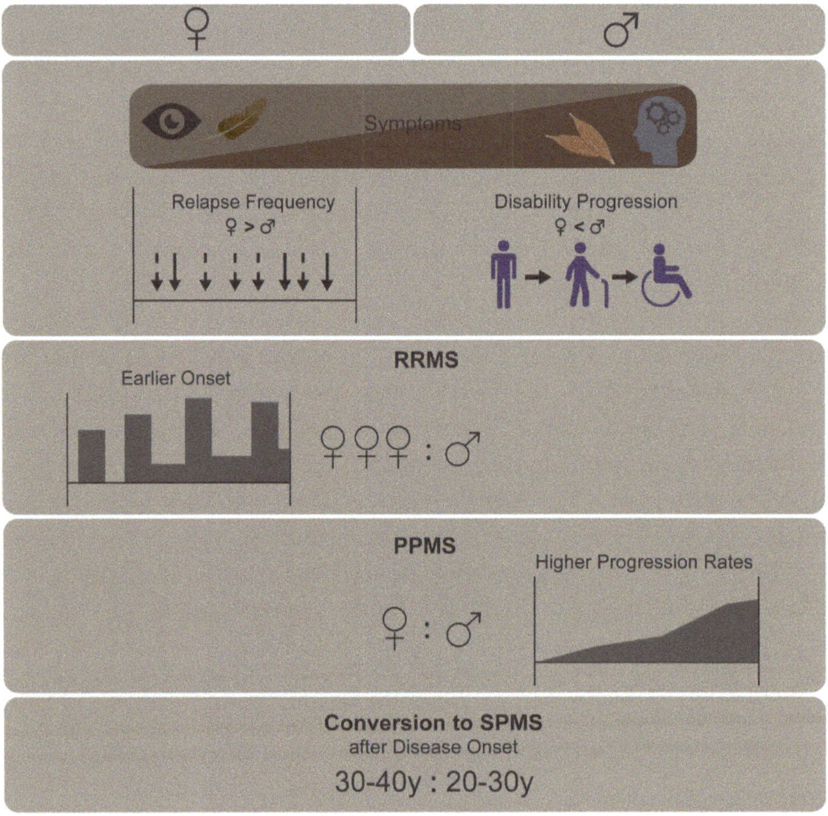

Fig. 1 Main differences in course of multiple sclerosis (MS) between women and men: Symptom frequencies differ between women (visual and sensory symptoms more common) and men (motor and cognitive symptoms more common). Whereas women are prone to higher relapse frequencies, men experience higher disability progression. RRMS has a female preponderance with earlier onset in women which is not detected in PPMS, again with higher progression rates in men. SPMS conversion occurs earlier in men compared to women. Abbreviations and symbols: >: higher than, ♀: female, ♂: male, PPMS: primary progressive multiple sclerosis; RRMS: relapsing-remitting multiple sclerosis; SPMS: secondary progressive multiple sclerosis; y: years.

1.3 Sex-specific hormonal aspects in autoimmunity and MS

Hormones are important constituents of sex differences in the immune system and in autoimmunity. No differences in the prevalence of MS are observed before puberty, but a female preponderance establishes shortly thereafter (Chitnis, 2013).

The different sex hormones have distinct roles in immunity and autoimmunity. Sex hormone receptors are expressed on immune cells (Cunningham & Gilkeson, 2011).

Estrogens are a major class of sex hormones and consist of Estrone (E1), Estradiol (E2) and Estriol (E3), in which E2 is the most prominent in pre-menopause. Estrogenes in (CNS-) autoimmunity may exert both pro- and anti-inflammatory effects, depending on cell and organ types, estrogene receptor (ER) subsets or timing and concentration of estrogenes. E2 in higher levels promotes an anti-inflammatory effect with reduction of pro- and increase in anti-inflammatory cytokine levels (Straub, 2007). In contrast, lower, pre-ovulatory levels of E2 cause a pro-inflammatory cytokine profile with increased tumor necrosis factor α (TNF-α) and IFN-γ expression (Gilmore, Weiner, & Correale, 1997). The ER-α receptor subtype mainly transfers these immunological effects. However, ER-β trans-repression reduces microglial activation (Saijo, Collier, Li, Katzenellenbogen, & Glass, 2011), and might have neuroprotective properties (Voskuhl & Gold, 2012). Moreover, ER-β trans-repression confers an anti-inflammatory phenotype of $CD11c^+$ macrophages and dendritic cells with reduced expression of pro-inflammatory inducible Nitric Oxide Synthetase (iNOS) and enhanced anti-inflammatory IL-10 secretion, resulting in a reduction of disease activity in active immunization EAE (Kim et al., 2018).

Progesterone is the main pro-gestational hormone and dominates in the second half of the female cycle and during pregnancy. It binds to its receptors expressed on many types of immune cells (Dosiou et al., 2008) and elicits an anti-inflammatory response. It favors a Th2 shift with a reduction in the expression of co-stimulatory receptors, reduction of inflammation of macrophages (TLR, NO-production), T_{reg} induction (Hughes, 2012) and natural killer- (NK-) cell apoptosis (Arruvito et al., 2008).

Prolactin is a pituitary gland hormone, but also secreted by different innate and adaptive immune cells (Freeman, Kanyicska, Lerant, & Nagy, 2000). Besides the regulation of lactation, prolactin influences the immune system via its receptor that belongs to the cytokine receptor superfamily (Ysrraelit & Correale, 2019).

Androgen levels, including testosterone and 5α-dihydrotestosterone as the main agonists on androgen receptors, are higher in men compared to women. They exert anti-inflammatory effects by shifting the Th1/Th2 ratio toward Th2, which correlates with the respective cytokine levels (IL-5, -10 and IL-33 higher, IFN-γ, TNF-α and IL-17 lower) and suppress immunoglobulin production (Russi et al., 2018).

Male (and female) MS patients express lower levels of androgens as age- and sex-matched controls. This was associated with an increased risk of MS-associated disability and cognitive decline in men (Chitnis, 2018). The higher age of MS onset in men compared to women might therefore in part be linked to an age-dependent decline in male androgen levels (andropause).

The role of different sex hormones has been studied in trial and model settings of MS.

Supplementation with E3 reduced disease severity in an active immunization EAE model (Jansson, Olsson, & Holmdahl, 1994), possibly by induction of an Th2-response with increased anti-inflammatory IL-10 (Kim, Liva, Dalal, Verity, & Voskuhl, 1999). These findings may partially explain a reduction of MS relapse activity during pregnancy (Confavreux, Hutchinson, Hours, Cortinovis-Tourniaire, & Moreau, 1998), where E3 is the predominant estrogen (Ali, Mangold, & Peiris, 2017). A small phase 2 clinical trial investigated the combination of Glatiramer acetate and E3. However, this study had methodological difficulties hampering a final evaluation of the results. The clinical and radiological outcomes of the study were not well compatible (Langer-Gould, 2016; Voskuhl et al., 2016).

Progesterone treatment also reduced disease severity in different active immunization EAE models (Hughes, 2012) and was attributed beneficial roles in myelin repair, neuroprotection and oligodendrocyte maturation in CNS injury (Labombarda et al., 2011; Ramien et al., 2016). Clinical implications of these findings remain thus far unclear. A randomized trial to limit post-partum relapses in MS using the synthetical gestagene nomegestrol (in conjunction with E2) did not show clinical or radiological effects (Vukusic et al., 2021).

The role of prolactin in MS is not well defined. Several studies propose an MS- (Moshirzadeh, Ghareghozli, Harandi, & Pakdaman, 2012; Wei, Liu, Cheng, & Hu, 2017) and EAE-aggravating effect (Riskind, Massacesi, Doolittle, & Hauser, 1991); still, prolactin might also have remyelinating features (Gregg et al., 2007).

In spontaneous SJL-mouse EAE models, testosterone supplementation reduced disease severity by a Th2-shift (Dalal, Kim, & Voskuhl, 1997). Androgens induced autoimmune regulator (AIRE) gene expression in thymus epithelial cells, which promotes T-cell tolerance by increasing the display of self-autoantigens in the thymus during T-cell development. In active immunization EAE, androgen and sex-dependent upregulation of AIRE protected against T-cell mediated autoimmunity (Zhu et al., 2016). Besides additional anti-inflammatory effects of androgens in active immunization EAE and *in vitro* (e.g., decrease in Th1 and Th17 infiltration and differentiation), decreasing neuronal survival under oxidative stress in cultured neurons was demonstrated as a potential surrogate of higher neurodegeneration (Massa et al., 2017). This may be another aspect helping to explain the phenotypical differences between MS in women, more prone to inflammation, and men, more prone to neurodegeneration. However, this experimental finding stands in contrast to epidemiological and clinical observations (Chitnis, 2018) underscoring the complex interplay of immune and neuronal functioning in the context of sex differences that are currently insufficiently understood and deserve further investigation.

2. Multiple sclerosis (MS)

MS is the most common chronic inflammatory disease of the CNS beginning mainly in young adulthood (Reich, Lucchinetti, & Calabresi, 2018). Clearly more common in women, prevalence in females has been rising, causing an increased sex bias (Milo & Kahana, 2010; Nussinovitch & Shoenfeld, 2012). Hormonal factors are involved in regulating the course of the disease, with sex hormones playing a major role. Here, we summarize sex and gender differences in MS and the influence of the life cycle and sexual hormones on the disease course.

In a first step, a PubMed literature search was performed (17-May-2021), revealing many data on sex referring to MS. With the search terms "multiple sclerosis AND sex," 4802 results have been detected of which only 31 were relevant to the topic after screening of headings and abstracts. Since the female reproductive cycle plays a role, a further search was carried out using the terms: "multiple sclerosis AND menarche"; "multiple sclerosis AND pregnancy" and "multiple sclerosis AND menopause." This search revealed 1724 hits with 19 being useful in this context. The search with the terms "multiple sclerosis AND gender" revealed 5878 hits of which only 10 were relevant to the topic after screening of headings and abstracts. The search for "multiple sclerosis AND andropause" displayed 3 hits, 2 being topic-related (one general and one MS-specific review).

2.1 Epidemiology

The prevalence of MS is unequally distributed between men and women (Milo & Kahana, 2010; Nussinovitch & Shoenfeld, 2012). In the beginning of the 20th century, MS was still considered more prevalent in men (Rolak, 2003). In the 1940s, the sex ratio shifted and a female preponderance is now universally reported, an aspect also known from other autoimmune diseases (Maghzi et al., 2010; Nussinovitch & Shoenfeld, 2012; Orton et al., 2006; Osoegawa et al., 2009; Pugliatti, Cossu, Sotgiu, Rosati, & Riise, 2009; Talley, 2005).

In recent decades, the sex ratio further increased toward women, and so did the incidence. In the 1970s, the risk of developing MS doubled among women in Denmark, while there was only a slight increase in men (Koch-Henriksen & Sorensen, 2010). Since the 1980s, the female-to-male prevalence increased from 2:1 to 3:1 (Koch-Henriksen & Sorensen, 2010). Interestingly, an increased incidence in women has been reported for relapsing remitting MS (RRMS), but not primary progressive MS (PPMS) (Celius & Smestad, 2009; Debouverie, Pittion-Vouyovitch, Louis, Roederer, & Guillemin, 2007; Miclea et al., 2019). An influence of behavioral (i.e., lifestyle changes predominantly affecting women) and environmental factors in disease mechanisms has been postulated as well as more complex gene-environment and gene-gene interactions (Koch-Henriksen & Sorensen, 2010). Numerous lifestyle changes in recent decades, especially in Western women, could contribute to the increased incidence of MS, such as cigarette smoking behavior, changes in diets with a tendency to obesity, earlier age of menarche, pregnancies at a higher age, use of oral contraceptives and hormone replacement therapy.

Overweight and obesity are a major current public health problem worldwide with a fourfold increase in the rate of obesity in adolescents reported over the past 30 years (Ogden, Carroll, Kit, & Flegal, 2014). Obesity in young adulthood, i.e., between 18 and 25 years of age, is associated with a higher susceptibility for MS (Gianfrancesco & Barcellos, 2016). The biological mechanisms linking obesity and MS are not entirely clear. Obesity is characterized by a chronic low-grade inflammatory response, associated with higher levels of pro-inflammatory C-reactive protein, IL-6 and leptin levels, which may influence the pathogenesis of MS (Sbarbati et al., 2006).

While smoking was more prevalent in men in the mid-20th century, a sex shift occurred toward a higher prevalence of smoking in women over the last 40 years (Koch-Henriksen & Sorensen, 2010). Although this has been suggested as an explanation for the increasing MS incidence in women (Harbo, Gold, & Tintore, 2013; Ysrraelit & Correale, 2019), the coincidental decrease in age at diagnosis at least challenges this hypothesis as a longer period of exposure to smoking preceding disease onset would be expected. Moreover, the sex difference with relative stability of male prevalence over seven decades renders smoking rather unlikely as a triggering factor almost exclusively affecting women.

Another major behavioral changes in women are later and fewer child births. The temporary immune tolerance during pregnancy may exert a protective effect against MS. Older age when first giving birth or a lower number of pregnancies may be, therefore, factors favoring the increasing incidence of MS in women (McCombe & Greer, 2013).

In summary, the many hypotheses surrounding the shift in MS incidence in women are yet to be clarified.

2.2 Disease course and phenotype

Clinical features, onset, and course of MS are not spared from sex differences: RRMS and secondary progressive MS (SPMS) have a 3:1 and 3:2 female-to-male ratio, respectively (Alonso & Hernan, 2008; Confavreux & Vukusic, 2006; Koch-Henriksen & Sorensen, 2010; Runmarker & Andersen, 1993; Thompson, 2004). On the contrary, in PPMS, the main demographic and clinical characteristics are male sex, older age at onset, and presence of spinal lesions (Thompson, 2004). Male sex, older age, and lesions predominantly in the spinal cord were additionally identified as the three predictors of evolution from radiologically isolated syndrome (RIS) to PPMS (Granberg, Martola, Kristoffersen-Wiberg, Aspelin, & Fredrikson, 2013). Female RRMS patients are younger than males at onset and it is the opposite in PPMS (Rommer et al., 2020).

Despite a clear definition of common MS symptoms, but not commonly differentiated by sex, symptoms differ between men and women and at different stages of the disease: Optic neuritis, visual symptoms and sensory disturbances at disease onset are more frequent in women, whereas motor symptoms are more frequent in men (Rommer et al., 2020). During the course of disease, woman experience more fatigue, depression, pain and cognitive impairment (Rommer et al., 2020).

A 17.37% higher annual relapse rate has been reported in women with a higher female-to-male ratio among the patients with higher annual relapse count. Within the initial 5 years after disease onset, this ratio was described with 2.3:1 in patients with no relapses vs 3.3:1 in patients with at least four relapses per year (Kalincik et al., 2013).

Contrarily, the risk of disability progression is higher in men. In fact, male sex is one of the early predictors of future permanent disability in MS. Different studies report a shorter time to disability progression in male RRMS patients as well as shorter time to conversion to SPMS (Ribbons et al., 2015; Savettieri et al., 2004; Weinshenker et al., 1991).

However, poorer prognosis of disability progression in males seems to be limited to relapsing onset MS. In PPMS, disease progression is similar in men and women (Ribbons et al., 2015). Within these studies, disease progression was not limited to motor disability progression, but also included faster cognitive decline in men, in part contradicting other studies (Rommer et al., 2020) and thus underscoring the need for more data particularly addressing this topic.

The notion of a "separation" of disability progression from relapse rates questions annualized relapse rate (ARR) as an adequate single primary outcome in clinical trials, as still usually done. It also raises the question of sex-specific trial outcome definitions.

Fig. 1 illustrates key epidemiological features and disease characteristics of MS.

2.3 Effects of sex hormones on MS

Apparent epidemiological sex differences in MS suggest that hormones play a key role in the development and course of the disease. As MS prevalence is higher in women and hormonal changes are mainly attributed to the female sex, their influence on the disease are by far better described.

2.3.1 Menarche and puberty

Over the 19th century, the age at menarche decreased considerably (Okasha, McCarron, McEwen, & Smith, 2001). Menarche is characterized by relevant hormonal changes including high production of estrogens, especially estradiol, and androgens. Age at menarche influences physiological and disease processes. In MS, many studies associate early menarche with an increased risk of MS and an earlier onset of the disease (Bove, Chua, et al., 2016; Sloka, Pryse-Phillips, & Stefanelli, 2006). The lower age at menarche in women with MS compared to women without MS suggests an inverse association between age at menarche and MS risk. For each year that menarche is delayed, MS risk was shown to decrease by 13% (Sloka et al., 2006).

Age at menarche has not only an effect on the risk of MS onset, but also on disease progression. Older age at menarche was associated with a reduced risk of reaching an Expanded Disability Status Scale (EDSS) score of 6.0 (i.e., requiring unilateral aid for a walking range of 100 m) in progressive MS. As a supporting notion for hormonal influences, increased relapse rates were observed in the peri-menarcheal period (D'Hooghe, Haentjens, Nagels, D'Hooghe, & De Keyser, 2012; Lulu, Graves, & Waubant, 2016).

2.3.2 Pregnancy

Positive effects of pregnancy on MS relapse rate are well studied. The decrease of the relapse rate during pregnancy is highest during the third trimester with a 70% reduction compared to the year before pregnancy. After delivery, a rebound effect can occur with increased relapse frequency (Confavreux et al., 1998).

These changes in disease activity are related to hormonal changes during pregnancy that affect the immune system. Pregnancy tends to suppress the immune system of the mother to prevent rejection of the fetus, which has paternal foreign antigens (Tafuri, Alferink, Moller, Hammerling, & Arnold, 1995). During late pregnancy estriol, estradiol and progesterone are increased. Estriol is a hormone unique to pregnancy and reaches high levels during the third trimester. In an animal model of MS, estriol treatment showed amelioration of disease symptoms, reduced inflammation and demyelination compared to vehicle treatment (Kim et al., 1999) suggesting a protective effect of estriol during pregnancy (see Section 1).

Protective effects on MS disease course during pregnancy are also attributed to progesterone. A remarkable increase in relapse rate is observed postpartum, when progesterone levels decrease dramatically (Confavreux et al., 1998). In experimental animal models, progesterone was associated with a shift of a Th1 pro-inflammatory response to a Th2 anti-inflammatory response, decreased inflammatory mediators and attenuated neuropathology and disease severity (Garay, Gonzalez Deniselle, Lima, Roig, & De Nicola, 2007) (see Section 1).

Despite well-established knowledge on pregnancy and MS relapse rate, the effects on disability levels remain controversial. Protection against relapse-related disability was demonstrated in a cohort-study of 2466 patients over 10 years (Jokubaitis et al., 2016) whereas another study did not show an effect on permanent disability scores (Vukusic et al., 2004).

2.3.3 Puerperium

The puerperium is associated with an abrupt drop in estrogen and progesterone levels, the hormones responsible for reducing the immune response during pregnancy. This change leads to an abrupt "normalization" of immune function and may favor an increase in disease activity. Indeed, relapse rates after childbirth significantly increased compared to the year preceding pregnancy (Finkelsztejn, Brooks, Paschoal Jr., & Fragoso, 2011).

Therefore, the post-partum period requires special attention by the treating neurologists and clearly discussed concepts for this phase with a benefit-risk assessment on how and when to re-install immunotherapy and how and how long breastfeeding can be facilitated in individual cases.

2.3.4 Breastfeeding

Studies on breastfeeding in MS remain controversial. A protection against relapses during exclusive breastfeeding and a lower risk of developing MS after prolonged breastfeeding were reported (Airas, Jalkanen, Alanen, Pirttila, & Marttila, 2010; Hellwig et al., 2015; Langer-Gould et al., 2020). However, other studies indicate that breastfeeding does not suppress disease susceptibility or activity (Airas et al., 2010; Portaccio et al., 2011). There is no clear evidence that breastfeeding provides sufficient protection from relapses to justify not resuming disease-modifying treatment after giving birth in women with (highly) active MS (Langer-Gould et al., 2020). Again, an individual benefit-risk assessment remains important.

2.3.5 Menopause

Only few information is available on a possible influence of menopause and hormone replacement therapy in MS. The effects of menopausal hormone changes may be confounded by age-related changes in MS disease activity and co-morbidities.

Two groups report a significant annual increase of EDSS with menopause, in one of these studies with an ARR reduction (Baroncini et al., 2019; Bove, Healy, et al., 2016). In contrast, a stable rate of EDSS progression across the menopausal transition with decreasing relapse rates (Ladeira et al., 2018) and a lacking association of menopause with increased risk of disability accumulation (Otero-Romero, 2017) have also been reported.

Fig. 2 illustrates the influence of the female life cycle from menarche to menopause on MS disease course.

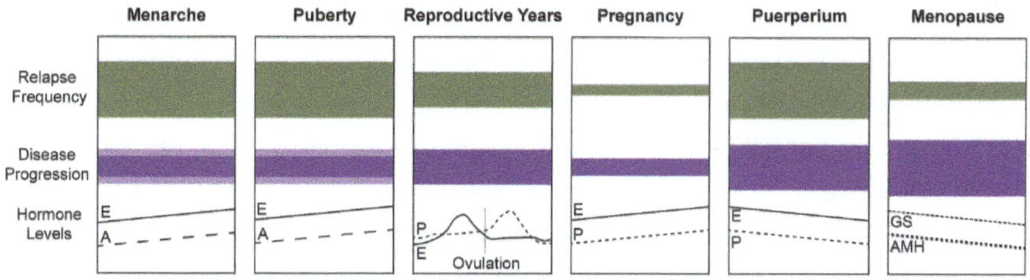

Fig. 2 Effects of the female life cycle from menarche to menopause on multiple sclerosis (MS) relapses and progression: Early menarche is associated with increased risk of MS and earlier onset of the disease. During pregnancy, a decrease in relapse rates is observed, particularly in the third trimester. After delivery, relapse rates may increase to almost three-times higher than pre-pregnancy levels during the first 3–6 months. There is little evidence on the possible influence of menopause and hormone replacement therapy on the course of MS. As an epidemiological observation, disability progression occurs more prominently with longer disease duration, and thus in later life. However, progression may occur independent of relapses in all phases of MS. Abbreviations: A: androgens; AMH: anti-Muellerian hormone; E: estriol; GS: gonadal steroids; P: progesterone.

2.3.6 Hormonal changes in men

The hormonal changes in men during life are deemed less pronounced than in women. However, testosterone levels change during life and testosterone plays an important role in the regulation of the immune system.

In experimental models, neuroprotective effects of testosterone on the development and progression of MS are described (see Section 1). Four pathophysiological mechanisms have been proposed: Testosterone (1) can cross the blood–brain barrier in its free form and directly influence neuronal cells, (2) has a protective effect against glutamate toxicity on spinal cord neurons and oxidative stress in neuronal cell lines, (3) leads to an increased expression of neurotrophic factors and (4) reduces microglial activation (Chitnis, 2018).

Lowered testosterone levels in up to 40% of men with early MS and an association of low total testosterone levels and increased disability and cognitive decline (Bove et al., 2014) suggest a clinical meaningfulness of the experimental findings, but need deeper investigation.

2.4 The role of gender in MS

The differences between men and women described so far are mainly based on sexual differences; however, differences are not limited to sex as a biological variable. Dealing with the diagnosis and living with the disease can significantly differ between men and women.

Gender differences are prominently apparent in the choice and handling of diagnosis and medication: Men may be willing to tolerate additional risks (efficacy weighted over side effects) when choosing therapies. However, this difference has no effect on compliance to medication (Clanet, Wolinsky, Ashton, Hartung, & Reingold, 2014; Jokubaitis et al., 2013). On the other hand, men are more skeptical about physical and occupational therapy than women (Bove, McHenry, et al., 2016) representing a major component of MS treatment. On a psychological level, women are more prone to anxiety disorders, while men are more prone to suicide upon MS diagnosis (Pompili et al., 2012). The effects on social life are also different: Men less frequently take advantage of the health care system, experience lower social support and leave their employment due to their MS more likely than woman (Bove, McHenry, et al., 2016).

To date, investigations of gender differences in MS are present and growing. However, they currently rather descriptively note differences. In a next step, addressing these differences with tailored communication, treatment and support strategies in trial settings would alleviate optimized and personalized counselling of patients with MS. An important future goal will be to address the needs of women and men with MS individually in the clinical routine setting, at best with evidence-based approaches.

3. Neuromyelitis optica spectrum disorder (NMOSD) and myelin-oligodendrocyte-glycoprotein antibody associated disorder (MOGAD)

Neuromyelitis optica spectrum disorder (NMOSD) and Myelin-Oligodendrocyte-Glycoprotein antibody associated disorder (MOGAD) are autoimmune demyelinating disorders of the central nervous system (CNS), mostly defined by the presence of specific autoantibodies involved in the disease. They are different entities from Multiple Sclerosis (MS), which harbors relevant implications for both differential diagnosis and acute and long-term treatment.

The majority of patients with NMOSD is positive for immunoglobulin G (IgG) antibodies against the water channel aquaporin-4 (AQP4) (Iorio & Papi, 2021). The diagnosis of MOGAD is by definition based on the detection of MOG-IgG (Jarius et al., 2018). As the clinical spectrum of antibody negative NMOSD is less well defined, we here only refer to AQP4-IgG positive NMOSD—from now on only named NMOSD—and MOGAD.

In a first step, a PubMed literature search was performed (12-Aug-2021), revealing sparse data on sex and gender referring to NMOSD and MOGAD. With the search terms "NMOSD AND sex" and "NMOSD AND gender," 232 results have been detected of which only twelve were relevant to the topic screening headings and abstracts. As the term "MOGAD" has just recently been introduced, we decided to use "MOG AND sex" and "MOG AND gender" as search terms in order not to miss potentially relevant results. This search revealed 214 hits with only eight being useful in this context. Overlapping hits between NMOSD and MOGAD were only counted once. The considerably lower number of search results as compared to MS is expected for different reasons: Both entities are considerably rarer than MS and their definition based on the reliable detection of a specific antibody in cell-based assays has only been set up in the early 2000s for AQP4-IgG in NMOSD (Lennon, Kryzer, Pittock, Verkman, & Hinson, 2005; Lennon et al., 2004), and in 2018 for MOG-IgG in MOGAD (Jarius et al., 2018).

The pathogenicity of AQP4-IgG is undoubted whereas in MOGAD, pathophysiological mechanisms are still less well defined. Just recently, as a distinct feature of MOGAD, a subset of patients was described to produce MOG-IgG intrathecally. This has not been detected in AQP4-IgG positive NMOSD patients (Akaishi et al., 2021).

3.1 Epidemiology

3.1.1 Neuromyelitis optica spectrum disorder (NMOSD)

NMOSD are a rare disease entity with a prevalence range of around 0.5–10 per 100,000 persons and an incidence range of below 0.5 to 7.3 per 1,000,000 persons (Hor et al., 2020; Papp et al., 2021). The age at onset of NMOSD tends to be higher than in MS with most data revealing a mean around 40 years (Hor et al., 2020; Papp et al., 2021). In children, NMOSD are even rarer, yet, childhood onset is possible. Several studies suggest a higher age at onset in men (Borisow et al., 2017; Kim et al., 2017; Li et al., 2021).

Due to substantial ethnic differences, it is important to dissect prevalence and incidence rates by different ethnic origin and region, which results in considerably different rates (Hor et al., 2020; Papp et al., 2021) (Fig. 3). These differences may also explain the variability of reported incidence rates in children with NMOSD (0.3–1.1 per 1,000,000) (Hor et al., 2020).

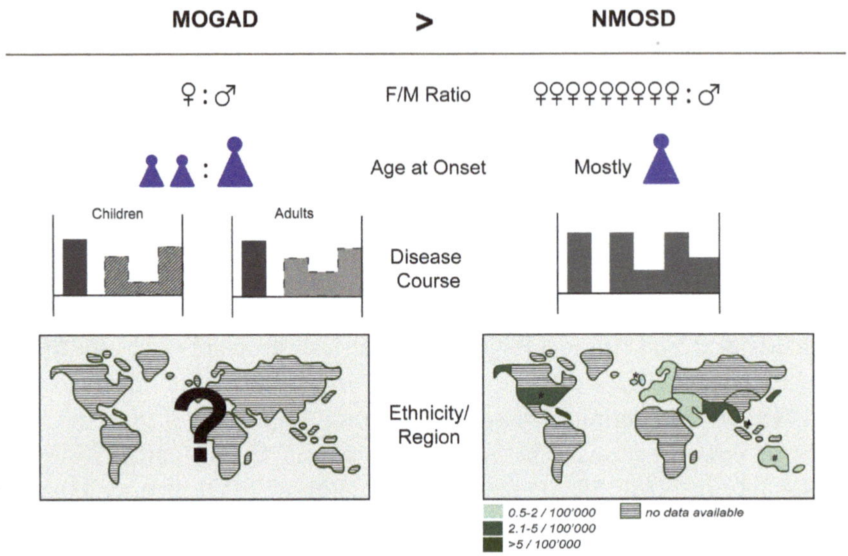

Fig. 3 Epidemiological characteristics of myelin-oligodendrocyte-glycoprotein antibody associated disorder (MOGAD) and neuromyelitis optica spectrum disorder (NMOSD): MOGAD is more frequent than NMOSD with a considerably different female-to-male ratio. Prevalence of MOGAD is twice as high in children compared to adults whereas NMOSD is rare in children. For both, NMOSD and MOGAD, disability mainly results from relapses, different from multiple sclerosis (MS, Section 1). Children are more likely to experience monophasic MOGAD, reported proportions of relapsing MOGAD in adults vary, but relapsing disease is common. NMOSD typically follows a relapsing course. Ethnic and regional differences are pronounced in NMOSD (Hor et al., 2020; Papp et al., 2021), but still not well determined for MOGAD. Prevalence is given per 100,000 persons. Abbreviations and symbols: >: higher than; ♀: female; ♂: male; F/M Ratio: female-to-male ratio; *: Blacks > Whites; #: Blacks/Asians/Maoris > Whites/Aborigines; ▲: Chinese > Malays.

The genetic and environmental determinants of ethnic and regional differences are thus far unclear. Methodological differences between studies or availability of adequate antibody testing methods may not fully explain this picture. Interestingly, the regional and racial differences do not overlap with the latitude gradients described for MS as African ethnicity is associated with the highest prevalence and incidence estimates (Papp et al., 2021).

Throughout all epidemiological studies, a female preponderance, substantially exceeding the reported female preponderance in MS, is reported, mostly summarized in a female-to-male ratio of around 9:1 (Hor et al., 2020). This high female-to-male ratio has relevant implications not only for treatment trials (Section 4), but on the interpretation of data in general as the male NMOSD population usually has a considerably lower size than the female NMOSD patients within a study limiting robust analyses for the male proportion.

Besides the general explanations for increased autoimmunity in women as summarized in Section 1, no sex-specific data on disease susceptibility for NMOSD have been reported, thus far. In line with this, women are more likely to be AQP4-IgG positive and to have co-existing anti-thyroid antibodies (Borisow et al., 2017; Sun et al., 2017). Strikingly, despite this higher proportion of women with AQP4-IgG which should alleviate NMOSD diagnosis, time between disease onset and diagnosis was significantly longer compared to men (Borisow et al., 2017). Misdiagnosis of MS may account for this diagnostic delay; however, a sex difference is not well explained by this.

3.1.2 Myelin-oligodendrocyte-glycoprotein antibody associated disorder (MOGAD)

With the recent definition of the specific antibody and varying diagnostic criteria, prevalence and incidence rates should be interpreted with caution, especially when comparing published data from different regions. In an additional literature search, no publications were identified answering the question of MOGAD prevalence thus far as studies investigated prevalence of MOG-IgG in NMOSD cohorts, and therefore not covering the full spectrum (as reviewed in Hor et al., 2020). A European study estimated the incidence rate of MOGAD to be 1.6 per 1,000,000 persons (de Mol et al., 2020). Of note, the incidence rate is more than double in children as compared to adults (de Mol et al., 2020). In fact, in acquired demyelinating syndromes (ADS) in children, MOG-IgG are found in 34% of patients (Bruijstens et al., 2020). As a clinical consequence, testing for MOG-IgG should be considered in all children presenting with ADS as this has prognostic and therapeutic implications.

Age at onset of MOGAD needs to be differentially investigated in children and adults. In children, an onset age of 9 years and, in adults, of 32–33 years, respectively, has been described (de Mol et al., 2020; Kitley et al., 2014). Interestingly, in three adult cohorts with predominantly optic neuritis as onset, a slightly older onset age has been found (39–41 years) (Ducloyer et al., 2020; Liu et al., 2021). Differences due to clinical phenotypes might thus be hypothesized.

Both in children and adults, an increasing number of studies reports an equal sex distribution in different clinical manifestations of MOGAD (Bruijstens et al., 2020; de Mol et al., 2020; Ducloyer et al., 2020; Hor et al., 2020; Kitley et al., 2014; Lechner et al., 2020; Zhang et al., 2020). Still, a German/Japanese study found a female preponderance comparable to MS (Liu et al., 2021) highlighting potential methodological differences between study settings in small cohorts of a rare disorder, but potentially also regional and ethnical differences of MOGAD.

Environmental and genetic factors contributing to NMOSD and MOGAD susceptibility are thus far puzzling. HLA haplotype association has been described robustly for NMOSD, but not consistently for pediatric-onset MOGAD, and not for adult-onset MOGAD, thus far (as reviewed in Ghafouri-Fard, Azimi, & Taheri, 2021; Lopez, Denkova, Ramanathan, Dale, & Brilot, 2021). Other risk loci mainly describing genes involved in immune function have been found in single studies, mostly without independent confirmation for NMOSD (Ghafouri-Fard et al., 2021). The role of an infectious prodrome, vitamin D, and UV exposure need further investigation in both disorders (Lopez et al., 2021).

3.2 Disease course and phenotype

3.2.1 Neuromyelitis optica spectrum disorder (NMOSD)

The clinical phenotype of NMOSD is summarized in the diagnostic criteria with the core features of optic neuritis, acute myelitis, area postrema and acute brainstem syndromes as well as acute diencephalic and cerebral syndromes (Wingerchuk et al., 2015).

NMOSD follows a relapsing disease course in the vast majority of patients (around 90%) with disability as a consequence of severe relapses rather than chronic progressive loss of function (Hinson, Lennon, & Pittock, 2016). A multicenter German study did neither report sex differences for this notion nor for relapse rate, relapse phenotype or outcome after different relapse treatment regimens (Borisow et al., 2017). By contrast, a Chinese study demonstrated sex differences for onset and relapse phenotypes more commonly presenting with optic neuritis over myelitis in men vs women, and even more importantly, a higher proportion of high-dose steroid non-responders in men (Li et al., 2021). A smaller Korean study detected just the opposite with men being prone to myelitis rather than optic neuritis (Kim et al., 2017). With the latter two studies including patients of

Asian origin, but presenting diverging results, again the issue of potential methodological problems arises in a rare disease with small sample sizes, especially for the underrepresented sex. Thus, it remains open whether these differences really reflect regional and racial differences.

3.2.2 Myelin-oligodendrocyte-glycoprotein antibody associated disorder (MOGAD)

The clinical spectrum of MOGAD is broader than the NMOSD spectrum, but may include all clinical presentations thereof as well as encephalitis not typically seen in NMOSD. Both phenotype and disease course vary with age. Whereas in children, the most common initial presentation is acute demyelinating encephalomyelitis (ADEM), optic neuritis and/or transverse myelitis are more commonly the onset syndrome in adults (Bruijstens et al., 2020; de Mol et al., 2020; Kitley et al., 2014). The Japanese adult cohort exhibited more cerebral syndromes as compared to the German cohort (Liu et al., 2021). Again, the reasons for these differences are hard to delineate.

As compared to children with MOGAD, functional recovery seems worse in adults (defined by an EDSS of $\geq$3.0 at last follow-up) (Cobo-Calvo et al., 2021). In addition to better recovery, children are more likely to have a monophasic disease course and MOG-IgG seems to be useful in relapse risk evaluation as children turning seronegative are more likely to be monophasic (Bruijstens et al., 2020; Cobo-Calvo et al., 2021). This observation was not found in adults (Cobo-Calvo et al., 2021). The reported frequency of a relapsing course in adults varies from 41% to 80%, mostly underscoring the need for more data (de Mol et al., 2020; Jarius et al., 2016).

Whereas in children, older age predisposes for sequential relapses, in the adult population, younger patients (18–39 years) are at higher risk for a relapsing disease course (Bruijstens et al., 2020; Cobo-Calvo et al., 2021). Thus far, a single retrospective analysis has demonstrated an increased risk of relapsing course in women vs men, an effect seemingly driven by the children included in this cohort (Cobo-Calvo et al., 2021). Another study did not detect differences in adults with regards to the frequency of a relapsing course in women and men (Jarius et al., 2016). However, descriptively this study has found higher annualized relapse rates in women which was not statistically significant in a cohort of 34 patients.

Taken together, the data are still incomplete and need confirmation in most aspects. There is thus far no clear picture on differences driven by sex for MOGAD.

Fig. 3 illustrates key epidemiological features and disease characteristics of NMOSD and MOGAD.

3.3 Effects of sex hormones on NMOSD and MOGAD

The scarce evidence on the influence of sex hormones on NMOSD and MOGAD is indirect as mostly sex and age categories as indicators of fertility are analyzed without an actual measurement of hormones or a specific individual evaluation of the hormonal status in study participants. This must be regarded as a caveat in the interpretation of data.

3.3.1 Neuromyelitis optica spectrum disorder (NMOSD)

3.3.1.1 Potential associations of fertility, menopause and hormonal contraception

A comparison of NMOSD characteristics between women of fertile age (defined as women within an age range of 15–40 years) and women beyond fertile age (defined as older than 40 years) demonstrated more favorable relapse recovery and relapse treatment response whereas relapse phenotype and relapse rates did not differ (Borisow et al., 2017). The same analysis in the (albeit considerably smaller) group of men separated by an age cut-off of 40 years revealed no differences for these parameters as a hint on effects attributed to sex, not age (Borisow et al., 2017).

Comparing non- and menopausal women in a Chinese cohort, unfortunately without a description on how this status was defined, onset and further relapse phenotype did differ with more myelitis events, worse disability outcomes and higher rates of high-dose steroid non-responders in menopausal women (Li et al., 2021). As these data are fairly consistent with a previous report of NMOSD relapse treatment outcome after myelitis (Kleiter et al., 2016), this seems plausible. Yet, again, a concluding answer whether true ethnic differences or methodological aspects may explain the diverging results cannot be given.

From a pathophysiological point of view, not only the influence of fertility on NMOSD, but also the influence of NMOSD on fertility is of interest: AQP4 and other aquaporin isoforms are expressed in the female and male reproductive tract and in CNS regions involved in the gonadotropic axis in mammals (Huang et al., 2006; Venero, Vizuete, Machado, & Cano, 2001). Human data is thus far lacking. However, an interaction of

AQP4-IgG with cellular homeostasis within the reproductive tract and hormonal regulation influencing male and female fertility and conception, especially the peri-implantation phase, may well be hypothesized. AQP4 deletion did in fact reduce fertility in female mice (Sun et al., 2009). A small study including 14 NMOSD patients ($n = 11$ AQP4-IgG positive, $n = 1$ MOG-IgG positive, $n = 2$ negative for both) detected lower anti-Muellerian hormone levels as compared to healthy controls, indicating a potential reduction of the ovarian reserve (Thone et al., 2018).

A questionnaire-based multicenter study on 217 subjects reported younger age at symptom onset in women with NMOSD with ever-use of hormonal contraception as compared to never-use (Bove et al., 2017). Yet, only NMOSD patients have been investigated in this study, impeding a comparison to other disease entities.

3.3.1.2 Pregnancy and breastfeeding

Especially the early post-partum period, namely 3–6 months after delivery, seems to be associated with a higher relapse risk in women with NMOSD (as reviewed in Altintas et al., 2020; Mao-Draayer et al., 2020). This phase was even shown to be associated with newly onset NMOSD almost 3-times higher than expected (Klawiter et al., 2017) and overshooting disease activity exceeding the pre-pregnancy level (Collongues et al., 2021). Different from MS where pregnancy is associated with lower relapse risks, especially during the later stages of pregnancy, the annualized relapse rate increased during the first trimester and returned to the baseline level thereafter until delivery in NMOSD patients (Klawiter et al., 2017). This finding was not confirmed in a recent study (Collongues et al., 2021).

The cessation of immunosuppressive treatments before or with conception is likely to have an influence on this and cessation behavior may have change within the interval between these two studies. However, the benefit-risk assessment in this situation is very difficult. Several agents used to treat NMOSD are contraindicated with pregnancy due to a teratogenic potential, e.g., methotrexate, mycophenolate mofetil, mitoxantrone or cyclophosphamide (Altintas et al., 2020). Data on pregnancy-related outcomes for most other agents is sparse (Mao-Draayer et al., 2020). The risk of a significant increase of permanent disability due to a relapse needs to be weighted with the wish to become pregnant and potential risks for the fetus. Azathioprine, rituximab or eculizumab might be potential options in this situation, as well as intravenous immunoglobulins or low-dose steroid treatment (Altintas et al., 2020; Mao-Draayer et al., 2020). However, robust data on a larger cohort level is lacking and these off-label options need to be discussed for each individual patient.

Continued immunosuppressive treatment (in general) reduced the risk of relapses during pregnancy in a mixed cohort including AQP4-IgG, MOG-IgG and antibody negative patients with an NMOSD phenotype (Collongues et al., 2021). This effect was driven by the AQP4-IgG positive subgroup. It needs to be mentioned that the three sub-cohorts were small and various immunotherapies have been used. A recommendation for a specific regimen thus cannot be derived from these data.

Treatment of relapses during pregnancy is possible with high-dose steroids or plasma exchange procedures including immunoadsorption (Altintas et al., 2020; Mao-Draayer et al., 2020).

Data on a potential protective effect of breastfeeding in NMOSD is thus far lacking. With the mostly relapsing disease course, higher disease activity post-partum and potential severe disability resulting from a single relapse, early re-initiation of immunotherapy after delivery seems reasonable and may be possible, again with an individualized benefit-risk assessment, and with some of the immunotherapeutic strategies (Altintas et al., 2020; Mao-Draayer et al., 2020).

As already pointed out above, it seems plausible that the presence of AQP4-IgG affects the reproductive cycle due to the expression of AQP4 in the respective tissues and hormonal axis. AQP4 is additionally expressed in the human placenta, experimental data indicate a potential role of AQP4-IgG in miscarriage (Saadoun et al., 2013). A higher risk of miscarriage has in fact been demonstrated in women after NMOSD onset, also after correction for maternal age and history of miscarriage (Nour et al., 2016).

Preeclampsia, as a potential risk for both the mother and the unborn child, has not been associated with NMOSD, but with a recent history of miscarriage and the presence of other co-existing autoimmune disorders (Nour et al., 2016). Whereas in other antibody-mediated autoimmune disorders, it is described that the newborn may experience symptoms due to a transplacental shift of pathogenic autoantibodies, NMOSD symptoms have not been described in newborns although AQP4-IgG crosses the placenta (Mao-Draayer et al., 2020).

3.3.2 Myelin-oligodendrocyte-glycoprotein antibody associated disorder (MOGAD)

Data addressing associations of sex hormones or male and female life cycles and MOGAD were not found. A systematic evaluation of pregnancy-related susceptibility or relapse risks for MOGAD is lacking. In a cohort that included ten women with documented pregnancies, five of them experienced relapses in association to pregnancy or the post-partum period, including first manifestations of MOGAD, whereas in the other five women, pregnancy and post-partum period were not in timely correlation to an event (Jarius et al., 2016). Similar to NMOSD and MS, an early post-partum peak of disease activity occurs in MOGAD, but seems to be less pronounced as compared to NMOSD (Collongues et al., 2021).

Distinct studies on the influence of hormonal states in both NMOSD and MOGAD are widely lacking for men and women arguing for further research in this field. Due to the rarity of both disorders, this calls for joint international efforts to investigate these issues, also with respect to assumed regional and ethnic differences.

3.4 The role of gender in NMOSD and MOGAD

All detected studies indicating the topic of «gender» as referred to the title of the publication, actually deal with sex as a biological variable, discussing epidemiological traits of NMOSD and MOGAD or potential hormonal influences as described above (Altintas et al., 2020; Kim et al., 2017; Sun et al., 2017; Wingerchuk, 2009). Distinct studies on gender effects were thus not identified for NMOSD and MOGAD. The general aspects as discussed for MS (Section 2) need to be considered, likewise. Further research investigating the role of gender in autoimmune demyelinating CNS disorders is necessary.

4. Immunotherapies in the context of sex differences

Despite a focus on sex differences in clinical and preclinical studies stipulated by the FDA (FDA, n.d.), sex differences of different aspects of MS and NMOSD immunotherapies are still mostly neglected. Phase 3 clinical trials are powered to detect differences between an investigational medical product and a control medication, but not to detect differences in clinical response and side effect patterns between sexes, not to name potential gender effects therein. Most studies in the post-marketing setting aim to investigate the effect of a recently licensed medication on safety and/or specific symptoms such as fatigue, cognition or patient-related outcome, e.g., satisfaction with the treatment. With regards to sex differences, post-hoc analyses of the pivotal phase 3 trials represent the mainstay of current practice to this end. Yet, these may only yield insufficient evidence as trials were not powered to prove significant differences within subgroups. This is of special relevance for the underrepresented sex due to a usually smaller sample size reflecting the female-to-male ratio in the indication tested. However, it may also be highly relevant in case of sex differences in the disease course and thus remarkably different effect sizes between sexes (Houtchens & Bove, 2018).

With these limitations, sex differences in MS and NMOSD immunotherapies are summarized for efficacy and safety aspects.

4.1 Search strategy

A PubMed search was performed (17-Nov-2021) for licensed MS and NMOSD medications according to the Swiss authority Swissmedic (swissmedic.ch) using the following search term combination: "[drug name] AND sex." Animal studies, case reports and case series were not considered.

Fig. 4 displays the timeline of PubMed hits as identified by our search strategy.

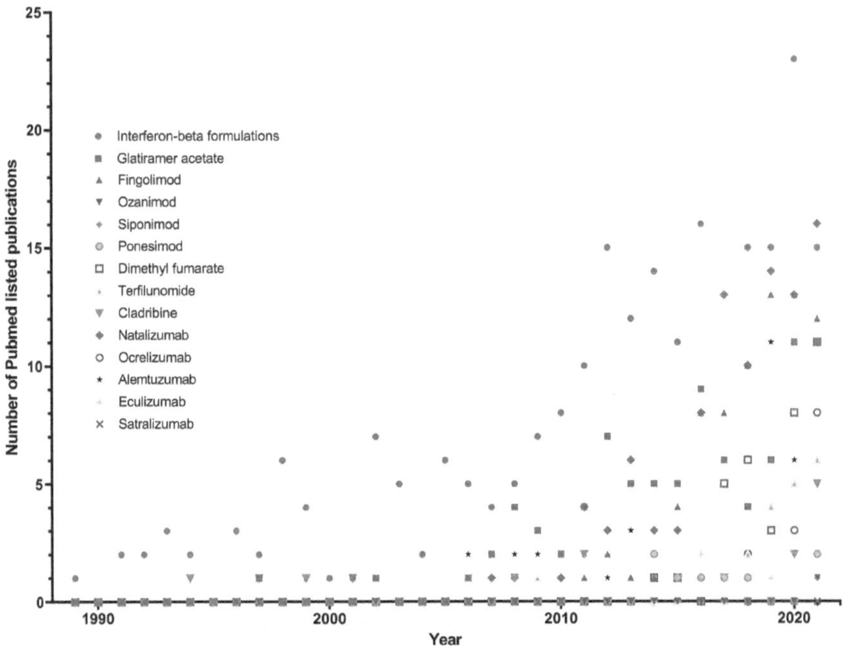

Fig. 4 Absolute number of PubMed listed publications per year using the search terms "[drug name] AND sex": All identified manuscripts are listed, irrespective of the content or further use in this review. The search has not been performed for the following recently licensed substances for the treatment of MS and are thus not part of this analysis: diroximel fumarate, natalizumab (subcutaneous application), ofatumumab.

In addition to the search strategy, important papers identified studying the data were also included even if they were not found by the initial search strategy. This review does thus not fulfill the criteria for a systematic review, but also differentially weighs aspects as deemed of importance by the authors.

4.2 Sex differences per substance

4.2.1 Sex-specific efficacy of MS treatments

Sex-specific data of efficacy parameters of disease-modifying MS treatments are summarized in Table 1.

As some of the data reveal, results between different analyses or even within analyses still do not paint a clear picture on sex differences with regards to treatment efficacy, as delineated for interferon-beta formulations or fingolimod. What factors drive these partially contradictory findings remains unclear.

Strikingly, even for longstanding available treatments, data are sparse.

4.2.2 Sex-specific efficacy of NMOSD treatments

Efficacy data of the pivotal trials are summarized in Table 2.

For both eculizumab (Soliris®) and satralizumab (Enspryng®), inclusion of patients followed an epidemiologically plausible female-to-male ratio (Section 3) resulting in around 90% of female trial participants. Thus, 13/143 male participants (8/96 treated with eculizumab, 5/47 treated with placebo) (Palace et al., 2021) were investigated for eculizumab. Preliminary data of the open-label extension studies of the pivotal trials of satralizumab have been presented including 11 men ($n = 111$ participants in total) (Kleiter et al., 2021).

4.2.3 Sex-specific safety of MS and NMOSD treatments

As the described literature search has only revealed sparse data on sex-specific safety aspects, findings are listed in Table 3.

Mostly, small independent studies have been detected, limiting interpretability due to size and focusing selected aspects of a specific treatment. For instance, occurrence of infusion-related reactions (IRR) under natalizumab do not represent a major safety issue in clinical practice (Sacco et al., 2020). However, the occurrence of progressive multifocal leukoencephalopathy (PML), a potentially life-threatening opportunistic infection, has not been investigated stratified by sex.

Table 1 Summary of sex-specific efficacy data of disease-modifying treatments of multiple sclerosis.

Substance	Mode of action	Indication	Efficacy data	Setting	Reference
Interferon beta formulations (Interferon beta-1a (Avonex®, Rebif®), PEG-ylated Interferon beta-1a (Plegridy®), Interferon beta-1b (Betaferon®/Betaseron®, Extavia®))	Not completely understood (regulation of innate and adaptive immune responses)	CIS, RRMS, SPMS (labels differ between formulations)	– Percentage of patients with relapse within the first year (female/male: 37%/37%, *P*-value 0.785) – ARR (female/male: 0.68/0.65, *P*-value 0.655, only available for 1/5 studies) – Percentage of patients with progression at year 2 (female/male: 27%/29%, *P*-value 0.776) – MRI without gadolinium enhancing lesions – at year 1 (female/male: 70%/79%, *P*-value 0.07) – at year 2 (female/male: 69%/69%, *P*-value 0.806)	Pooled analysis of 5 studies on intramuscular interferon beta-1a, 1406 MS patients (1027/1406 (73%) female)	Rudick et al. (2011)
			– No sex difference for conversion from CIS to MS, defined by McDonald 2005 criteria	Subgroup analysis of the pivotal REFLEX trial on interferon beta-1a	Freedman et al. (2014)
			– Significant risk reduction for male MS patients for time to first relapse (HR 0.87 (0.77–0.97), *P*-value 0.0164) – Trend for increased risk for male MS patients for 1-point EDSS progression (HR 1.17 (95%-CI 0.98–1.40), *P*-value 0.0882)	Real-world data of 2570 MS patients (1774/2570 (69%) female), Italian MS centers over up to 7 years, propensity score matched Cox model	Trojano et al. (2009)

			– Relapse ratio under treatment (female/male) 1.21 (95%-CI 1.04–1.40, *P*-value 0.013), not explained by other covariates (neutralizing antibodies against interferon beta, age, disease activity (relapses)/disability level (EDSS) before treatment start, time on medication)	Nationwide Danish Multiple Sclerosis registry, 2034 MS patients (1395/2034 (68.6%) female), unadjusted statistical model	Magyari, Koch-Henriksen, Laursen, and Sorensen (2014)
Glatiramer acetate (different formulations of Copaxone®, biosimilar Glatiramyl®/CLIFT®)	Binding through the major histocompatibility complex molecules leading to glatiramer acetate induced immune responses toward immunotolerance	RRMS	– Male sex favoring the use of glatiramer acetate over interferon beta-1a in RRMS	2 cohorts: – CombiRX trial (glatiramer acetate: 182/255 (71.4%) female; interferon beta-1a: 172/248 (69.4%) female) – MSBase with validation and generalization extracts (validation: glatiramer acetate: 859/1156 (74.3%) female; interferon beta-1a: 865/1156 (74.8%) female; generalization: glatiramer acetate: 1955/2660 (73.5%) female; interferon beta-1a: 1938/2660 (72.9%) female), treatment response score to predict patient-specific benefit of glatiramer acetate over interferon beta-1a	Bovis et al. (2021)

Continued

Table 1 Summary of sex-specific efficacy data of disease-modifying treatments of multiple sclerosis.—cont'd

Substance	Mode of action	Indication	Efficacy data	Setting	Reference
			– no effect on disability progression in total cohort (primary endpoint) – male participants: significant reduction of probability of progression (HR: 0.71, 95%-CI 0.53–0.95; $P=0.0193$)	Phase 3 study, 943 **PPMS** patients (483/943 (51.2%) female), randomized to glatiramer acetate or placebo (2:1), post-hoc analysis	Wolinsky et al. (2007)
Sphingosin-1-phosphate receptor modulators (Fingolimod (Gilenya®), Ozanimod (Zeposia®), Ponesimod (Ponvory®) and Siponimod (Mayzent®))	Group of drugs targeting S1PR with different selectivity for the receptor subclasses 1–5, mainly influencing immune cell trafficking by internalization and degradation of the $S1PR_1$	Label depending on S1PRM type and regulatory authority, ranging from CIS to SPMS, not PPMS			
– Fingolimod			– ARR ratio for fingolimod 0.5 mg vs. placebo: – men: 0.33 (95%-CI 0.22–0.50) – women: 0.50 (95%-CI 0.39–0.65) – Freedom from disability progression: – fingolimod-treated men: 86.7% (95%-CI 80.7–92.8) – fingolimod-treated women: 80.6% (95%-CI 76.0–85.1)	Subgroup analysis of the pivotal FREEDOMS trial, 842 RRMS patients (594/842 (70.5%) female), no formal analysis comparing fingolimod-treated men and women	Devonshire et al. (2012)

			– ARR reduction: – men: higher reduction in fingolimod- vs. placebo-treated, only borderline significant reduction vs. interferon beta-1a – women: higher reduction in fingolimod- vs. placebo- and interferon beta-1a treated	Pooled post-hoc analysis of the phase 3 studies (FREEDOMS, FREEDOMS II, and TRANSFORMS), 2416 RRMS patients (686 men (360 fingolimod, 187 placebo, 139 interferon beta.1a), 1730 women (852 fingolimod, 586 placebo, 292 interferon beta-1a))	Derfuss, Ontaneda, Nicholas, Meng, and Hawker (2016)
Ozanimod, Ponesimod, Siponimod			No data on sex differences found		
Dimethyl fumarate (Tecfidera®)	Pleiotropic immunological mechanisms including reduction of oxidative stress	RRMS	– HR for relapses of dimethyl fumarate-treated patients compared to placebo: – men: 0.47 (95%-CI 0.28–0.79) – women: 0.53 (95%-CI 0.28–0.79)	Subgroup analysis of the pivotal DEFINE trial (placebo: 306/408 (75.0%) female; dimethyl fumarate: 296/410 (72.2%) female)	Li et al. (2017)
Teriflunomide (Aubagio®)	Reversible inhibition of mitochondrial enzyme dihydro-orotate dehydrogenase (DHODH) needed for pyrimidine biosynthesis	RRMS	Addition of testosterone compared to E2 led to stronger downregulation of inflammatory cytokines	*In vitro* study of the active leflunomide metabolite A771726, representing teriflunomide, on inflammatory cytokine production of cultured human monocytes and primary cultures of human synovial macrophages from patients with rheumatoid arthritis	Cutolo et al. (2009)

Continued

Table 1 Summary of sex-specific efficacy data of disease-modifying treatments of multiple sclerosis.—cont'd

Substance	Mode of action	Indication	Efficacy data	Setting	Reference
Cladribine (2-chlorodeoxyadenosine, Mavenclad®)	Deoxyadenosine analogue, activated by phosphorylation by intracellular deoxycytidine kinase, differentially expressed on different immune cells, preferentially depleting lymphocytes	Highly active RRMS		For pivotal trials in relapsing MS, no data on sex differences found.	
			– Rate of conversion to RRMS: – men: HR 0.277 (95%-CI 0.129–0.593) – women: HR 0.357 (95%-CI 0.207–0.616)	Subgroup analysis of phase 3 study ORACLE-MS in CIS (according to McDonald 2005 criteria), cladribine (262/410 (63.9%) female) vs. placebo (138/206 (67.0%) female)	Cree et al. (2021)
Natalizumab (Tysabri®)	Monoclonal antibody against very late antigen (VLA)-4 on an IgG4 backbone, blocking immune cell trafficking from the blood into the central nervous system	Highly active RRMS, active RRMS in anti-JCV antibody negative patients	– Rate ratios for relapse rate (favoring natalizumab over placebo): – men: AFFIRM: 0.38 (95%-CI 0.25–0.59); SENTINEL: 0.36 (95%-CI 0.26–0.51) – women: AFFIRM: 0.32 (95%-CI 0.25–0.42); SENTINEL: 0.48 (95%-CI 0.40–0.57) – HRs of disability progression (favoring natalizumab over placebo): – men: AFFIRM: 0.68 (95%-CI 0.40–1.17); SENTINEL: 0.90 (95%-CI 0.59–1.37) – women: AFFIRM: 0.54 (95%-CI 0.38–0.75); SENTINEL: 0.73 (95%-CI 0.55–0.97) – effect on disability progression only significant in women	Subgroup analysis of phase 3 trials AFFIRM and SENTINEL in RRMS, natalizumab (449/627 (71.6%) female), placebo (211/315 (67.0%) female), interferon beta-1a + placebo (420/582 (72.2%) female), interferon beta-1a + natalizumab (442/589 (75.0%) female)	Hutchinson et al. (2009)

			– men: similar incidence rate ratio for on-treatment relapse, likelihood of first relapse, probability of disability improvement for natalizumab and fingolimod – women: lower incidence rate ratio for on-treatment relapse (0.76 (95%-CI 0.65–0.88)), lower HR of first relapse (0.79 (95%-CI 0.63–0.98)), higher probability of disability improvement (1.40 (95%-CI 1.13–1.73)) for treatment with natalizumab compared to fingolimod	Combined analysis of three MS registries (MSBase, OFSEP registry, Danish registry), 5148 patients (3698/5148 (71.8%) female) treated with natalizumab ($n = 1989$) or fingolimod ($n = 3159$)	Sharmin et al. (2021)
Ocrelizumab (Ocrevus®)	Monoclonal antibody against CD20 leading to depletion of CD20-positive immune cells (mostly B-cells)	Relapsing and primary progressive MS	– Rate ratios of ARR (favoring ocrelizumab): – men: 0.56 (95%-CI 0.40–0.80) – women: 0.53 (95%-CI 0.41–0.68) – Number of T1 gadolinium enhancing lesions (favoring ocrelizumab): – men: 0.04 (95%-CI 0.02–0.07) – women: 0.07 (95%-CI 0.04–0.11) – HR of 12-week confirmed disability progression (favoring ocrelizumab): – men: 0.56 (95%-CI 0.34–0.91) – women: 0.64 (95%-CI 0.44–0.93)	Subgroup analysis of phase 3 studies in relapsing MS (OPERA I, II), ocrelizumab (541/827 (65.4%) female) vs interferon beta-1a (552/829 (66.6%) female)	Turner et al. (2019)

Continued

Table 1 Summary of sex-specific efficacy data of disease-modifying treatments of multiple sclerosis.—cont'd

Substance	Mode of action	Indication	Efficacy data	Setting	Reference
			– No difference in the rate of patients with confirmed disability progression (primary endpoint) for women between ocrelizumab (36%) and placebo (35.5%)	Subgroup analysis of the phase 3 study in PPMS (ORATORIO), ocrelizumab (237/488 (48.6%) female) vs placebo (124/244 (50.9%) female)	Houtchens and Bove, 2018, Roche (2021)
Alemtuzumab (Lemtrada®)	Monoclonal antibody against CD52 with IgG1 backbone causing lysis of CD52 positive immune cells (mostly T- and B-lymphocytes)	Highly active RRMS after failure of at least one immunotherapy	– HR of sustained accumulation of disability: – men: 0.198 – women: 0.399 (95%-CI not given) – Rate ratios of relapse rates: – men: 0.285 (95%-CI not given) – women: 0.274 (95%-CI not given)	Post-hoc analysis of a phase 2 study, 334 treatment-naïve MS patients, randomized 1:1:1 (interferon beta-1a: 70/107 (65.4%) female, alemtuzumab 12 mg/d: 69/107 (64.5%) female, alemtuzumab 24 mg/d: 69/108 (63.9%) female)	Coles et al. (2011)

Labelling refers to Swissmedic if not indicated differently.
Abbreviations: 95%-CI—95% confidence interval, ARR—annualized relapse rate, CD20 / CD52—cluster of differentiation 20 / 52, CIS—clinically isolated syndrome, E2—estradiol, EDSS—expanded disability status scale, HR—hazard ratio, MRI—magnetic resonance imaging, MS—multiple sclerosis, PPMS—primary progressive multiple sclerosis, RRMS—relapsing-remitting multiple sclerosis, S1PR(M)—sphingosin-1-phosphate receptor (modulator), SPMS—secondary progressive multiple sclerosis.

Table 2 Summary of sex-specific efficacy data of disease-modifying treatments of neuromyelitis optica spectrum disorder.

Substance	Mode of action	Indication	Efficacy data	Setting	Reference
Eculizumab (Soliris®)	Monoclonal antibody against C5 of the complement cascade	AQP4-IgG positive NMOSD	Significant reduction of relapses in women and men for eculizumab vs placebo	Subgroup analysis of phase 3 trial	Palace et al. (2021)
Satralizumab (Enspryng®)	Monoclonal antibody against the IL-6 receptor	AQP4-IgG positive NMOSD	No data on sex differences	Preliminary data of OLE	Kleiter et al. (2021)

Abbreviations: AQP4-IgG—aquaporin 4 immunoglobulin G, IL-6—interleukin 6, NMOSD—neuromyelitis optica spectrum disorder, OLE—open label extension, vs—versus.

Table 3 Summary of sex-specific safety data of disease-modifying treatments of multiple sclerosis and neuromyelitis optica spectrum disorder.

Substance	Available safety data	Reference

Interferon beta formulations	– No sex differences in frequencies of overall AE and SAE (frequencies not given)
	– Specific AE:
	  – Headache (female/male: 51%/39%, *P*-value not given)
	  – Urinary tract infections (female/male: 23%/12%, *P*-value not given)
	  – Flu-like symptoms (female/male: 69%/74%, *P*-value not given)
	  – Elevated liver enzymes (female/male: 5%/11%, *P*-value not given)
		Rudick et al. (2011)

	– Elevated liver enzymes (802 MS patients, 590/802 (74%) female): – higher risk in men, independent of dosage (OR 2.1, 95%-CI 1.3–3.3)	Chan, Kingwell, Oger, Yoshida, and Tremlett (2011)
Glatiramer acetate	No data on sex differences found	

Continued

Table 3 Summary of sex-specific safety data of disease-modifying treatments of multiple sclerosis and neuromyelitis optica spectrum disorder.—cont'd

Substance	Available safety data	Reference
Sphingosin-1-phosphate receptor modulators		
– Fingolimod	– Abnormalities of white blood cell counts (300 RRMS patients, 221/300 (73.6%) female): – no sex difference for leukopenia grade 2 and 3 – Lymphopenia grade 4 less frequent in male patients (OR 0.309, 95%-CI 0.11–0.71, *P*-value 0.011)	Baharnoori et al. (2018)
	– Lymphopenia (418 German patients (286/418 (68.4%) female); 438 Swedish patients (295/438 (67.4%) female) – German: no sex difference for lymphopenia grade 4 – Swedish: higher risk of lymphopenia grade 4 for female patients (OR: 2.56, 95%-CI 1.32–4.98, *P*-value 0.005)	Warnke et al. (2014)
	– Infections/liver enzymes (225 RRMS patients, 154/225 (68.4%) female): – increased frequency of infections in the first year of treatment in women; not observed after first year – Elevated liver enzymes: more frequent increase in men	Manni et al. (2017)
– Siponimod	– Interaction with oral contraceptives (23 healthy women): – no impact of siponimod on pharmacokinetics/–dynamics of oral contraceptives	Biswal et al. (2014)
– Ozanimod and Ponesimod	No data on sex differences found	
Dimethyl fumarate	– Interaction with oral contraceptives (32 healthy women): – No interaction of with orgestimate 250 μg/ethinylestradiol 35 μg	Zhu et al. (2017)
Teriflunomide	– Teratogenicity in animal experiments	Sanofi-Aventis (2021)
Cladribine	– Teratogenicity for male and female patients	(Merck, 2021)

Table 3 Summary of sex-specific safety data of disease-modifying treatments of multiple sclerosis and neuromyelitis optica spectrum disorder.—cont'd

Substance	Available safety data	Reference
Natalizumab	– IRR: 11133 natalizumab infusions (302 MS patients, 208/302 (68.9%) female): – no sex effect on IRR probability	Sacco et al. (2020)
	– PML: – no sex-specific data on PML occurrence or outcome – Risk of seroconversion for anti-JCV antibodies (1001 RRMS patients, 742/1001 (74.2%) female): higher adjusted HR in men	Dwyer et al. (2021)
Ocrelizumab	– Breast cancer (only female patients of pivotal trials): more frequently in ocrelizumab vs comparator (low rates in comparator groups (0/668 treated with interferon beta-1a or placebo	Roche (2021)
Alemtuzumab	No data on sex differences found	
Eculizumab	No sex-specific data reported (sex not defined as pre-specified subgroup)	Palace et al. (2021)
Satralizumab	No data on sex differences found	

Abbreviations: 95%-CI—95% confidence interval, AE—adverse event, HR—hazard ratio, IRR—infusion-related reaction, JCV—John Cunningham virus, MS—multiple sclerosis, OR—Odds ratio, PML—progressive multifocal leukoencephalopathy, RRMS—relapsing-remitting multiple sclerosis, SAE—serious adverse event.

5. Conclusions

Sex effects of MS and NMOSD treatments are still not sufficiently addressed in clinical studies although the first trial of a modern MS immunotherapy has already been published nearly 50 years ago (Abramsky, Teitelbaum, & Arnon, 1977). To date, sex differences are mainly addressed in post-hoc analyses of pivotal phase 2 and 3 trials or analyses of real-world cohorts, bearing the inherent limitations of such investigations with a lack of power and different sources of bias. Not a single prospective phase 3 study was powered to identify sex differences in treatment response or side effects. The reporting of sex-specific safety and efficacy aspects is also lacking a standardized approach limiting data interpretation (Tables 1–3). These factors represent a shortcoming of immunotherapies in MS and NMOSD that are widely used in clinical practice in both sexes under the assumption of equal efficacy and safety.

Strikingly, sex-specific safety aspects have not been analyzed in pooled evaluations from large phase 2 or 3 trials and open-label extensions for all substances discussed here, but partially only in small independent studies with the respective limitations.

Our summary of different risks for liver injury in patients treated with interferon beta formulations or fingolimod being higher for men (Manni et al., 2017; Rudick et al., 2011), and potential interactions with contraceptive measures and pregnancy outcomes for women (Biswal et al., 2014; Varyte, Arlauskiene, & Ramasauskaite, 2021; Zhu et al., 2017) underpins that future registry studies using routinely available clinical data in a large scale are urgently needed to address sex specific drug safety. The sparse examples and partially inconclusive data that we were able to find delineate the lack of evidence (Baharnoori et al., 2018; Warnke et al., 2014).

Due to this data gap, an inter-dependence of possible sex-specific side effects remain likewise unclear: It remains speculative whether increased infection rates could be associated with a preponderance of pronounced lymphopenia in women under S1PRM. On the level of a Danish Nation Wide registry of the general population investigating 98,344 individuals (54,163/98344 female, 55.1%), an association of overall lymphopenia with the risk of any infection had been demonstrated (HR: 1.31, 95% CI 1.14–1.52) (Warny, Helby, Nordestgaard, Birgens, & Bojesen, 2018).

With the female preponderance of MS and the disease onset mostly within the reproductive phase, hormonal contraception is widely used in this patient population, but very sparsely investigated as a sex-specific safety aspect. For S1PRM, this is of special relevance as they are associated with a potential poor pregnancy outcome (Varyte et al., 2021).

Likewise, post-marketing surveillance to further determine the risk of breast cancer under anti-CD20 treatment will be important in clinical practice, but also collected in a standardized fashion, and published to increase the amount of evidence.

Sex differences in efficacy should likewise be investigated more in depth respecting differentially presenting disease in men and women (Section 2, Fig. 1; Section 3, Fig. 3). Examples include a lack of a significant effect of ocrelizumab in female PPMS patients on disability progression (Houtchens & Bove, 2018; Roche, 2021). The interpretation of these findings remains not well-determined as it may result from different baseline characteristics that were not investigated (Houtchens & Bove, 2018) or truly different disease characteristics of PPMS in men and women lowering the chance to detect differences due to slower progression in women

(Section 2). This phenomenon has been described more than a decade before in a trial of glatiramer acetate in PPMS with no significant treatment effects in the whole cohort, but reduced progression probability in men treated with the active substance (Wolinsky et al., 2007). Similarly, the very low number of male cases included in NMOSD studies (Kleiter et al., 2021; Palace et al., 2021) will create difficulties in interpreting results for male participants.

However, all analyses presented in our review do not fulfill the level of evidence of phase 3 clinical trials. This may inspire to rethink clinical studies and to include the investigation of sex differences already during the planning phase.

References

Abdullah, M., Chai, P. S., Chong, M. Y., Tohit, E. R., Ramasamy, R., Pei, C. P., et al. (2012). Gender effect on in vitro lymphocyte subset levels of healthy individuals. *Cellular Immunology*, *272*(2), 214–219. https://doi.org/10.1016/j.cellimm.2011.10.009.

Abramsky, O., Teitelbaum, D., & Arnon, R. (1977). Effect of a synthetic polypeptide (COP 1) on patients with multiple sclerosis and with acute disseminated encephalomeylitis. Preliminary report. *Journal of the Neurological Sciences*, *31*(3), 433–438. https://doi.org/10.1016/0022-510x(77)90220-9.

Afshan, G., Afzal, N., & Qureshi, S. (2012). CD4+CD25(hi) regulatory T cells in healthy males and females mediate gender difference in the prevalence of autoimmune diseases. *Clinical Laboratory*, *58*(5–6), 567–571.

Airas, L., Jalkanen, A., Alanen, A., Pirttila, T., & Marttila, R. J. (2010). Breast-feeding, postpartum and prepregnancy disease activity in multiple sclerosis. *Neurology*, *75*(5), 474–476. https://doi.org/10.1212/WNL.0b013e3181eb5860.

Akaishi, T., Takahashi, T., Misu, T., Kaneko, K., Takai, Y., Nishiyama, S., et al. (2021). Difference in the Source of anti-AQP4-IgG and anti-MOG-IgG antibodies in CSF in Patients with neuromyelitis optica spectrum disorder. *Neurology*, *97*(1), e1–e12. https://doi.org/10.1212/WNL.0000000000012175.

Ali, E. S., Mangold, C., & Peiris, A. N. (2017). Estriol: Emerging clinical benefits. *Menopause*, *24*, 1081–1085. https://doi.org/10.1097/GME.0000000000000855.

Alonso, A., & Hernan, M. A. (2008). Temporal trends in the incidence of multiple sclerosis: a systematic review. *Neurology*, *71*(2), 129–135. https://doi.org/10.1212/01.wnl.0000316802.35974.34.

Altintas, A., Dargvainiene, J., Schneider-Gold, C., Asgari, N., Ayzenberg, I., Ciplea, A. I., et al. (2020). Gender issues of antibody-mediated diseases in neurology: (NMOSD/autoimmune encephalitis/MG). *Therapeutic Advances in Neurological Disorders*, *13*. https://doi.org/10.1177/1756286420949808. 1756286420949808.

Arruvito, L., Giulianelli, S., Flores, A. C., Paladino, N., Barboza, M., Lanari, C., et al. (2008). NK cells expressing a progesterone receptor are susceptible to progesterone-induced apoptosis. *Journal of Immunology*, *180*, 5746–5753. https://doi.org/10.4049/jimmunol.180.8.5746.

Baharnoori, M., Gonzalez, C. T., Chua, A., Diaz-Cruz, C., Healy, B. C., Stankiewicz, J., et al. (2018). Predictors of hematological abnormalities in multiple sclerosis patients treated with fingolimod and dimethyl fumarate and impact of treatment switch on lymphocyte and leukocyte count. *Multiple Sclerosis and Related Disorders*, *20*, 51–57. https://doi.org/10.1016/j.msard.2017.12.003.

Baroncini, D., Annovazzi, P. O., De Rossi, N., Mallucci, G., Torri Clerici, V., Tonietti, S., et al. (2019). Impact of natural menopause on multiple sclerosis: A multicentre study. *Journal of Neurology, Neurosurgery, and Psychiatry, 90*(11), 1201–1206. https://doi.org/10.1136/jnnp-2019-320587.

Bianchi, I., Lleo, A., Gershwin, M. E., & Invernizzi, P. (2012). The X chromosome and immune associated genes. *Journal of Autoimmunity, 38*(2–3), J187–J192. https://doi.org/10.1016/j.jaut.2011.11.012.

Biswal, S., Veldandi, U. K., Derne, C., Golla, G., Muhsen, N., & Legangneux, E. (2014). Effect of oral siponimod (BAF312) on the pharmacokinetics and pharmacodynamics of a monophasic oral contraceptive in healthy female subjects. *International Journal of Clinical Pharmacology and Therapeutics, 52*(11), 996–1004. https://doi.org/10.5414/CP202142.

Borisow, N., Kleiter, I., Gahlen, A., Fischer, K., Wernecke, K. D., Pache, F., et al. (2017). Influence of female sex and fertile age on neuromyelitis optica spectrum disorders. *Multiple Sclerosis, 23*(8), 1092–1103. https://doi.org/10.1177/1352458516671203.

Bouman, A., Schipper, M., Heineman, M. J., & Faas, M. M. (2004). Gender difference in the non-specific and specific immune response in humans. *American Journal of Reproductive Immunology, 52*(1), 19–26. https://doi.org/10.1111/j.1600-0897.2004.00177.x.

Bove, R., Chua, A. S., Xia, Z., Chibnik, L., De Jager, P. L., & Chitnis, T. (2016). Complex relation of HLA-DRB1*1501, age at menarche, and age at multiple sclerosis onset. *Neurology Genetics, 2*(4), e88. https://doi.org/10.1212/NXG.0000000000000088.

Bove, R., Elsone, L., Alvarez, E., Borisow, N., Cortez, M. M., Mateen, F. J., et al. (2017). Female hormonal exposures and neuromyelitis optica symptom onset in a multicenter study. *Neurology Neuroimmunology & Neuroinflammation, 4*(3), e339. https://doi.org/10.1212/NXI.0000000000000339.

Bove, R., Healy, B. C., Musallam, A., Glanz, B. I., De Jager, P. L., & Chitnis, T. (2016). Exploration of changes in disability after menopause in a longitudinal multiple sclerosis cohort. *Multiple Sclerosis, 22*(7), 935–943. https://doi.org/10.1177/1352458515606211.

Bove, R., McHenry, A., Hellwig, K., Houtchens, M., Razaz, N., Smyth, P., et al. (2016). Multiple sclerosis in men: Management considerations. *Journal of Neurology, 263*(7), 1263–1273. https://doi.org/10.1007/s00415-015-8005-z.

Bove, R., Musallam, A., Healy, B. C., Raghavan, K., Glanz, B. I., Bakshi, R., et al. (2014). Low testosterone is associated with disability in men with multiple sclerosis. *Multiple Sclerosis, 20*(12), 1584–1592. https://doi.org/10.1177/1352458514527864.

Bovis, F., Kalincik, T., Lublin, F., Cutter, G., Malpas, C., Horakova, D., et al. (2021). Treatment response score to glatiramer acetate or interferon beta-1a. *Neurology, 96*(2), e214–e227. https://doi.org/10.1212/WNL.0000000000010991.

Bruijstens, A. L., Lechner, C., Flet-Berliac, L., Deiva, K., Neuteboom, R. F., Hemingway, C., et al. (2020). E.U. paediatric MOG consortium consensus: Part 1—Classification of clinical phenotypes of paediatric myelin oligodendrocyte glycoprotein antibody-associated disorders. *European Journal of Paediatric Neurology, 29*, 2–13. https://doi.org/10.1016/j.ejpn.2020.10.006.

Celius, E. G., & Smestad, C. (2009). Change in sex ratio, disease course and age at diagnosis in Oslo MS patients through seven decades. *Acta Neurologica Scandinavica. Supplementam, 2009*(189), 27–29. https://doi.org/10.1111/j.1600-0404.2009.01208.x.

Chan, S., Kingwell, E., Oger, J., Yoshida, E., & Tremlett, H. (2011). High-dose frequency beta-interferons increase the risk of liver test abnormalities in multiple sclerosis: A longitudinal study. *Multiple Sclerosis, 17*(3), 361–367. https://doi.org/10.1177/1352458510388823.

Chitnis, T. (2013). Role of puberty in multiple sclerosis risk and course. *Clinical Immunology, 149*(2), 192–200. https://doi.org/10.1016/j.clim.2013.03.014.

Chitnis, T. (2018). The role of testosterone in MS risk and course. *Multiple Sclerosis, 24*(1), 36–41. https://doi.org/10.1177/1352458517737395.

Clanet, M. C., Wolinsky, J. S., Ashton, R. J., Hartung, H. P., & Reingold, S. C. (2014). Risk evaluation and monitoring in multiple sclerosis therapeutics. *Multiple Sclerosis*, *20*(10), 1306–1311. https://doi.org/10.1177/1352458513513207.

Cobo-Calvo, A., Ruiz, A., Rollot, F., Arrambide, G., Deschamps, R., Maillart, E., et al. (2021). Clinical features and risk of relapse in children and adults with myelin oligodendrocyte glycoprotein antibody-associated disease. *Annals of Neurology*, *89*(1), 30–41. https://doi.org/10.1002/ana.25909.

Coles, A. J., Fox, E., Vladic, A., Gazda, S. K., Brinar, V., Selmaj, K. W., et al. (2011). Alemtuzumab versus interferon beta-1a in early relapsing-remitting multiple sclerosis: Post-hoc and subset analyses of clinical efficacy outcomes. *Lancet Neurology*, *10*(4), 338–348. https://doi.org/10.1016/S1474-4422(11)70020-5.

Collongues, N., Alves Do Rego, C., Bourre, B., Biotti, D., Marignier, R., da Silva, A. M., et al. (2021). Pregnancy in patients with AQP4-Ab, MOG-Ab, or double-negative neuromyelitis optica disorder. *Neurology*, *96*(15), e2006–e2015. https://doi.org/10.1212/WNL.0000000000011744.

Confavreux, C., Aimard, G., & Devic, M. (1980). Course and prognosis of multiple sclerosis assessed by the computerized data processing of 349 patients. *Brain*, *103*(2), 281–300. https://doi.org/10.1093/brain/103.2.281.

Confavreux, C., Hutchinson, M., Hours, M. M., Cortinovis-Tourniaire, P., & Moreau, T. (1998). Rate of pregnancy-related relapse in multiple sclerosis. Pregnancy in Multiple Sclerosis Group. *The New England Journal of Medicine*, *339*(5), 285–291. https://doi.org/10.1056/NEJM199807303390501.

Confavreux, C., & Vukusic, S. (2006). Natural history of multiple sclerosis: A unifying concept. *Brain*, *129*(Pt. 3), 606–616. https://doi.org/10.1093/brain/awl007.

Cook, I. F. (2008). Sexual dimorphism of humoral immunity with human vaccines. *Vaccine*, *26*(29–30), 3551–3555. https://doi.org/10.1016/j.vaccine.2008.04.054.

Cree, B. A. C., Bowen, J. D., Hartung, H. P., Vermersch, P., Hughes, B., Damian, D., et al. (2021). Subgroup analysis of clinical and MRI outcomes in participants with a first clinical demyelinating event at risk of multiple sclerosis in the ORACLE-MS study. *Multiple Sclerosis and Related Disorders*, *49*(102), 695. https://doi.org/10.1016/j.msard.2020.102695.

Cunningham, M., & Gilkeson, G. (2011). Estrogen receptors in immunity and autoimmunity. *Clinical Reviews in Allergy and Immunology*, *40*(1), 66–73. https://doi.org/10.1007/s12016-010-8203-5.

Cutolo, M., Montagna, P., Brizzolara, R., Sulli, A., Seriolo, B., Villaggio, B., et al. (2009). Sex hormones modulate the effects of Leflunomide on cytokine production by cultures of differentiated monocyte/macrophages and synovial macrophages from rheumatoid arthritis patients. *Journal of Autoimmunity*, *32*(3–4), 254–260. https://doi.org/10.1016/j.jaut.2009.02.016.

Dalal, M., Kim, S., & Voskuhl, R. R. (1997). Testosterone therapy ameliorates experimental autoimmune encephalomyelitis and induces a T helper 2 bias in the autoantigen-specific T lymphocyte response. *Journal of Immunology (Baltimore, Md. 1950)*, *159*, 3–6.

de Mol, C. L., Wong, Y., van Pelt, E. D., Wokke, B., Siepman, T., Neuteboom, R. F., et al. (2020). The clinical spectrum and incidence of anti-MOG-associated acquired demyelinating syndromes in children and adults. *Multiple Sclerosis*, *26*(7), 806–814. https://doi.org/10.1177/1352458519845112.

Debouverie, M., Pittion-Vouyovitch, S., Louis, S., Roederer, T., & Guillemin, F. (2007). Increasing incidence of multiple sclerosis among women in Lorraine, Eastern France. *Multiple Sclerosis*, *13*(8), 962–967. https://doi.org/10.1177/1352458507077938.

Derfuss, T., Ontaneda, D., Nicholas, J., Meng, X., & Hawker, K. (2016). Relapse rates in patients with multiple sclerosis treated with fingolimod: Subgroup analyses of pooled data from three phase 3 trials. *Multiple Sclerosis and Related Disorders*, *8*, 124–130. https://doi.org/10.1016/j.msard.2016.05.015.

Devonshire, V., Havrdova, E., Radue, E. W., O'Connor, P., Zhang-Auberson, L., Agoropoulou, C., et al. (2012). Relapse and disability outcomes in patients with multiple sclerosis treated with fingolimod: Subgroup analyses of the double-blind, randomised, placebo-controlled FREEDOMS study. *Lancet Neurol*, *11*(5), 420–428. https://doi.org/10.1016/S1474-4422(12)70056-X.

D'Hooghe, M. B., Haentjens, P., Nagels, G., D'Hooghe, T., & De Keyser, J. (2012). Menarche, oral contraceptives, pregnancy and progression of disability in relapsing onset and progressive onset multiple sclerosis. *Journal of Neurology*, *259*(5), 855–861. https://doi.org/10.1007/s00415-011-6267-7.

Dosiou, C., Hamilton, A. E., Pang, Y., Overgaard, M. T., Tulac, S., Dong, J., et al. (2008). Expression of membrane progesterone receptors on human T lymphocytes and Jurkat cells and activation of G-proteins by progesterone. *The Journal of Endocrinology*, *196*, 67–77. https://doi.org/10.1677/JOE-07-0317.

Du, S., Itoh, N., Askarinam, S., Hill, H., Arnold, A. P., & Voskuhl, R. R. (2014). XY sex chromosome complement, compared with XX, in the CNS confers greater neurodegeneration during experimental autoimmune encephalomyelitis. *Proceedings of the National Academy of Sciences of the United States of America*, *111*(7), 2806–2811. https://doi.org/10.1073/pnas.1307091111.

Ducloyer, J. B., Caignard, A., Aidaoui, R., Ollivier, Y., Plubeau, G., Santos-Moskalyk, S., et al. (2020). MOG-Ab prevalence in optic neuritis and clinical predictive factors for diagnosis. *The British Journal of Ophthalmology*, *104*(6), 842–845. https://doi.org/10.1136/bjophthalmol-2019-314845.

Dwyer, C. M., Jokubaitis, V. G., Stankovich, J., Baker, J., Haartsen, J., Butzkueven, H., et al. (2021). High rates of JCV seroconversion in a large international cohort of natalizumab-treated patients. *Therapeutic Advances in Neurological Disorders*, *14*. https://doi.org/10.1177/17562864211998915. 17562864211998915.

FDA. n.d. www.fda.gov/science-research/womens-health-research/understanding-sex-differences-fda.

Feldman, C. H., Hiraki, L. T., Liu, J., Fischer, M. A., Solomon, D. H., Alarcon, G. S., et al. (2013). Epidemiology and sociodemographics of systemic lupus erythematosus and lupus nephritis among US adults with Medicaid coverage, 2000-2004. *Arthritis and Rheumatism*, *65*(3), 753–763. https://doi.org/10.1002/art.37795.

Finkelsztejn, A., Brooks, J. B., Paschoal, F. M., Jr., & Fragoso, Y. D. (2011). What can we really tell women with multiple sclerosis regarding pregnancy? A systematic review and meta-analysis of the literature. *BJOG*, *118*(7), 790–797. https://doi.org/10.1111/j.1471-0528.2011.02931.x.

Freedman, M. S., De Stefano, N., Barkhof, F., Polman, C. H., Comi, G., Uitdehaag, B. M., et al. (2014). Patient subgroup analyses of the treatment effect of subcutaneous interferon beta-1a on development of multiple sclerosis in the randomized controlled REFLEX study. *Journal of Neurology*, *261*(3), 490–499. https://doi.org/10.1007/s00415-013-7222-6.

Freeman, M. E., Kanyicska, B., Lerant, A., & Nagy, G. (2000). Prolactin: Structure, function, and regulation of secretion. *Physiological Reviews*, *80*, 1523–1631. https://doi.org/10.1152/physrev.2000.80.4.1523.

Furman, D., Hejblum, B. P., Simon, N., Jojic, V., Dekker, C. L., Thiebaut, R., et al. (2014). Systems analysis of sex differences reveals an immunosuppressive role for testosterone in the response to influenza vaccination. *Proceedings of the National Academy of Sciences of the United States of America*, *111*, 869–874. https://doi.org/10.1073/pnas.1321060111.

Garay, L., Gonzalez Deniselle, M. C., Lima, A., Roig, P., & De Nicola, A. F. (2007). Effects of progesterone in the spinal cord of a mouse model of multiple sclerosis. *The Journal of Steroid Biochemistry and Molecular Biology*, *107*(3–5), 228–237. https://doi.org/10.1016/j.jsbmb.2007.03.040.

Ghafouri-Fard, S., Azimi, T., & Taheri, M. (2021). A comprehensive review on the role of genetic factors in neuromyelitis optica spectrum disorder. *Frontiers in Immunology*, *12*, 737673. https://doi.org/10.3389/fimmu.2021.737673.

Gianfrancesco, M. A., & Barcellos, L. F. (2016). Obesity and multiple sclerosis susceptibility: A review. *Journal of Neurology and Neuromedicine*, *1*(7), 1–5. https://doi.org/10.29245/2572.942x/2016/7.1064.

Gilmore, W., Weiner, L. P., & Correale, J. (1997). Effect of estradiol on cytokine secretion by proteolipid protein-specific T cell clones isolated from multiple sclerosis patients and normal control subjects. *Journal of iImunology*, *158*, 446–451.

Granberg, T., Martola, J., Kristoffersen-Wiberg, M., Aspelin, P., & Fredrikson, S. (2013). Radiologically isolated syndrome—Incidental magnetic resonance imaging findings suggestive of multiple sclerosis, a systematic review. *Multiple Sclerosis*, *19*(3), 271–280. https://doi.org/10.1177/1352458512451943.

Gregg, C., Shikar, V., Larsen, P., Mak, G., Chojnacki, A., Yong, V. W., et al. (2007). White matter plasticity and enhanced remyelination in the maternal CNS. *Journal of Neuroscience*, *27*, 1812–1823. https://doi.org/10.1523/JNEUROSCI.4441-06.2007.

Harbo, H. F., Gold, R., & Tintore, M. (2013). Sex and gender issues in multiple sclerosis. *Therapeutic Advances in Neurological Disorders*, *6*(4), 237–248. https://doi.org/10.1177/1756285613488434.

Hellwig, K., Rockhoff, M., Herbstritt, S., Borisow, N., Haghikia, A., Elias-Hamp, B., et al. (2015). Exclusive breastfeeding and the effect on postpartum multiple sclerosis relapses. *JAMA Neurology*, *72*(10), 1132–1138. https://doi.org/10.1001/jamaneurol.2015.1806.

Hewagama, A., Patel, D., Yarlagadda, S., Strickland, F. M., & Richardson, B. C. (2009). Stronger inflammatory/cytotoxic T-cell response in women identified by microarray analysis. *Genes and Immunity*, *10*(5), 509–516. https://doi.org/10.1038/gene.2009.12.

Hinson, S. R., Lennon, V. A., & Pittock, S. J. (2016). Autoimmune AQP4 channelopathies and neuromyelitis optica spectrum disorders. *Handbook of Clinical Neurology*, *133*, 377–403. https://doi.org/10.1016/B978-0-444-63432-0.00021-9.

Hor, J. Y., Asgari, N., Nakashima, I., Broadley, S. A., Leite, M. I., Kissani, N., et al. (2020). Epidemiology of neuromyelitis optica spectrum disorder and its prevalence and incidence worldwide. *Frontiers in Neurology*, *11*, 501. https://doi.org/10.3389/fneur.2020.00501.

Houtchens, M. K., & Bove, R. (2018). A case for gender-based approach to multiple sclerosis therapeutics. *Frontiers in Neuroendocrinology*, *50*, 123–134. https://doi.org/10.1016/j.yfrne.2018.07.001.

Huang, H. F., He, R. H., Sun, C. C., Zhang, Y., Meng, Q. X., & Ma, Y. Y. (2006). Function of aquaporins in female and male reproductive systems. *Human Reproduction Update*, *12*(6), 785–795. https://doi.org/10.1093/humupd/dml035.

Hughes, G. C. (2012). Progesterone and autoimmune disease. *Autoimmunity Reviews*, *11*, A502–A514. https://doi.org/10.1016/j.autrev.2011.12.003.

Hutchinson, M., Kappos, L., Calabresi, P. A., Confavreux, C., Giovannoni, G., Galetta, S. L., et al. (2009). The efficacy of natalizumab in patients with relapsing multiple sclerosis: Subgroup analyses of AFFIRM and SENTINEL. *Journal of Neurology*, *256*(3), 405–415. https://doi.org/10.1007/s00415-009-0093-1.

Iorio, R., & Papi, C. (2021). Neuromyelitis optica, aquaporin-4 antibodies, and neuroendocrine disorders. *Handbook of Clinical Neurology*, *181*, 173–186. https://doi.org/10.1016/B978-0-12-820683-6.00013-0.

Jacobson, D. L., Gange, S. J., Rose, N. R., & Graham, N. M. (1997). Epidemiology and estimated population burden of selected autoimmune diseases in the United States. *Clinical Immunology and Immunopathology*, *84*, 223–243. https://doi.org/10.1006/clin.1997.4412.

Jansson, L., Olsson, T., & Holmdahl, R. (1994). Estrogen induces a potent suppression of experimental autoimmune encephalomyelitis and collagen-induced arthritis in mice. *Journal of Neuroimmunology*, *53*, 203–207. https://doi.org/10.1016/0165-5728(94)90030-2.

Jarius, S., Paul, F., Aktas, O., Asgari, N., Dale, R. C., de Seze, J., et al. (2018). MOG encephalomyelitis: International recommendations on diagnosis and antibody testing. *Journal of Neuroinflammation*, *15*(1), 134. https://doi.org/10.1186/s12974-018-1144-2.

Jarius, S., Ruprecht, K., Kleiter, I., Borisow, N., Asgari, N., Pitarokoili, K., et al. (2016). MOG-IgG in NMO and related disorders: A multicenter study of 50 patients. Part 2: Epidemiology, clinical presentation, radiological and laboratory features, treatment responses, and long-term outcome. *Journal of Neuroinflammation*, *13*(1), 280. https://doi.org/10.1186/s12974-016-0718-0.

Jokubaitis, V. G., Spelman, T., Kalincik, T., Lorscheider, J., Havrdova, E., Horakova, D., et al. (2016). Predictors of long-term disability accrual in relapse-onset multiple sclerosis. *Annals of Neurology*, *80*(1), 89–100. https://doi.org/10.1002/ana.24682.

Jokubaitis, V. G., Spelman, T., Lechner-Scott, J., Barnett, M., Shaw, C., Vucic, S., et al. (2013). The Australian Multiple Sclerosis (MS) immunotherapy study: A prospective, multicentre study of drug utilisation using the MSBase platform. *PLoS One*, *8*(3), e59694. https://doi.org/10.1371/journal.pone.0059694.

Kalincik, T., Vivek, V., Jokubaitis, V., Lechner-Scott, J., Trojano, M., Izquierdo, G., et al. (2013). Sex as a determinant of relapse incidence and progressive course of multiple sclerosis. *Brain*, *136*(Pt. 12), 3609–3617. https://doi.org/10.1093/brain/awt281.

Kim, S., Liva, S. M., Dalal, M. A., Verity, M. A., & Voskuhl, R. R. (1999). Estriol ameliorates autoimmune demyelinating disease: Implications for multiple sclerosis. *Neurology*, *52*(6), 1230–1238. https://doi.org/10.1212/wnl.52.6.1230.

Kim, R. Y., Mangu, D., Hoffman, A. S., Kovash, R., Jung, E., Itoh, N., et al. (2018). Oestrogen receptor beta ligand acts on CD11c+ cells to mediate protection in experimental autoimmune encephalomyelitis. *Brain*, *141*, 132–147. https://doi.org/10.1093/brain/awx315.

Kim, S. M., Waters, P., Woodhall, M., Kim, Y. J., Kim, J. A., Cheon, S. Y., et al. (2017). Gender effect on neuromyelitis optica spectrum disorder with aquaporin4-immunoglobulin G. *Multiple Sclerosis*, *23*(8), 1104–1111. https://doi.org/10.1177/1352458516674366.

Kitley, J., Waters, P., Woodhall, M., Leite, M. I., Murchison, A., George, J., et al. (2014). Neuromyelitis optica spectrum disorders with aquaporin-4 and myelin-oligodendrocyte glycoprotein antibodies: A comparative study. *JAMA Neurology*, *71*(3), 276–283. https://doi.org/10.1001/jamaneurol.2013.5857.

Klawiter, E. C., Bove, R., Elsone, L., Alvarez, E., Borisow, N., Cortez, M., et al. (2017). High risk of postpartum relapses in neuromyelitis optica spectrum disorder. *Neurology*, *89*(22), 2238–2244. https://doi.org/10.1212/WNL.0000000000004681.

Klein, S. L., & Flanagan, K. L. (2016). Sex differences in immune responses. *Nature Reviews. Immunology*, *16*(10), 626–638. https://doi.org/10.1038/nri.2016.90.

Klein, S. L., Jedlicka, A., & Pekosz, A. (2010). The Xs and Y of immune responses to viral vaccines. *The Lancet Infectious Diseases*, *10*(5), 338–349. https://doi.org/10.1016/s1473-3099(10)70049-9.

Klein, S. L., Pekosz, A., Passaretti, C., Anker, M., & Olukoya, P. (2010). *Sex, gender and influenza* (pp. 1–58). Geneva: World Health Organization.

Kleiter, I., Gahlen, A., Borisow, N., Fischer, K., Wernecke, K. D., Wegner, B., et al. (2016). Neuromyelitis optica: Evaluation of 871 attacks and 1,153 treatment courses. *Annals of Neurology*, *79*(2), 206–216. https://doi.org/10.1002/ana.24554.

Kleiter, I., Traboulsee, A., Palace, J., Yamamura, T., Fujihara, K., Saiz, A., et al. (2021). Long-term efficacy of satralizumab in aquaporin-4-IgGseropositive neuromyelitis optica spectrum disorder (NMOSD): Results from the open-label extension periods of SAkuraSky and SAkuraStar (P024), ECTRIMS 2021. *Multiple Sclerosis Journal, 27*(2S), 134–740. https://doi.org/10.1177/13524585211044667.

Koch-Henriksen, N., & Sorensen, P. S. (2010). The changing demographic pattern of multiple sclerosis epidemiology. *Lancet Neurology, 9*(5), 520–532. https://doi.org/10.1016/S1474-4422(10)70064-8.

Labombarda, F., González, S., Lima, A., Roig, P., Guennoun, R., Schumacher, M., et al. (2011). Progesterone attenuates astro- and microgliosis and enhances oligodendrocyte differentiation following spinal cord injury. *Experimental Neurology, 231*, 135–146. https://doi.org/10.1016/j.expneurol.2011.06.001.

Ladeira, F., Salavisa, M., Caetano, A., Barbosa, R., Sa, F., & Correia, A. S. (2018). The influence of menopause in multiple sclerosis course: A longitudinal cohort study. *European Neurology, 80*(3–4), 223–227. https://doi.org/10.1159/000496374.

Langer-Gould, A. (2016). Sex hormones and multiple sclerosis: Another informative failure. *Lancet Neurology, 15*(1), 22–23. https://doi.org/10.1016/S1474-4422(15)00348-8.

Langer-Gould, A., Smith, J. B., Albers, K. B., Xiang, A. H., Wu, J., Kerezsi, E. H., et al. (2020). Pregnancy-related relapses and breastfeeding in a contemporary multiple sclerosis cohort. *Neurology, 94*(18), e1939–e1949. https://doi.org/10.1212/WNL.0000000000009374.

Lechner, C., Breu, M., Wendel, E. M., Kornek, B., Schanda, K., Baumann, M., et al. (2020). Epidemiology of pediatric NMOSD in Germany and Austria. *Frontiers in Neurology, 11*, 415. https://doi.org/10.3389/fneur.2020.00415.

Lehmann, S. M., Krüger, C., Park, B., Derkow, K., Rosenberger, K., Baumgart, J., et al. (2012). An unconventional role for miRNA: let-7 activates Toll-like receptor 7 and causes neurodegeneration. *Nature Neuroscience, 15*, 827–835. https://doi.org/10.1038/nn.3113.

Lennon, V. A., Kryzer, T. J., Pittock, S. J., Verkman, A. S., & Hinson, S. R. (2005). IgG marker of optic-spinal multiple sclerosis binds to the aquaporin-4 water channel. *The Journal of Experimental Medicine, 202*(4), 473–477. https://doi.org/10.1084/jem.20050304.

Lennon, V. A., Wingerchuk, D. M., Kryzer, T. J., Pittock, S. J., Lucchinetti, C. F., Fujihara, K., et al. (2004). A serum autoantibody marker of neuromyelitis optica: Distinction from multiple sclerosis. *Lancet, 364*(9451), 2106–2112. https://doi.org/10.1016/S0140-6736(04)17551-X.

Li, R., Li, C., Mao, Z., Huang, Q., Shu, Y., Chang, Y., et al. (2021). Male patients with neuromyelitis optica spectrum disorders: Different clinical characteristics and worse steroid treatment response. *Neurological Sciences, 42*(8), 3267–3274. https://doi.org/10.1007/s10072-020-04933-1.

Li, R., Sun, X., Shu, Y., Mao, Z., Xiao, L., Qiu, W., et al. (2017). Sex differences in outcomes of disease-modifying treatments for multiple sclerosis: A systematic review. *Multiple Sclerosis and Related Disorders, 12*, 23–28. https://doi.org/10.1016/j.msard.2017.01.001.

Libert, C., Dejager, L., & Pinheiro, I. (2010). The X chromosome in immune functions: When a chromosome makes the difference. *Nature Reviews. Immunology, 10*, 594–604. https://doi.org/10.1038/nri2815.

Liu, J., Mori, M., Zimmermann, H., Brandt, A., Havla, J., Tanaka, S., et al. (2021). Anti-MOG antibody-associated disorders: Differences in clinical profiles and prognosis in Japan and Germany. *Journal of Neurology, Neurosurgery, and Psychiatry, 92*, 377–383. https://doi.org/10.1136/jnnp-2020-324422.

Lopez, J. A., Denkova, M., Ramanathan, S., Dale, R. C., & Brilot, F. (2021). Pathogenesis of autoimmune demyelination: From multiple sclerosis to neuromyelitis optica spectrum disorders and myelin oligodendrocyte glycoprotein antibody-associated disease. *Clinical & Translational Immunology, 10*(7), e1316. https://doi.org/10.1002/cti2.1316.

Lulu, S., Graves, J., & Waubant, E. (2016). Menarche increases relapse risk in pediatric multiple sclerosis. *Multiple Sclerosis*, *22*(2), 193–200. https://doi.org/10.1177/1352458515581873.

Maghzi, A. H., Ghazavi, H., Ahsan, M., Etemadifar, M., Mousavi, S., Khorvash, F., et al. (2010). Increasing female preponderance of multiple sclerosis in Isfahan, Iran: A population-based study. *Multiple Sclerosis*, *16*(3), 359–361. https://doi.org/10.1177/1352458509358092.

Magyari, M., Koch-Henriksen, N., Laursen, B., & Sorensen, P. S. (2014). Gender effects on treatment response to interferon-beta in multiple sclerosis. *Acta Neurologica Scandinavica*, *130*(6), 374–379. https://doi.org/10.1111/ane.12277.

Manni, A., Direnzo, V., Iaffaldano, A., Di Lecce, V., Tortorella, C., Zoccolella, S., et al. (2017). Gender differences in safety issues during Fingolimod therapy: Evidence from a real-life relapsing multiple sclerosis cohort. *Brain and Behavior: A Cognitive Neuroscience Perspective*, 7(10), e00804. https://doi.org/10.1002/brb3.804.

Mao-Draayer, Y., Thiel, S., Mills, E. A., Chitnis, T., Fabian, M., Katz Sand, I., et al. (2020). Neuromyelitis optica spectrum disorders and pregnancy: Therapeutic considerations. *Nature Reviews. Neurology*, *16*(3), 154–170. https://doi.org/10.1038/s41582-020-0313-y.

Massa, M. G., David, C., Jorg, S., Berg, J., Gisevius, B., Hirschberg, S., et al. (2017). Testosterone differentially affects T cells and neurons in murine and human models of neuroinflammation and neurodegeneration. *The American Journal of Pathology*, *187*(7), 1613–1622. https://doi.org/10.1016/j.ajpath.2017.03.006.

McCombe, P. A., & Greer, J. M. (2013). Female reproductive issues in multiple sclerosis. *Multiple Sclerosis*, *19*(4), 392–402. https://doi.org/10.1177/1352458512452331.

Merck. (2021). *Summary of product characteristics—Mavenclad*. https://www.ema.europa.eu/en/documents/product-information/mavenclad-epar-product-information_en.pdf.

Miclea, A., Salmen, A., Zoehner, G., Diem, L., Kamm, C. P., Chaloulos-Iakovidis, P., et al. (2019). Age-dependent variation of female preponderance across different phenotypes of multiple sclerosis: A retrospective cross-sectional study. *CNS Neuroscience & Therapeutics*, *25*(4), 527–531. https://doi.org/10.1111/cns.13083.

Milo, R., & Kahana, E. (2010). Multiple sclerosis: Geoepidemiology, genetics and the environment. *Autoimmunity Reviews*, *9*(5), A387–A394. https://doi.org/10.1016/j.autrev.2009.11.010.

Moldovan, I. R., Cotleur, A. C., Zamor, N., Butler, R. S., & Pelfrey, C. M. (2008). Multiple sclerosis patients show sexual dimorphism in cytokine responses to myelin antigens. *Journal of Neuroimmunology*, *193*(1–2), 161–169. https://doi.org/10.1016/j.jneuroim.2007.10.010.

Moshirzadeh, S., Ghareghozli, K., Harandi, A. A., & Pakdaman, H. (2012). Serum prolactin level in patients with relapsing-remitting multiple sclerosis during relapse. *Journal of Clinical Neurology*, *19*, 622–623. https://doi.org/10.1016/j.jocn.2011.07.032.

Nour, M. M., Nakashima, I., Coutinho, E., Woodhall, M., Sousa, F., Revis, J., et al. (2016). Pregnancy outcomes in aquaporin-4-positive neuromyelitis optica spectrum disorder. *Neurology*, *86*(1), 79–87. https://doi.org/10.1212/WNL.0000000000002208.

Nussinovitch, U., & Shoenfeld, Y. (2012). The role of gender and organ specific autoimmunity. *Autoimmunity Reviews*, *11*(6–7), A377–A385. https://doi.org/10.1016/j.autrev.2011.11.001.

Ogden, C. L., Carroll, M. D., Kit, B. K., & Flegal, K. M. (2014). Prevalence of childhood and adult obesity in the United States, 2011-2012. *JAMA*, *311*(8), 806–814. https://doi.org/10.1001/jama.2014.732.

Okasha, M., McCarron, P., McEwen, J., & Smith, G. D. (2001). Age at menarche: Secular trends and association with adult anthropometric measures. *Annals of Human Biology*, *28*(1), 68–78. https://doi.org/10.1080/03014460150201896.

Orton, S. M., Herrera, B. M., Yee, I. M., Valdar, W., Ramagopalan, S. V., Sadovnick, A. D., et al. (2006). Sex ratio of multiple sclerosis in Canada: A longitudinal study. *Lancet Neurology*, *5*(11), 932–936. https://doi.org/10.1016/S1474-4422(06)70581-6.

Osoegawa, M., Kira, J., Fukazawa, T., Fujihara, K., Kikuchi, S., Matsui, M., et al. (2009). Temporal changes and geographical differences in multiple sclerosis phenotypes in Japanese: Nationwide survey results over 30 years. *Multiple Sclerosis*, *15*(2), 159–173. https://doi.org/10.1177/1352458508098372.

Otero-Romero, S. (2017). *Impact of oral contraceptives and menopause on MS risk and prognosis: Results from the Barcelona CIS cohort.* ECTRIMS.

Palace, J., Wingerchuk, D. M., Fujihara, K., Berthele, A., Oreja-Guevara, C., Kim, H. J., et al. (2021). Benefits of eculizumab in AQP4+ neuromyelitis optica spectrum disorder: Subgroup analyses of the randomized controlled phase 3 PREVENT trial. *Multiple Sclerosis and Related Disorders*, *47*, 102641. https://doi.org/10.1016/j.msard.2020.102641.

Papp, V., Magyari, M., Aktas, O., Berger, T., Broadley, S. A., Cabre, P., et al. (2021). Worldwide incidence and prevalence of neuromyelitis optica: A systematic review. *Neurology*, *96*(2), 59–77. https://doi.org/10.1212/WNL.0000000000011153.

Pelfrey, C. M., Cotleur, A. C., Lee, J. C., & Rudick, R. A. (2002). Sex differences in cytokine responses to myelin peptides in multiple sclerosis. *Journal of Neuroimmunology*, *130*(1–2), 211–223. https://doi.org/10.1016/s0165-5728(02)00224-2.

Pompili, M., Forte, A., Palermo, M., Stefani, H., Lamis, D. A., Serafini, G., et al. (2012). Suicide risk in multiple sclerosis: A systematic review of current literature. *Journal of Psychosomatic Research*, *73*(6), 411–417. https://doi.org/10.1016/j.jpsychores.2012.09.011.

Portaccio, E., Ghezzi, A., Hakiki, B., Martinelli, V., Moiola, L., Patti, F., et al. (2011). Breastfeeding is not related to postpartum relapses in multiple sclerosis. *Neurology*, *77*(2), 145–150. https://doi.org/10.1212/WNL.0b013e318224afc9.

Pugliatti, M., Cossu, P., Sotgiu, S., Rosati, G., & Riise, T. (2009). Clustering of multiple sclerosis, age of onset and gender in Sardinia. *Journal of the Neurological Sciences*, *286*(1–2), 6–13. https://doi.org/10.1016/j.jns.2009.07.013.

Ramagopalan, S. V., Goldacre, R., Skingsley, A., Conlon, C., & Goldacre, M. J. (2013). Associations between selected immune-mediated diseases and tuberculosis: Record-linkage studies. *BMC Medicine*, *11*, 97. https://doi.org/10.1186/1741-7015-11-97.

Ramien, C., Taenzer, A., Lupu, A., Heckmann, N., Engler, J. B., Patas, K., et al. (2016). Sex effects on inflammatory and neurodegenerative processes in multiple sclerosis. *Neuroscience & Biobehavioral Reviews*, *67*, 137–146. https://doi.org/10.1016/j.neubiorev.2015.12.015.

Reich, D. S., Lucchinetti, C. F., & Calabresi, P. A. (2018). Multiple sclerosis. *The New England Journal of Medicine*, *378*(2), 169–180. https://doi.org/10.1056/NEJMra1401483.

Ribbons, K. A., McElduff, P., Boz, C., Trojano, M., Izquierdo, G., Duquette, P., et al. (2015). Male sex is independently associated with faster disability accumulation in relapse-onset MS but not in primary progressive MS. *PLoS One*, *10*(6), e0122686. https://doi.org/10.1371/journal.pone.0122686.

Riskind, P. N., Massacesi, L., Doolittle, T. H., & Hauser, S. L. (1991). The role of prolactin in autoimmune demyelination: Suppression of experimental allergic encephalomyelitis by bromocriptine. *Annals of Neurology*, *29*, 542–547. https://doi.org/10.1002/ana.410290514.

Roberts, C. W., Walker, W., & Alexander, J. (2001). Sex-associated hormones and immunity to protozoan parasites. *Clinical Microbiology Reviews*, *14*, 476–488. https://doi.org/10.1128/CMR.14.3.476-488.2001.

Roche. (2021). *Labeling ocrevus (FDA)*. https://www.accessdata.fda.gov/drugsatfda_docs/nda/2017/761053Orig1s000Lbl.pdf.
Rolak, L. A. (2003). MS: The basic facts. *Clinical Medicine & Research*, *1*(1), 61–62. https://doi.org/10.3121/cmr.1.1.61.
Rommer, P. S., Ellenberger, D., Hellwig, K., Haas, J., Pohlau, D., Stahmann, A., et al. (2020). Relapsing and progressive MS: The sex-specific perspective. *Therapeutic Advances in Neurological Disorders*, *13*, 1756286420956495. https://doi.org/10.1177/1756286420956495.
Rudick, R. A., Kappos, L., Kinkel, R., Clanet, M., Phillips, J. T., Herndon, R. M., et al. (2011). Gender effects on intramuscular interferon beta-1a in relapsing-remitting multiple sclerosis: Analysis of 1406 patients. *Multiple Sclerosis*, *17*(3), 353–360. https://doi.org/10.1177/1352458510384605.
Runmarker, B., & Andersen, O. (1993). Prognostic factors in a multiple sclerosis incidence cohort with twenty-five years of follow-up. *Brain*, *116*(Pt. 1), 117–134. https://doi.org/10.1093/brain/116.1.117.
Russi, A. E., Ebel, M. E., Yang, Y., & Brown, M. A. (2018). Male-specific IL-33 expression regulates sex-dimorphic EAE susceptibility. *Proceedings of the National Academy of Sciences of the United States of America*, *115*(7), E1520–e1529. https://doi.org/10.1073/pnas.1710401115.
Saadoun, S., Waters, P., Leite, M. I., Bennett, J. L., Vincent, A., & Papadopoulos, M. C. (2013). Neuromyelitis optica IgG causes placental inflammation and fetal death. *Journal of Immunology*, *191*(6), 2999–3005. https://doi.org/10.4049/jimmunol.1301483.
Sacco, R., Disanto, G., Maraffi, I., Candrian, U., Kamm, C. P., Rossi, S., et al. (2020). Infusion-related reactions during Natalizumab treatment: Do we still need a post-infusion observation period? *Multiple Sclerosis and Related Disorders*, *38*, 101523. https://doi.org/10.1016/j.msard.2019.101523.
Saijo, K., Collier, J. G., Li, A. C., Katzenellenbogen, J. A., & Glass, C. K. (2011). An ADIOL-ERβ-CtBP transrepression pathway negatively regulates microglia-mediated inflammation. *Cell*, *145*, 584–595. https://doi.org/10.1016/j.cell.2011.03.050.
Sanofi-Aventis. (2021). *Summary of product characteristics—Aubagio*. https://www.ema.europa.eu/en/documents/product-information/aubagio-epar-product-information_en.pdf.
Savettieri, G., Messina, D., Andreoli, V., Bonavita, S., Caltagirone, C., Cittadella, R., et al. (2004). Gender-related effect of clinical and genetic variables on the cognitive impairment in multiple sclerosis. *Journal of Neurology*, *251*(10), 1208–1214. https://doi.org/10.1007/s00415-004-0508-y.
Sbarbati, A., Osculati, F., Silvagni, D., Benati, D., Galie, M., Camoglio, F. S., et al. (2006). Obesity and inflammation: Evidence for an elementary lesion. *Pediatrics*, *117*(1), 220–223. https://doi.org/10.1542/peds.2004-2854.
Sharmin, S., Lefort, M., Andersen, J. B., Leray, E., Horakova, D., Havrdova, E. K., et al. (2021). Natalizumab versus fingolimod in patients with relapsing-remitting multiple sclerosis: A subgroup analysis from three international cohorts. *CNS Drugs*, *35*(11), 1217–1232. https://doi.org/10.1007/s40263-021-00860-7.
Sloka, J. S., Pryse-Phillips, W. E., & Stefanelli, M. (2006). The relation between menarche and the age of first symptoms in a multiple sclerosis cohort. *Multiple Sclerosis*, *12*(3), 333–339. https://doi.org/10.1191/135248506ms1267oa.
Souyris, M., Cenac, C., Azar, P., Daviaud, D., Canivet, A., Grunenwald, S., et al. (2018). TLR7 escapes X chromosome inactivation in immune cells. *Science Immunology*, *3*-(19). https://doi.org/10.1126/sciimmunol.aap8855.
Straub, R. H. (2007). The complex role of estrogens in inflammation. *Endocrine Reviews*, *28*, 521–574. https://doi.org/10.1210/er.2007-0001.
Sun, H., Sun, X., Li, J., Huo, Y., Wu, L., Huang, D., et al. (2017). Gender differences among Chinese patients with neuromyelitis optica spectrum disorders. *Multiple Sclerosis and Related Disorders*, *17*, 5–8. https://doi.org/10.1016/j.msard.2017.06.008.

Sun, X. L., Zhang, J., Fan, Y., Ding, J. H., Sha, J. H., & Hu, G. (2009). Aquaporin-4 deficiency induces subfertility in female mice. *Fertility and Sterility*, *92*(5), 1736–1743. https://doi.org/10.1016/j.fertnstert.2008.07.1785.

Tafuri, A., Alferink, J., Moller, P., Hammerling, G. J., & Arnold, B. (1995). T cell awareness of paternal alloantigens during pregnancy. *Science*, *270*(5236), 630–633. https://doi.org/10.1126/science.270.5236.630.

Talley, C. L. (2005). The emergence of multiple sclerosis, 1870-1950: A puzzle of historical epidemiology. *Perspectives in Biology and Medicine*, *48*(3), 383–395. https://doi.org/10.1353/pbm.2005.0079.

Thompson, A. (2004). Overview of primary progressive multiple sclerosis (PPMS): Similarities and differences from other forms of MS, diagnostic criteria, pros and cons of progressive diagnosis. *Multiple Sclerosis*, *10*(Suppl. 1), S2–S7. https://doi.org/10.1191/1352458504ms1024oa.

Thone, J., Lichtenberg, S., Stahl, A., Pache, F., Kleiter, I., Ruprecht, K., et al. (2018). Ovarian reserve in women with neuromyelitis optica spectrum disorder. *Frontiers in Neurology*, *9*, 446. https://doi.org/10.3389/fneur.2018.00446.

Trojano, M., Pellegrini, F., Paolicelli, D., Fuiani, A., Zimatore, G. B., Tortorella, C., et al. (2009). Post-marketing of disease modifying drugs in multiple sclerosis: An exploratory analysis of gender effect in interferon beta treatment. *Journal of the Neurological Sciences*, *286*(1–2), 109–113. https://doi.org/10.1016/j.jns.2009.06.036.

Turner, B., Cree, B. A. C., Kappos, L., Montalban, X., Papeix, C., Wolinsky, J. S., et al. (2019). Ocrelizumab efficacy in subgroups of patients with relapsing multiple sclerosis. *Journal of Neurology*, *266*(5), 1182–1193. https://doi.org/10.1007/s00415-019-09248-6.

Varyte, G., Arlauskiene, A., & Ramasauskaite, D. (2021). Pregnancy and multiple sclerosis: An update. *Current Opinion in Obstetrics & Gynecology*, *33*(5), 378–383. https://doi.org/10.1097/GCO.0000000000000731.

Venero, J. L., Vizuete, M. L., Machado, A., & Cano, J. (2001). Aquaporins in the central nervous system. *Progress in Neurobiology*, *63*(3), 321–336. https://doi.org/10.1016/s0301-0082(00)00035-6.

vom Steeg, L. G., & Klein, S. L. (2016). SeXX matters in infectious disease pathogenesis. *PLoS Pathogens*, *12*(2), e1005374. https://doi.org/10.1371/journal.ppat.1005374.

Voskuhl, R. R., & Gold, S. M. (2012). Sex-related factors in multiple sclerosis susceptibility and progression. *Nature Reviews. Neurology*, *8*, 255–263. https://doi.org/10.1038/nrneurol.2012.43.

Voskuhl, R. R., Sawalha, A. H., & Itoh, Y. (2018). Sex chromosome contributions to sex differences in multiple sclerosis susceptibility and progression. *Multiple Sclerosis*, *24*(1), 22–31. https://doi.org/10.1177/1352458517737394.

Voskuhl, R. R., Wang, H., Wu, T. C., Sicotte, N. L., Nakamura, K., Kurth, F., et al. (2016). Estriol combined with glatiramer acetate for women with relapsing-remitting multiple sclerosis: A randomised, placebo-controlled, phase 2 trial. *Lancet Neurology*, *15*(1), 35–46. https://doi.org/10.1016/S1474-4422(15)00322-1.

Vukusic, S., Hutchinson, M., Hours, M., Moreau, T., Cortinovis-Tourniaire, P., Adeleine, P., et al. (2004). Pregnancy and multiple sclerosis (the PRIMS study): Clinical predictors of post-partum relapse. *Brain*, *127*(Pt. 6), 1353–1360. https://doi.org/10.1093/brain/awh152.

Vukusic, S., Ionescu, I., Cornu, C., Bossard, N., Durand-Dubief, F., Cotton, F., et al. (2021). Oral nomegestrol acetate and transdermal 17-beta-estradiol for preventing post-partum relapses in multiple sclerosis: The POPARTMUS study. *Multiple Sclerosis Journal*, *27*, 1458–1463. https://doi.org/10.1177/1352458520978218.

Warnke, C., Dehmel, T., Ramanujam, R., Holmen, C., Nordin, N., Wolfram, K., et al. (2014). Initial lymphocyte count and low BMI may affect fingolimod-induced lymphopenia. *Neurology*, *83*(23), 2153–2157. https://doi.org/10.1212/WNL.0000000000001049.

Warny, M., Helby, J., Nordestgaard, B. G., Birgens, H., & Bojesen, S. E. (2018). Lymphopenia and risk of infection and infection-related death in 98,344 individuals from a prospective Danish population-based study. *PLoS Medicine*, *15*(11), e1002685. https://doi.org/10.1371/journal.pmed.1002685.

Wei, W., Liu, L., Cheng, Z.-L., & Hu, B. (2017). Increased plasma/serum levels of prolactin in multiple sclerosis: A meta-analysis. *Postgraduate Medicine*, *129*, 605–610. https://doi.org/10.1080/00325481.2017.1282297.

Weinshenker, B. G., Rice, G. P., Noseworthy, J. H., Carriere, W., Baskerville, J., & Ebers, G. C. (1991). The natural history of multiple sclerosis: A geographically based study. 3. Multivariate analysis of predictive factors and models of outcome. *Brain*, *114*(Pt. 2), 1045–1056. https://doi.org/10.1093/brain/114.2.1045.

Wingerchuk, D. M. (2009). Neuromyelitis optica: Effect of gender. *Journal of the Neurological Sciences*, *286*(1–2), 18–23. https://doi.org/10.1016/j.jns.2009.08.045.

Wingerchuk, D. M., Banwell, B., Bennett, J. L., Cabre, P., Carroll, W., Chitnis, T., et al. (2015). International consensus diagnostic criteria for neuromyelitis optica spectrum disorders. *Neurology*, *85*(2), 177–189. https://doi.org/10.1212/WNL.0000000000001729.

Wolinsky, J. S., Narayana, P. A., O'Connor, P., Coyle, P. K., Ford, C., Johnson, K., et al. (2007). Glatiramer acetate in primary progressive multiple sclerosis: Results of a multinational, multicenter, double-blind, placebo-controlled trial. *Annals of Neurology*, *61*(1), 14–24. https://doi.org/10.1002/ana.21079.

Ysrraelit, M. C., & Correale, J. (2019). Impact of sex hormones on immune function and multiple sclerosis development. *Immunology*, *156*(1), 9–22. https://doi.org/10.1111/imm.13004.

Zhang, M., Shen, J., Zhou, S., Du, X., Li, W., Yu, L., et al. (2020). Clinical and neuroimaging characteristics of pediatric acute disseminating encephalomyelitis with and without antibodies to myelin oligodendrocyte glycoprotein. *Frontiers in Neurology*, *11*, 593287. https://doi.org/10.3389/fneur.2020.593287.

Zhu, M.-L., Bakhru, P., Conley, B., Nelson, J. S., Free, M., Martin, A., et al. (2016). Sex bias in CNS autoimmune disease mediated by androgen control of autoimmune regulator. *Nature Communications*, 7, 11350. https://doi.org/10.1038/ncomms11350.

Zhu, B., Nestorov, I., Zhao, G., Meka, V., Leahy, M., Kam, J., et al. (2017). Evaluation of potential drug-drug interaction between delayed-release dimethyl fumarate and a commonly used oral contraceptive (norgestimate/ethinyl estradiol) in healthy women. *Clinical Pharmacology in Drug Development*, *6*(6), 604–613. https://doi.org/10.1002/cpdd.377.

CHAPTER SIX

Sex and gender differences in dementia

Patrick Salwierz[a], Carly Davenport[a], Vishaal Sumra[a], M. Florencia Iulita[b,c,d], Maria Teresa Ferretti[d], and Maria Carmela Tartaglia[a,e,*]

[a]Tanz Centre for Research in Neurodegenerative Diseases, University of Toronto, Toronto, ON, Canada
[b]Sant Pau Memory Unit, Department of Neurology, Hospital de la Santa Creu i Sant Pau, Biomedical Research Institute Sant Pau, Universitat Autònoma de Barcelona, Barcelona, Spain
[c]Center of Biomedical Investigation Network for Neurodegenerative Diseases (CIBERNED), Madrid, Spain
[d]Women's Brain Project, Guntershausen, Switzerland
[e]Memory Clinic, Krembil Brain Institute, University Health Network, Toronto, ON, Canada
*Corresponding author: e-mail address: carmela.tartaglia@uhn.ca

Contents

International Review of Neurobiology, Volume 164
ISSN 0074-7742
https://doi.org/10.1016/bs.irn.2022.07.002

Abstract

The dementia landscape has undergone a striking paradigm shift. The advances in understanding of neurodegeneration and proteinopathies has changed our approach to patients with cognitive impairment. Firstly, it has recently been shown that the various proteinopathies that are the cause of the dementia begin to build up long before the appearance of any obvious symptoms. This has cemented the idea that there is an urgency in diagnosis as it occurs very late in the pathophysiology of these diseases. Secondly, that accurate diagnosis is required to deliver targeted therapies, that is precision medicine. With this latter point, the realization that various factors of a person need to be considered as they may impact the presentation and progression of disease has risen to the forefront. Two of these factors aside from race and age are biological sex and gender (social construct), as both can have tremendous impact on manifestation of disease. This chapter will cover what is known and remains to be known on the interaction of sex and gender with some of the major causes of dementia.

1. Sex and gender differences in Alzheimer's disease

1.1 Overview

Alzheimer's disease (AD) is the most common cause of dementia accounting for 50–75% of cases (Nichols, Szoeke, Vollset, et al., 2019), and appears to disproportionately affect females (Mielke, Vemuri, & Rocca, 2014; Podcasy & Epperson, 2016). According to recent estimates based on biomarker data, approximately 413 million individuals globally are on the AD continuum (Gustavsson, Norton, Fast, et al., 2022). Of the 413 million, approximately 315 million have preclinical AD, 69 million have mild cognitive impairment (MCI) due to AD, and 30 million have AD dementia (Gustavsson et al., 2022). In the US, one in nine individuals over the age of 65 (11.3%) live with AD, and almost two-thirds are females (Nichols et al., 2019). AD is commonly stratified into early onset AD (EOAD) and late onset AD (LOAD) based on an age cut-off of 65 years (Wingo, Lah, Levey, & Cutler, 2012). EOAD accounts for 5–10% of all cases in the US (Lambert, Bickel, Prince, et al., 2014). The underlying pathology in EOAD and LOAD, namely amyloid-β (Aβ) plaques (also known as senile plaques) and neurofibrillary tau tangles, is similar, however, important differences between the two forms of AD exist. Individuals with EOAD present with more severe grey matter atrophy (Frisoni et al., 2005), a greater burden of senile plaques and neurofibrillary tangles, greater synaptic loss (Buerger, Ewers, Pirttilä, et al., 2006), and greater deficits of acetylcholine. The cognitive profiles of EOAD and LOAD also differ where patients with

EOAD more often present with a non-amnestic profile (Koss, Edland, Fillenbaum, et al., 1996). Finally, the clinical trajectory of EOAD is often more aggressive and patients with EOAD have a shorter life expectancy (Gumus, Multani, Mack, & Tartaglia, 2021; Jacobs, Sano, Marder, et al., 1994; Wattmo & Wallin, 2017). Due to the limited availability of large autopsy cohort studies of patients with EOAD, there is a paucity of data regarding sex differences. However, there is some evidence to suggest that females may experience a greater burden of tau pathology compared with males (Barnes et al., 2005; Liesinger, Graff-Radford, Duara, et al., 2018; Oveisgharan et al., 2018; Spina, la Joie, Petersen, et al., 2021).

While the majority of AD cases are sporadic, 1% of all AD cases are due to autosomal dominant genetic mutations, such as those of the genes *PSEN1, PSEN2* and *APP* (Potter III, Long, & Lichtenstein, 2020). In addition, other genetically-determined forms of AD are related to duplications or triplications of the *APP* gene. The latter occurs in individuals with Down syndrome (chromosome 21 trisomy), who account for the largest population at genetic risk of AD (Fortea et al., 2021). The *APP* gene encoding the amyloid precursor protein resides on chromosome 21, and thus nearly all adults with Down syndrome will develop full-blown amyloid and tau pathology consistent with AD by the age of 40 years (Fagan, Henson, Li, et al., 2021; MANN DMA, 1988). Down syndrome, together with autosomal dominant AD, constitute the main evidence for the amyloid cascade hypothesis in which genetic mutations or gene dose loading result in increased amyloid production, leading to tau hyperphosphorylation, neurodegeneration and dementia. Indeed, the temporality of AD biomarkers in Down syndrome is similar to the sporadic forms and identical to autosomal dominant AD (Fortea, Vilaplana, Carmona-Iragui, et al., 2020; Wu, Yeh, Hsieh, et al., 2021).

While sex differences have been mainly explored in sporadic AD, some literature is starting to emerge on sex differences in Down Syndrome and autosomal dominant AD (Fortea et al., 2020). Studies of AD risk by sex in people with Down syndrome are conflicting (Lai et al., 1999, 2020; Mhatre, Lee, Pang, et al., 2021; Schupf et al., 1998; Sinai, Mokrysz, Bernal, et al., 2018) and there are yet no studies about sex differences across a range of biomarker modalities in this population. Some have found greater risk and earlier age at diagnosis in men (Mhatre et al., 2021; Schupf et al., 1998), while others have found greater risk in women (Lai et al., 1999) or no sex differences (Lai et al., 2020; Sinai et al., 2018). The reasons for these inconsistencies might be related to differences in age between the cohorts, differences in sample size, differences in menopause status or estrogen bioavailability in women, which are known to influence risk and age at diagnosis of dementia in Down syndrome (Coppus, Evenhuis, Verberne, et al., 2010; Cosgrave, Tyrrell, McCarron, Gill, & Lawlor, 1999; Schupf, Pang, Patel, et al., 2003; Schupf, Winsten, Patel, et al., 2006).

Age is the greatest risk factor for AD, and sex differences in prevalence have been largely attributed to the longer lifespan of females (Chêne, Beiser, Au, et al., 2015; Hebert, Scherr, McCann, Beckett, & Evans, 2001). However, studies examining the incidence of AD have been mixed (Hebert et al., 2001; Kawas & Zonderman, 2000). European studies have more consistently shown sex differences in the incidence of AD. For example, according to a recent meta-analysis, the incidence of AD in Europe was 7.02 per 1000 person-years in males and 13.25 per 1000 person-years in females (Niu, Álvarez-Álvarez, Guillén-Grima, & Aguinaga-Ontoso, 2017). Similarly, a 2018 study of 16, 926 men and women in the Swedish Twin Registry found women to have higher incidence of AD, but only starting after the age of 80. Studies of Asian cohorts have also observed a higher incidence in women after the age of 80–85 (Liu et al., 1998; Yoshitake, Kiyohara, Kato, et al., 1995). In contrast, many US studies have not observed differences between the sexes in the rates of AD (Bachman, Wolf, Linn, et al., 1993; Edland, Rocca, Petersen, Cha, & Kokmen, 2002; Ganguli, Dodge, Chen, Belle, & DeKosky, 2000; Hebert et al., 2001; Kawas & Zonderman, 2000; Kukull, Higdon, Bowen, et al., 2002; Mielke et al., 2014; Rocca, Cha, Waring, & Kokmen, 1998), suggesting that geographic, cultural, and social factors could also play a role (Mielke et al., 2014; Mielke, Teresa Ferretti, Florencia Iulita, Hayden, & Khachaturian, 2018).

1.2 Biomarkers

Historically, defining AD consisted of a clinico-pathological diagnosis characterized by the insidious onset and progressive impairment of memory and other cognitive functions, and was only confirmed at autopsy by the presence of extracellular amyloid plaques and intracellular tau neurofibrillary tangles (McKhann et al., 1984; Tiraboschi, Hansen, Thal, & Corey-Bloom, 2004). The discovery of biological markers that allowed the monitoring of these pathological hallmarks *in vivo* revealed that the neuropathological changes of AD begin decades before symptom onset, in a phase known as preclinical. Thus, the typical clinical disease progression of AD moves from a preclinical, asymptomatic phase to mild cognitive impairment (MCI) and finally, dementia (Dubois, Feldman, Jacova, et al., 2007; Sperling, Aisen, Beckett, et al., 2011).

In recent years, the field has navigated towards a biological definition of AD to aid in achieving an understanding of the underlying mechanisms and obtaining biological targets for disease-modifying interventions (Jack, Bennett, Blennow, et al., 2018). The current framework proposed by the National Institute on Aging—Alzheimer's Association (NIA-AA) defines AD by biomarkers of: Aβ plaques (cortical amyloid positron emission tomography (PET) ligand binding and low CSF $A\beta_{42}$) (Fagan et al., 2007; Mattsson, Zetterberg, Hansson, et al., 2009; Visser, Verhey, Knol, et al., 2009), fibrillar tau (cortical tau PET ligand binding and elevated CSF phosphorylated tau (P-tau)) (Brier, Gordon, Friedrichsen, et al., 2016; Buerger et al., 2006; Mattsson et al., 2009; Visser et al., 2009), and neurodegeneration/neuronal injury (elevated CSF total tau (T-tau), FDG PET hypometabolism and atrophy on MRI) (Besson, Joie, Doeuvre, et al., 2015; Dickerson, Bakkour, Salat, et al., 2009; Fox et al., 2001; Minoshima et al., 1997; Seab et al., 1988). Biomarkers of Aβ and tau are considered core to AD pathogenesis, whereas biomarkers of neuronal injury are non-specific, but contribute to disease progression and severity (Dubois, Feldman, Jacova, et al., 2014; Jack et al., 2018). These biomarkers manifest changes 15–20 years prior to the onset of clinical symptoms and follow a temporal ordering whereby Aβ begins to accumulate first followed by tau and markers of neuronal injury/neurodegeneration (Buchhave et al., 2012; Jack, Knopman, Jagust, et al., 2013).

Until recently, there was a paucity of research exploring sex difference in AD biomarkers, which has been further hindered by the statistical adjustments most studies make to control for sex (Ferretti, Iulita, Cavedo, et al., 2018). Studies investigating sex differences in Aβ burden in the preclinical/asymptomatic stage of the disease have revealed mixed results, with some showing males had increased Aβ burden (Scheinin, Wikman, Jula, et al., 2014), while others the opposite (Gottesman, Schneider, Zhou, et al., 2016; Vemuri, Knopman, Lesnick, et al., 2017) or no differences at all (Bouter, Vogelgsang, & Wiltfang, 2019; Buckley, Mormino, Chhatwal, et al., 2019; Jack, Wiste, Weigand, et al., 2015; Li, Shofer, Petrie, et al., 2017; Schoonenboom, Pijnenburg, Mulder, et al., 2004). In MCI and AD, post-mortem studies have found no differences in the spatial distribution or extent of Aβ plaques between males and females (Barnes et al., 2005). Similarly, sex does not appear to impact PET amyloid positivity

(Jansen, Ossenkoppele, Knol, et al., 2015) or CSF concentrations of CSF $A\beta_{42}$ in individuals living with AD (Mattsson, Lönneborg, Boccardi, Blennow, & Hansson, 2017). However, evidence suggests that among clinically-normal individuals, females with increased Aβ burden showed faster rates of cognitive decline compared to males with similar levels of Aβ (Buckley, Mormino, Rabin, et al., 2019). The mechanism explaining the interaction between Aβ and sex remains to be elucidated.

Growing evidence suggests that females experience more widespread tau pathology compared with males (Oveisgharan et al., 2018). Studies have also observed region-specific differences in cortical areas such as the midfrontal, superior temporal, entorhinal and inferior parietal cortices in which females with AD show greater burden than males (Barnes et al., 2005; Edwards, la Joie, Iaccarino, et al., 2021). Similarly, among cognitively unimpaired individuals, females showed higher retention of PET tau in the medial temporal lobes, however only among participants with increased amyloid burden (Buckley, Mormino, Rabin, et al., 2019; Smith, Strandberg, Mattsson-Carlgren, et al., 2020). There are indications that these differences are influenced by the presence of *APOE4,* which is discussed below.

Studies suggest that the same amount of accumulated pathology may have different effects in males and females. Work by Barnes and colleagues found that each additional unit of AD pathology (i.e., global measures of neuritic plaques, diffuse plaques, and neurofibrillary tangles) resulted in a 3-fold increase in the odds of clinical AD in males and a 20-fold increase in females (Barnes et al., 2005). The authors controlled for the number of cerebral infarcts suggesting this result is unlikely to have been confounded by a disproportionate cerebrovascular disease burden among females (Barnes et al., 2005). A more recent study by Koran, Wagener, and Hohman found that females with low levels of CSF Aβ and high total tau (suggesting higher deposition in the brain) had greater hippocampal atrophy and faster cognitive decline, especially those with lower education (Koran, Wagener, & Hohman, 2017). Further studies are required to determine whether sex-specific atrophy patterns exist in individuals with AD as greater losses of hippocampal volume have been reported in both males (Sundermann, Biegon, Rubin, et al., 2016) and females (Apostolova, Dinov, Dutton, et al., 2006).

The *APOE4* allele is present in >50% of patients with AD, but only in 15% of healthy individuals (Ward, Crean, Mercaldi, et al., 2012), making it among the strongest genetic risk factors for sporadic AD. Indeed, *APOE4* carriers show earlier AD symptomatology and greater pathology, both in the general population and in people with Down syndrome (Bejanin, Iulita, Vilaplana, et al., 2021; Corder, Saunders, Strittmatter, et al., 1993; Morris, Roe, Xiong, et al., 2010; Reiman, Caselli, Yun, et al., 1996; Strittmatter, Saunders, Schmechel, et al., 1993). Studies suggest that female carriers of *APOE4* have a higher incidence of MCI and AD, as well as higher rates of conversion from MCI to AD (Altmann, Tian, Henderson, & Greicius, 2014; Kim, Kim, Kim, et al., 2015; Neu, Pa, Kukull, et al., 2017). However, this long-standing finding was called into question by a recent large meta-analysis by Neu and colleagues which found that females had higher risk of AD only at younger ages (65–75 years), whereas between the ages of 55–85 years, males and females had nearly the same odds (Neu et al., 2017). In Down syndrome, female APOE4 carriers show a 3-year earlier diagnosis of symptomatic AD, a difference which was not seen in men (manuscript under review). Furthermore, female carriers have shown higher levels of CSF tau compared to male carriers (Hohman, Dumitrescu, Barnes, et al., 2018). Taken together, these results suggest that APOE may modulate risk for neurodegeneration in a sex-specific manner. In addition to *APOE4,* the largest genome-wide association study (GWAS) meta-analysis to date, The International Genomics of Alzheimer's Project, identified 20 genetic loci associated with AD (Lambert, Ibrahim-Verbaas, Harold, et al., 2013; Stocker, Möllers, Perna, & Brenner, 2018). Many of the loci identified reinforce the importance of already established AD pathways such as *APP* and tau, as well as pathways involved in immune response, inflammation, cell migration, lipid transport and endocytosis (Lambert et al., 2013). However, new pathways have also been implicated such as those involved with hippocampal synaptic function, cytoskeletal function, axonal transport, and myeloid cell function (Lambert et al., 2013). A recent study by Fan and colleagues found that if GWAS are analyzed separately for males and females, then sex-specific polygenic scores can be produced, and sex-specific gene effects may also be revealed (Fan, Banks, Thompson, et al., 2020). However, such studies are currently scarce.

In the transcriptomics space, a recent study by Mathys et al., which profiled and analysed single-nucleus transcriptomes from the prefrontal cortex of individuals with AD found cell-type specific changes at early stages of pathology, whereas in later stages the genes upregulated tended to be common across all cell types. Furthermore, myelination related genes were consistently perturbed across most major cell types suggesting a potential reactionary response to maintain appropriate myelin function and integrity in response to pathology. The study also found that cells from females were over-represented among the AD-associated cell subpopulations, and that sex differences were also present in the marker genes of these subpopulations. The authors interpreted this finding as potentially suggesting a sex-specific differential transcriptional response to AD pathology (Stocker et al., 2018).

1.3 Neuropsychological markers/clinical progression/ disease onset

Sex differences in the cognitive performance of healthy individuals have been well-documented. Females tend to perform better on verbal tasks (Gale, Baxter, & Thompson, 2016; Jack et al., 2015; Li & Singh, 2014; McCarrey, An, Kitner-Triolo, Ferrucci, & Resnick, 2016) and show slower cognitive decline at all ages (Laws, Irvine, & Gale, 2016; Li & Singh, 2014). In contrast, males perform better on visuospatial and motor coordination tasks (Laws et al., 2016; Li & Singh, 2014; McCarrey et al., 2016). These differences persist into prodromal AD (Sundermann et al., 2016; Sundermann, Biegon, et al., 2016), however in AD, females perform worse in verbal memory and fluency tasks (Benke, Delazer, & Sanin, 2013; Gale et al., 2016; Lehrner, Hagmann, & Marksteiner, 2015). Sundermann and colleagues have shown that among patients with amnestic MCI, the female advantage in verbal memory persists even when levels of disease burden are similar between sexes (Sundermann et al., 2017), suggesting that females may possess greater verbal memory-specific cognitive reserve, which allows them to sustain performance further into disease course (Stern et al., 1994, 2003; Sundermann, Maki, Biegon, et al., 2019).

Several studies have shown faster rates of cognitive decline in female patients with cognitive impairment (Holland, Desikan, Dale, & Mcevoy, 2013; Gamberger, Lavrač, Srivatsa, Tanzi, & Doraiswamy, 2017; Lin et al., 2015). A recent pooled cohort study of non-demented individuals showed that despite higher baseline global, executive and memory scores, females showed faster rates of decline in global and executive scores, but not in memory, after controlling for age, race, education and mean systolic blood pressure (Levine, Gross, Briceño, et al., 2021). Females also show rates of progression from MCI to AD that are 2–3% greater per annum than males (Tifratene, Robert, Metelkina, Pradier, & Dartigues, 2015). The faster rates of cognitive decline may result from later diagnosis of AD in women (Ferretti et al., 2018). Pradier and colleagues showed that females had lower Mini-Mental State Examination (MMSE) scores at initial diagnosis suggesting they were further along in their disease course (Pradier, Sakarovitch, le Duff, et al., 2014). Similarly, Sundermann and colleagues have shown that given the female advantage in verbal memory noted above, sex-specific cut offs should be utilized on tests such as the Rey Auditory Verbal Learning Test (Sundermann et al., 2019). Indeed, after applying sex-specific cut offs, one recent study by Sundermann et al found amnestic MCI to be underdiagnosed by 10% in females and over-diagnosed by 10% in males (Sundermann et al., 2019).

Neuropsychiatric symptoms (NPS) such as delusions, hallucinations, depression, anxiety, and apathy are very common in AD to the extent that more than 80% of patients will experience at least one NPS throughout the course of the disease (Lyketsos et al., 2002). Furthermore, NPSs can present in advance of cognitive deficits (Lyketsos et al., 2002). Among cognitively normal individuals, depression, anxiety, irritability, and apathy can predict more rapid cognitive decline (Donovan, Amariglio, Zoller, et al., 2014; Duara, Loewenstein, Greig, et al., 2011; Lyketsos et al., 2002). There also appears to be differences in NPS between EOAD and LOAD such that patients with EOAD have a greater prevalence and severity of anxiety, depression, disinhibition, hallucinations, euphoria, and appetite change compared to patients with LOAD (Gumus et al., 2021). Furthermore, the most common NPS in EOAD were anxiety and depression whereas in LOAD, it was depression and irritability (Gumus et al., 2021; Kaiser et al., 2014; Panegyres & Chen, 2014; Porter, Buxton, Fairbanks, et al., 2003). Regarding sex differences in NPS, a recent meta-analysis by Eikelboom and colleagues found that females experience greater prevalence and severity of depressive symptoms, aberrant motor behavior, and psychotic symptoms, whereas males experienced greater severity of apathy; suggesting females incur a greater burden of NPS (Eikelboom, Pan, Ossenkoppele, et al., 2022). These results are concordant with other studies which also found females to possess a greater burden of NPS (Cohen, Eisdorfer, Gorelick, et al., 1993; Eikelboom et al., 2022; Nagata, Nakajima, Shinagawa, et al., 2017; Tao, Peters, Drye, et al., 2018).

In 2020, Livingston et al put forth the following 12 modifiable risk factors for dementia: less education, hypertension, hearing impairment, smoking, obesity, depression, physical inactivity, diabetes, low social contact, excessive alcohol consumption, traumatic brain injury (TBI), and air pollution (Livingston, Huntley, Sommerlad, et al., 2020). These risk factors can be broadly grouped as medical/psychiatric comorbidities (obesity, hypertension, diabetes, hearing impairment, TBI, and depression) and social/environmental (less education, low social contact, physical inactivity, excessive alcohol consumption, smoking, and air pollution). The sex differences associated with these 12 modifiable risk factors are shown in Table 1 below.

Table 1 Sex differences associated with modifiable risk factors of dementia.

Modifiable risk factors	Reported sex differences
Medical/psychiatric co-morbidities	*Obesity:* Higher incidence in females (Gannon, Robison, Custozzo, & Zuloaga, 2019; Kautzky-Willer, Harreiter, & Pacini, 2016)
	Hypertension: Higher incidence among males when compared to pre-menopausal females, however it is higher among post-menopausal females (Benjamin, Virani, Callaway, et al., 2018; Lima, Wofford, & Reckelhoff, 2012)
	Diabetes: The incidence is equal between sexes, however, males are more likely to develop Type II diabetes at similar body weight and females have a higher incidence of obesity, a major risk factor for developing diabetes (Gannon et al., 2019; Kautzky-Willer et al., 2016)
	Hearing impairment: Hearing ability at most ages and frequencies was found to decline more than twice as fast in males (Pearson, Morrell, & Gordon-Salant, 1995; Villavisanis, Berson, Lauer, Cosetti, & Schrode, 2020)
	TBI: The incidence of TBI is higher among males (Nguyen, Fiest, McChesney, et al., 2016). However, females are at an increased of TBI secondary to domestic violence (Corrigan, Wolfe, Mysiw, Jackson, & Bogner, 2003)
	Depression: More prevalent and more severe among females (NIMH, n.d.)
Social/environmental	*Education*: Historically, males have had more opportunities for higher education, however, the most recent US census showed that educational attainment was equal among males and females (Ryan & Bauman, 2015)
	Social contact: Females generally have larger social networks and exhibit higher participation in social activities (Cacioppo, Hawkley, & Thisted, 2010)

Table 1 Sex differences associated with modifiable risk factors of dementia.—cont'd

Modifiable risk factors	Reported sex differences
	Physical activity: Males engage in more physical activity over the lifespan even after accounting for unequal opportunity (Rosenfeld, 2017)
	Alcohol consumption: Females generally drink less and are less likely to develop alcohol-related disorders
	Smoking: Males tend to use all tobacco products at higher rates than females (Erol & Karpyak, 2015), however, the number of women who smoke is increasing (Akhter, Persaud, Zaokari, Zhao, & Zhu, 2021)
	Air Pollution: Females appear to be more vulnerable to air pollution (Liu, Sun, Yu, et al., 2020), however, prospective cohort studies are lacking

In addition, a growing literature indicates the occurrence of sex-specific risk factors, many of which are related to the female reproductive life. For instance, early menopause has been linked to risk of cognitive decline and AD (Rocca et al., 2021). For males, the use of anti-androgen drugs for prostate cancer has been associated with higher risk of dementia (Huang, Liu, Pang, et al., 2020), however more studies are needed to elucidate this.

1.4 Treatment

The currently available pharmacological treatments for AD are cholinesterase inhibitors (donepezil, rivastigmine, and galantamine) and the *N*-methyl-D-asparate receptor antagonist memantine. A systematic review of randomized controlled studies for pharmacological interventions in AD published by Canevelli et al examined over 1200 articles and found no studies where sex or gender differences were reported separately (Canevelli, Quarata, Remiddi, et al., 2017). Canevelli et al proceeded to review studies which performed ad hoc, secondary, or post-hoc analyses accounting for sex and gender differences and only found data on donepezil, which showed comparable functional and cognitive efficacy between the sexes (Greenberg, Tennis, Brown, et al., 2000; Winblad, Engedal, Soininen, et al., 2001). Similarly, a recent systematic review and meta-analysis by Martinkova et al found that only 12.5% of clinical trials of experimental drugs for AD reported sex-stratified results, however this proportion appeared to increase over time (Martinkova, Quevenco, Karcher, et al., 2021). Regarding sex differences in the adverse effects of cholinesterase inhibitors, a systematic review by Mehta and colleagues found an almost complete lack of reporting on sex differences among the 16 trials included in their analysis (Mehta, Rodrigues, Lamba, et al., 2017).

1.5 Gender

Many of the relevant factors related to gender identity (e.g., education, occupation, exercise, diet, etc.) can be tied to Stern's model of cognitive reserve, which posits that individuals with greater cognitive reserve are more resistant to cognitive decline due to compensatory mechanisms (i.e., differential synaptic organization and utilization of specific brain regions) (Stern, 2009). Thus, gender-related social and cultural differences will impact the attainment of cognitive reserve, which may manifest as differences in rates of cognitive decline (Mielke et al., 2014). Indeed, besides biological sex, gender is recognized as an important determinant of health by the World Health Organization (World Health Organization, 2011). Historically, women occupied a caretaker role of the household and engaged in less physically demanding work than men, resulting in lower cardiovascular fitness, and consequently, a higher incidence of cardiovascular disease (Hörder, Johansson, Guo, et al., 2018). Thus, it has been posited that the increased risk of AD among women may be at least partially explained by the increased incidence of cardiovascular disease given that there is robust evidence to suggest cardiovascular disease may contribute to AD progression (Leszek, Mikhaylenko, Belousov, et al., 2021). In addition, smoking, another risk factor for both dementia and cardiovascular disease (Livingston et al., 2020), was viewed as less socially acceptable for women; however, in recent years, the number of women who smoke has increased (Akhter et al., 2021).

Historically, men have had more opportunities for higher educational and occupational attainment, although the most recent US census showed that educational attainment was equal among men and women (Ryan & Bauman, 2015). A recent study by Bloomberg and colleagues found that women in later birth cohorts had increasingly better memory scores, and in fluency, women in later birth cohorts or with higher educational attainment outperformed men. The authors posit that the reduction of disparities in educational attainment due to secular increases in access to education could mitigate the excess risk of AD observed in women in the future. Similarly, other studies have found that these changes in educational attainment have translated to slower rates of cognitive decline when comparing younger and older cohorts (Gerstorf, Ram, Hoppmann, Willis, & Schaie, 2011; Schaie & Willis, 2010).

Table 1 summarizes the sex differences associated with social/environmental factors; however, much work is needed to further elucidate the relationship between gender and AD.

2. Sex and gender differences in vascular cognitive impairment/vascular dementia

2.1 Overview

Vascular dementia (VaD) is caused by a heterogenous group of cerebrovascular disorders characterized by the reduction or occlusion of blood flow to the brain (Vijayan & Reddy, 2016). These disorders include, but are not limited to stroke, cerebral amyloid angiopathy (CAA), subarachnoid hemorrhage, vasculitis, silent infarcts, and small vessel disease (Portegies, Koudstaal, & Ikram, 2016). Cerebrovascular integrity is fundamental for the maintenance of the brain's high metabolic demand and to prevent disease. Vascular health depends on genetic, anatomical, functional, and hormonal mechanisms that can differ between sexes. Typically, individuals with VaD will show deficits in attention, information processing, and executive functioning, however, cognitive profiles are variable and dependent upon the neuroanatomical structures affected (O'Brien & Thomas, 2015).

Over the years, VaD as a concept has undergone considerable change. Building on the early works of Otto Binswanger, Alois Alzheimer, and Emilie Kraeplin (Román, 2022; Yang, Kim, Lee, & Young, 2016), Hachinski, Potter and Merskey (1987) demonstrated that infarcts were relatively rare, and diffuse white matter lesions were more frequently associated with cognitive decline in VaD (Hachinski, Potter, & Merskey, 2022). It is now widely accepted that progressive changes in the brain related to small vessel disease (SVD) are the major cause of, and the most common pathology in VaD (Hachinski, 2015; Rosenberg, 2017).

During the 1990s, the term vascular cognitive impairment (VCI) was introduced to better account for the full spectrum of deficits ranging from MCI to dementia (Hachinski, Bowler, & Loeb, 1993). As a concept, VCI has been widely adopted and is currently defined as "a syndrome with evidence of clinical stroke or subclinical vascular brain injury and cognitive impairment affecting at least one cognitive domain" (Black, DeCarli, Greenberg, et al., 2011; Iadecola, 2013).

2.2 Frequency

VCI is the second most common cause of dementia after AD accounting for roughly 20% of cases (Korczyn, Vakhapova, & Grinberg, 2012; Román, 2003). Approximately 10% of dementia cases can be considered pure VCI (i.e., no other pathologies), whereas 50–70% are mixed (2019 Alzheimer's disease facts and figures, 2019). In population-based studies, the absolute risk of VCI is estimated to be 7% (Pendlebury & Rothwell, 2009). In contrast, hospital-based studies estimate the absolute risk to be over 40% (Pendlebury & Rothwell, 2009). The dramatic discrepancy between these cohorts can be attributed to the impact of stroke on precipitating dementia, wherein dementia occurs in 15–30% of individuals 3 months post-stroke (Pendlebury & Rothwell, 2009).

Sex and gender differences exist in VCI. Although the overall prevalence of dementia is higher among women, the prevalence of VCI/VaD appears to occur more frequently in males, however, the results are mixed (Copeland, Mccracken, Dewey, et al., 1999; Fujishima & Kiyohara, 2002; Imfeld, Brauchli Pernus, Jick, & Meier, 2013). Data from the Rotterdam Study found that the overall incidence of VaD was lower in females (rate ratio 0.57, 95% CI: 0.34–0.97) (Ruitenberg, Ott, van Swieten, Hofman, & Breteler, 2001). Similarly, a study by Di Carlo and colleagues found a significantly higher risk of VaD among males (RR = 2.23) (di Carlo, Baldereschi, Amaducci, et al., 2002). Other studies show trends for higher incidence among males, but none meet statistical significance (Copeland et al., 1999; Fujishima & Kiyohara, 2002; Imfeld et al., 2013), and some studies show no difference at all (Andersen, Launer, Dewey, et al., 1999). Further complicating the picture, studies examining the prevalence of VCI/VaD among different age groups have observed the prevalence to be higher among men before the age of 79, and higher in women thereafter (Dong et al., 2007; Ruitenberg et al., 2001). Given that women tend to live longer, these results should be interpreted with caution (Gannon et al., 2019). Further studies are needed to elucidate sex differences in the prevalence of VCI.

2.3 Biomarkers

Given the heterogeneity in the underlying causes of VCI, the frequent coexistence of degenerative and vascular pathology, and the paucity of pathognomonic radiological features, neuroimaging serves to describe rather than diagnose the brains of individuals with VCI (Hachinski, Iadecola, Petersen, et al., 2006). Nevertheless, the National Institute of Neurological Disorders and Stroke–Canadian Stroke Network Vascular Cognitive Impairment Harmonization Standards proposed the following Magnetic Resonance Imaging (MRI) measures to guide prospective studies on VCI: brain atrophy, white matter hyperintensities (WMH), infarction, hemorrhage, and other (mass lesions, arteriovenous malformations, extra-axial fluid collections, other malformations, dysplasia, etc.) (Hachinski et al., 2006). Neuroimaging can provide valuable information about the underlying cerebrovascular pathology and can be divided into three main groups: (1) the end-stage tissue damages, (2) disrupted vessels and (3) vascular malfunctioning (Andersen et al., 1999; Lang, Kindy, Kozel, Schultz, & Taheri, 2018).

Typically, end-stage tissue damage due to ischemic and non-ischemic cell death appears as infarcts (lacunar or micro) and WMH (periventricular or subcortical). The incidence of stroke is higher among males compared to pre-menopausal women (Benjamin et al., 2018), however this trend begins to reverse post-menopause to the extent that females over the age of 85 years have a higher incidence (Ahangar, Saadat, Heidari, Taheri, & Alijanpour, 2018; Appelros, Stegmayr, & Terent, 2009; Petrea et al., 2009), and the prevalence of stroke is also higher among females (Benjamin et al., 2018). In addition, evidence suggests that females experience worse functional outcomes post-stoke compared to males (Bushnell, Chaturvedi, Gage, et al., 2018) in terms of physical, cognitive, and functional impairment, often accompanied by a higher risk for depression following stroke. Likewise, cerebrovascular disease has been shown as the first cardiovascular manifestation in women (versus coronary heart disease for men) (Leening, Ferket, Steyerberg, et al., 2014), an additional example of how vascular insults could affect brain homeostasis differently between sexes.

The higher prevalence among females is largely attributed to increased life expectancy (Benjamin et al., 2018) and the reduced vulnerability to cerebrovascular injuries due to the protective effect of estrogen (Girouard, Lessard, Capone, Milner, & Iadecola, 2008). Recent studies have observed that although overall incidence of stroke is declining in both males and females the rate decline is slower in females (Madsen, Khoury, Alwell, et al., 2017). As described below, the higher incidence of stroke among males is likely related to the higher incidence of hypertension in males (Gillis & Sullivan, 2016) in mid-life and the neuroprotective role of estrogen among females before menopause (Robison, Gannon, Salinero, & Zuloaga, 2019). Regarding WMH, several studies suggest that females have higher volumes of WMH, as a proportion of the total white matter volume, and faster rates of WMH accumulation compared to males (de Leeuw, de Groot, Achten, et al., 2001; DeCarli, Massaro, Harvey, et al., 2005; Fatemi, Kantarci, Graff-Radford, et al., 2018; Sachdev, Parslow, Wen, Anstey, & Easteal, 2009; van den Heuvel, Admiraal-Behloul, ten Dam, et al., 2004). Interestingly, vascular risk factors such as hypertension, atherosclerosis, and diabetes are associated with the development of WMH, however this association appears to be stronger in males than females suggesting other pathophysiologic mechanisms may be responsible for the sexual dimorphism (Alqarni, Jiang, Crawford, et al., 2021). Several studies have suggested that female-specific factors such as the reduction of estrogen post-menopause, bilateral oophorectomy, and hypertensive pregnancy may play a role (Cook, Morgan, Dunkin, et al., 2002; Liu, Hu, Ji, et al., 2009; Miller et al., 2020; Miller, Miller, & Barnes, 2019).

The second group of causes, disrupted vessels, manifest as cerebral hemorrhages and microbleeds on MRI (Lang et al., 2018). It is well-established that the prevalence of aneurysms, a common precursor to hemorrhagic stroke, and subarachnoid hemorrhages are more common among females (Cras, Bos, Arfan Ikram, et al., 2020; Muehlschlegel, 2018; Turan et al., 2016).

The third group of causes, vascular malfunctioning, consists of blood-brain barrier (BBB) leakage, decreased cerebrovascular reactivity, and enlarged perivascular spaces (Lang et al., 2018). The breakdown of the BBB has been implicated in the pathogenesis of neurodegeneration (Blair, Frauen, Zhang, et al., 2015; Montagne, Zhao, Zlokovic, & v., 2017; Sepehrband, Barisano, Sheikh-Bahaei, et al., 2021). The dynamics of the BBB are complex and multifaceted; however, sex hormones appear to play a role. For example, estrogen has been shown to protect the BBB under inflammatory conditions by enhancing BBB functionality and decreasing leukocyte extravasation across the BBB (Maggioli, McArthur, Mauro, et al., 2016; Weber & Clyne, 2021). The loss of estrogen's protective role on the BBB following menopause may help to explain the increased risk of cerebrovascular events in females in late life (Montagne et al., 2017; Pepine & Nichols, 2006). Cerebrovascular reactivity refers to the ability of brain blood vessels to increase blood flow in response to vasodilatory stimuli and reflects microstructural integrity, which is disrupted in cerebrovascular disease (CVD) (Black, Gao, & Bilbao, 2009). Increased arterial stiffness has been associated with decreased cerebrovascular reactivity, however the results are mixed (Sabra, Intzandt, Desjardins-Crepeau, et al., 2021).

Estrogen and testosterone play protective roles against arterial stiffness (Maas & Appelman, 2010; Pepine & Nichols, 2006; Shaw, Bairey Merz, Pepine, et al., 2006; Towfighi, Zheng, & Ovbiagele, 2009) and it has been suggested that females experience increased AS following the post-menopausal drop of estrogen leading to a greater prevalence of CVD in later life (Sabra et al., 2021). In addition, the association between increased arterial stiffness and mortality is almost threefold higher in women than in men (Regnault, Thomas, Safar, et al., 2012). Finally, the relationship between perivascular enlargement and VCI has not been well-elucidated. However, a recent study by Sepehrband and colleagues showed that patients with MCI had regional differences in perivascular volumes compared to healthy controls, and this effect was driven primarily by females (Sepehrband et al., 2021).

2.4 Neuropsychological markers/clinical progression/ disease onset

Cognitive changes are more variable in VCI compared to neurodegenerative diseases, as impairments are dependent upon the neuroanatomical structures affected. However, subcortical pathology resulting in disruptions of frontostriatal circuits is common, and as such, deficits in attention, information processing, and executive functioning are frequently observed (O'Brien & Thomas, 2015).

Over much of the life span, females appear protected from CVD, however, this trend is seen to reverse following menopause and the associated reduction in estrogen production (Petrea et al., 2009; Robison et al., 2019). The sexual dimorphisms related to CVD appear to be related to the neuroprotective effects of estrogen on cerebrovasculature (Robison et al., 2019). Work by Girouard and colleagues found that females mice were less susceptible to the cerebrovascular actions of hypertension, an effect that was mediated by estrogen (Girouard et al., 2008). Similarly, *in vivo* treatment with estrogen was shown to be protective against Aβ deposition and Aβ-induced toxicity associated with CAA pathology (Rhodin, Thomas, Clark, Garces, & Bryant, 2003). The increased risk and incidence of CVD post-menopause may be mediated by an increased risk of cardiovascular disease (i.e., hypertension) and metabolic disease (i.e., diabetes, obesity, dyslipidemia). The sex differences related to cardiovascular and metabolic disease can be found in Table 2 below.

Females also experience sex-specific risk factors related to oral contraceptive use, pregnancy, and preeclampsia/eclampsia. Oral contraceptives

Table 2 Sex differences in incidence/prevalence rates of cardiovascular and metabolic diseases stratified by pre- and post-menopause.

	Pre-menopause	Post-menopause
Cardiovascular disease		
Stroke	Higher incidence of brain infarction and intracerebral hemorrhage in males, higher incidence of subarachnoid hemorrhage in females (Appelros et al., 2009)	Incidence of stroke begins to equalize between males and women (Ahangar et al., 2018; Appelros et al., 2009; Petrea et al., 2009), and becomes greater in females age 85+ (Petrea et al., 2009)
Hypertension	Higher incidence in males (Gillis & Sullivan, 2016), however, midlife hypertension increases the risk of dementia among females only	Higher incidence among females after 60 years of age (Gillis & Sullivan, 2016; Gilsanz, Mayeda, Glymour, et al., 2017; Yoon et al., 2015)
Metabolic disease		
Type II diabetes	No sex differences (Gannon et al., 2019; Kautzky-Willer et al., 2016)	No sex differences (Gannon et al., 2019; Kautzky-Willer et al., 2016)
	Note: Males are more likely to develop Type II diabetes at similar body weight, however, females have a higher incidence of obesity, a major risk factor for developing diabetes (Gannon et al., 2019; Kautzky-Willer et al., 2016). In addition, males Type II diabetes experience an increased risk of MCI compared to females (Roberts, Knopman, Geda, et al., 2014)	
Obesity	Higher incidence in females (di Cesare, Bentham, Stevens, et al., 2016; Gannon et al., 2019)	Higher incidence in females (di Cesare et al., 2016; Gannon et al., 2019)
	Note: Although incidence of obesity is higher among females, one study found that obesity was a specific predictor of conversion from normal cognition to MCI (within a 5-year follow up period) (Pankratz, Roberts, Mielke, et al., 2015), and a specific risk factor for dementia (within 3-year follow up period) (Hayden, Zandi, Lyketsos, et al., 2006) in males	
Dyslipidemia	Higher incidence in males (O'Meara et al., 2004)	Higher incidence in females (Azad, & al Bugami M, Loy-English I., 2007; Cignarella, Kratz, & Bolego, 2010; Wildman, Colvin, Powell, et al., 2008)

have been associated with increased risk of thrombotic events, potentially contributing to cognitive impairment and VCI (Dulicek, Ivanova, Kostal, et al., 2018; Gannon et al., 2019). Studies investigating the relationship between pregnancy and dementia have yielded mixed results. Some found that a greater number of pregnancies or months spent pregnant correlated with decreased risk of dementia (Gannon et al., 2019), while others found higher risk (Bae, Lipnicki, Han, et al., 2020; Song, Wu, Zhou, et al., 2020) or no difference (Gemmill & Weiss, 2022; Najar, Östling, Waern, et al., 2020; Shimizu, Sawada, Iwasaki, et al., 2019). However, further investigations are needed to validate these results as well as the specific relationship between VCI/VaD and pregnancy. Finally, preeclampsia/eclampsia can cause changes to the brain which can affect CVD. For example, a study by Aukes and colleagues found that females with eclampsia were 3.3 times more likely to have WMH compared to normotensive pregnancies (Aukes, de Groot, Aarnoudse, & Zeeman, 2009).

Although females show a lower risk of CVD throughout most of the lifespan, evidence suggests that the outcomes (e.g., stroke severity and resulting morbidity and mortality) are significantly worse for females (Bouter et al., 2019; Morris et al., 2010; Strittmatter et al., 1993). Furthermore, the effects of cardiovascular and metabolic risk factors on VCI/VaD risk show sex differences. For example, a recent study by Huo and colleagues found that specific cardiovascular conditions and risk factors, such as dyslipidemia, diabetes, and coronary artery, have stronger associations with cognitive decline in midlife for males than females despite the higher prevalence of those conditions in males (Huo, Vemuri, Graff-Radford, et al., 2022). In addition, a recent meta-analysis found that compared to males with diabetes, females have a 19% greater risk of developing VaD (Chatterjee, Peters, Woodward, et al., 2016).

Much work is needed given the complex relationship between hormones, CVD and VCI, and the paucity of data on the impact of sex on clinical manifestation and trajectory of VCI.

2.5 Treatment

The primary approach to treating VCI has been the management of vascular risks factors (e.g., hypertension, diabetes, and dyslipidemia) (Bocti, Black, & Frank, 2007). The effects of pharmacological management of risk factors on VaD has not been well-established, and the effect on dementia more broadly has returned mixed results (Gorelick, Scuteri, Black, et al., 2011). For example, according to the 2011 scientific statement on vascular contributions to cognitive impairment and dementia put forth by the American Heart Association and the American Stroke Association, only one out of six large, randomized control trials of anti-hypertensives which included an evaluation of dementia and cognition, reported a reduction in the incidence of dementia with control of hypertension (Gorelick et al., 2011). However, in 2016, Iulita and Girouard reviewed and discussed several methodological considerations like patient heterogeneity, choice of anti-hypertensive drug, study duration and sensitivity of neuropsychological tests used, which may have negatively biased the outcomes of these trials (Iulita & Girouard, 2017).

Many of the treatments currently being used for AD have been applied to VaD due to clinical and pathological evidence suggesting compromise of the cholinergic system in VaD (Behl, Bocti, Swartz, et al., 2007; Bocti et al., 2005; Erkinjuntti, Román, & Gauthier, 2004), as well as the frequent co-occurrence of VaD with AD (Chui & Ramirez-Gomez, 2015). These treatments include cholinesterase inhibitors donepezil, galantamine, rivastigmine, and the NMDA receptor antagonist memantine. Data from clinical trials support the use of donepezil and galantamine to improve cognition in patients with VaD, whereas support for the use of rivastigmine and memantine has not been well-established (Gorelick et al., 2011). Unfortunately, these studies did not examine the effects of sex.

Studies investigating sex differences in cholinesterase inhibitor treatment responses in patients with VaD are scarce and inconsistent (Edwards et al., 2021). Most of the studies examining the utility of cholinesterase inhibitors included individuals with AD, not VaD. Much work is needed to elucidate how these differences manifest in VaD, as very little data exists at present.

2.6 Gender

Similar to AD, the major risk factors for VaD include cardiovascular disease, smoking, and obesity, which may be influenced by social factors (Akhter et al., 2021). Historical differences in gender roles were reviewed in the previous section on AD (*Section: Gender*). Future studies should aim to further investigate the complex interplay between sex and gender in VaD specifically, as there is currently a paucity of research.

3. Sex and gender differences in Parkinson's disease/dementia with Lewy bodies

Alpha-synucleinopathies are the second most common form of neurodegenerative disorder after Alzheimer's disease (Outeiro, Koss, Erskine, et al., 2019). Within alpha-synucleinopathies, the accumulation of aggregated alpha-synuclein into Lewy bodies or Lewy neurites results in Lewy body diseases (LBD), consisting of the phenotypes of Parkinson's disease (PD), Parkinson's disease dementia (PDD), and dementia with Lewy bodies (DLB) (Turcano & Savica, 2020). The focus of this chapter will be on sex and gender differences in DLB and PDD.

3.1 Clinical

There is considerable overlap of clinical symptoms in DLB and PDD, and differentiation is based on whether motor symptoms are preceded or followed by the onset of dementia symptoms; these cases are classified as DLB or PDD, respectively. The 2017 revised consensus criteria for the diagnosis of probable and possible DLB dictate that when dementia develops before or concurrently with parkinsonism, the diagnosis should be DLB. When dementia appears in the context of well-established PD, the diagnosis should be PDD (McKeith, Boeve, et al., 2017). In DLB cases, the majority will develop parkinsonism over time, and increasingly, PDD is thought of as being an eventual common consequence of PD, representing a convergence of clinical symptoms at later disease stages that fuels the ongoing debate on the nosological relationship of LBDs (Jellinger, 2018).

The clinical features of DLB include fluctuating cognition, recurrent visual hallucinations, rapid eye movement (REM) sleep behavior disorder (RBD), and one or more spontaneous cardinal features of parkinsonism (bradykinesia, resting tremor, or rigidity) (McKeith et al., 2017). DLB fluctuations are typically delirium-like, spontaneous alterations in cognition, attention and arousal (McKeith et al., 2017). Fluctuations best predict DLB when present during the early stages of dementia (McKeith et al., 2017). Recurrent visual hallucinations occur in up to 80% of patients with DLB (McKeith et al., 2017). These hallucinations are fully-formed, featuring people or animals, and patients are typically able to report these experiences (McKeith et al., 2017). RBD is the absence of normal REM sleep atonia and manifests with dream enactment behavior (McKeith et al., 2017). RBD often begins many years before other symptoms and is considered a core clinical feature because it frequently occurs in autopsy-confirmed cases of DLB compared to non-DLB cases (McKeith et al., 2017). Parkinsonism features eventually occur in over 85% of DLB cases (McKeith et al., 2017).

The revised 2017 criteria for diagnosis of DLB relies on the essential feature of the presence of dementia as well as core and supportive clinical features and indicative and supportive biomarkers, which are summarized in Table 3 (McKeith et al., 2017). Probable diagnosis relies on two or more

Table 3 Revised criteria for clinical diagnosis of probable and possible dementia with Lewy bodies (DLB) (McKeith et al., 2017).

Essential feature
Dementia; deficits of attention, executive function and visuoperceptual ability
Core clinical features
Fluctuation cognition
Visual hallucinations
RBD
One or more spontaneous cardinal feature of parkinsonism (bradykinesia, resting tremor, or rigidity)
Supportive clinical features
Severe sensitivity to antipsychotic agents
Postural instability
Repeated falls
Syncope or other transient episodes of unresponsiveness
Severe autonomic dysfunction (constipation, orthostatic hypotension, urinary incontinence)
Hypersomnia
Hyposmia
Hallucinations in other modalities
Systematized delusions
Apathy, anxiety, and depression
Indicative biomarkers
Reduced DaT uptake in basal ganglia demonstrated by SPECT or PET
Low uptake iodine-MIBG myocardial scintigraphy
PSG-confirmed RBD
Supportive biomarkers
Relative preservation of medial temporal lobe structures on CT/MRI scan
Generalized low uptake on SPECT/PET perfusion/metabolism scan with reduced occipital activity 6 the cingulate island sign on FDG-PET imaging
Prominent posterior slow-wave activity on EEG with periodic fluctuations in the pre-alpha/theta range

core clinical features or only one core clinical feature accompanied by one or more indicative biomarkers (McKeith et al., 2017). At this time, the diagnosis of probable DLB should not be made solely relying on biomarkers (McKeith et al., 2017). Possible DLB can be diagnosed if one core feature is present or one or more indicative biomarkers are present in the absence of any core clinical features (McKeith et al., 2017).

According to the Movement Disorder Society (MDS) task force for PDD, the defining feature is that dementia develops in the context of established PD; probable or possible diagnosis of PDD relies on the presence of deficits in at least two of the cognitive domains and for them to be severe enough to affect normal functioning (Emre, Aarsland, Brown, et al., 2007). The diagnosis is broken into two levels of disease severity and the diagnostic criteria are outlined in Table 4. The prevalence of dementia in PD is close to 30% and is more often associated with the postural instability and gait disorder (PIGD) motor phenotype and older age (Emre et al., 2007). The onset of dementia is insidious and progresses slowly; it involves deficits in attention, memory, visuo-spatial, constructional and executive function domains, with preserved core language function (Emre et al., 2007). PDD is also associated with behavioral and neuropsychiatric symptoms such as hallucinations, delusions, mood disturbance and apathy. Hallucinations are not quite as common in PDD as in DLB, but nevertheless are present in PDD cases with a prevalence range of 60–80% (Emre et al., 2007).

Worldwide estimates of PD incidence range from 5 to >35 new cases per 100,000 individuals each year (Poewe, Seppi, Tanner, et al., 2017). A population-based study from Minnesota, USA, with pathologically-validated clinical diagnosis, reported incidence of PD at 14.2 cases per 100,000 person-years (Savica et al., 2013). Epidemiological data on DLB and PDD remain limited, but using the Minnesota population-based study data, the overall incidence rate of DLB was 3.5 cases per 100,000 person-years (Savica et al., 2013; Walker, Possin, Boeve, & Aarsland, 2015). The incidence rate of PDD was lower, with 2.5 cases per 100,000 person-years (Savica et al., 2013). However, due to the limitation of ascertainment bias in the Minnesota study, these rates should be interpreted as minimum estimates (Savica et al., 2013). The more recent DIAMOND-Lewy research programme carried out over 5 years at secondary care National Health Service (NHS) memory and movement disorder services in England was created to address the lack of knowledge on the accurate prevalence of DLB and PDD in the country and improve the diagnosis and

Table 4 Guideline for diagnosis of MCI in PD (MDS) (Litvan, Goldman, Tröster, et al., 2012; Vasconcellos & Pereira, 2015).

Inclusion criteria	
	Diagnosis of PD by UKPDBB
	Gradual cognitive decline
	Cognitive impairment documented by neuropsychological testing or a global cognitive scale
Exclusion criteria	
	Diagnosis of dementia based on MDS criteria
	Other changes that might explain cognitive deficits delirium, stroke, severe depression, metabolic abnormalities, adverse effects of medications or electroconvulsive therapy
	Other comorbidities related to PD
Level 1	
	Impairment on a global cognitive scale, or
	Impairment on at least two neuropsychological brief assessments
Level 2	
	Two neuropsychological tests in each of the five cognitive domains (attention and working memory, executive function, language, memory, and visuospatial function)
	Impairment of at least two tests represented by two changes in one cognitive domain (1 or 2 changes in different cognitive domains)
	Changes in neuropsychological test scores can be demonstrated by 1 or 2 SD below or significant decline demonstrated by serial evaluation or Premorbid significant decline in relation to previous status
Guidelines for diagnosis on PDD (MDS) (Emre et al., 2007; Vasconcellos & Pereira, 2015)	
Level 1	Fill criteria for PD proposed by UKPDBB
	Parkinsonism signs precede the development of dementia
	PD associated with global impairment of cognition
	Cognitive impairment interferes with activities of daily living
	Involvement of one or more cognitive domain (attention, executive function, visuo-constructive skill or memory)
Level 2	All of the above
	Evaluate four domains of cognition (global cognition, subcortical frontal function, instrumental function, and neuropsychiatric function)

management of these cases. The programme found evidence of delays in diagnosis, identified barriers to diagnosis and reported lower frequency of diagnosis than expected (O'brien, Taylor, Thomas, et al., 2022). A systematic review published in 2016 included 22 studies and found DLB incidence rates ranging from 0.5 to 1.6 per 1000 person-years and that prevalence rates of DLB were 0.3–24.4% of all cases of dementia. This paper also states they are the third systematic review on the topic and that the prevalence of DLB is chronically understudied (Hogan, Fiest, Roberts, et al., 2016).

Sex differences in PD are well-known, wherein incidence in males ranges from 1.3 to 2x more often than in females (Meoni, Macerollo, & Moro, 2020; Poewe et al., 2017). Since epidemiological data on PDD and DLB are so limited, there is consequentially poor understanding of the prevalence and incidence of DLB and PDD in the sexes. The Minnesota population-based study showed a predominance of DLB in males, with a higher incidence of DLB in males across all age groups, and during the peak incidence decade (ages 70–79 years), there was a fourfold higher incidence rate (Hely, Reid, Adena, Halliday, & Morris, 2008; Savica et al., 2013). Sex differences in the incidence of PDD appear marked because PD is approximately 1.5 times more common in males than females, and females often develop the more benign tremor dominant (TD) phenotype that is less associated with cognitive dysfunction (Reekes et al., 2020). A higher prevalence of dementia in PD is reported in males, and the male sex has been found to be a significant predictor of conversion from PD-MCI to PDD (Cereda, Cilia, Klersy, et al., 2016; Cholerton, Johnson, Fish, et al., 2018). A large study from China, however, going over 8405 dementia medical records reported patients with DLB comprised 5.6% and patients with PDD comprised 1.7% and the study reported only slightly more males than females in DLB (50.9%) and in PDD (57.9%) (Gan, Liu, Wang, et al., 2021). The literature to date largely attributes the sex differences found in PD to the protective effects of estrogen on dopaminergic pathways and sex chromosome determined dimorphism in the development of the dopaminergic system (Cholerton et al., 2018).

A study using neuropathologically-confirmed DLB cases in the National Alzheimer's Coordinating Center (NACC) database found that despite older age, lower education and more severe tau co-pathology, females with DLB were less likely to have a diagnosis of dementia (Bayram, Coughlin, Banks, & Litvan, 2021). This study also found that females were less likely to exhibit some of the core DLB features like hallucinations despite underlying LB pathology and were therefore commonly misdiagnosed with AD (Bayram et al., 2021). Similar to AD, wherein females bear a higher tau-burden before developing symptoms, females with PD appear to bear a higher LB-pathological burden before developing symptoms. Still, symptom severity can meet or surpass the level in males after a certain level of pathological burden (Bayram et al., 2021).

Despite most PD cases being diagnosed at an older age, there are still a small proportion of cases which are diagnosed before the age of 40 (Subramanian et al., 2022). For females diagnosed during ongoing menses, there are no specific treatment options tailored to fluctuating hormonal levels, even though in the literature there are reports that females experience a worsening of motor symptoms during the premenstrual phase (Quinn & Marsden, 1986). For females considering pregnancy after diagnosis, there are currently no evidence-based guidelines for the management of PD during pregnancy. L-dopa monotherapy is the most common treatment used during pregnancy and is not associated with an increased risk to the fetus but amantadine, for example, is a known teratogen (Seier & Hiller, 2017). Literature on maternal safety surrounding C-sections and obstetric anesthesiology for patients with PD is severely lacking (Subramanian et al., 2022).

In totality, these findings suggest the importance of age, sex and education-matched neuropathology-confirmed studies and the need for future sex-specific diagnostic criteria as these sex-related phenotypic differences are better-characterized (Bayram et al., 2021).

3.2 Biomarkers

PD, PDD and DLB are all associated with nigrostriatal dopaminergic neuron loss, and reduced dopamine transporter (DaT) uptake in basal ganglia can be detected using single-photon emission computed tomography (SPECT) or positron emission tomography (PET) imaging with a ligand binding to the presynaptic DaT (Outeiro et al., 2019). For example, I-ioflupane SPECT, also known as DaTscan (GE Healthcare), is approved for routine clinical use and can differentiate PD, PDD, and DLB from mimics such as AD (Outeiro et al., 2019; Poewe et al., 2017). Previous studies have shown higher DaT availability in females versus males. This is thought to be attributed to either higher density in the female striatum or differences in overall striatal volume between sexes (Varrone, Dickson, Tossici-Bolt, et al., 2013). However there is no evidence to suggest that correction for sex is necessary or beneficial for the diagnostic performance of I-ioflupane SPECT (Schmitz-Steinkrüger, Lange, Apostolova, et al., 2021).

There is very little published on sex differences in structural brain metrics relating to PD (Salminen et al., 2022). In the limited literature available, expected sex differences are often regressed out or confounded by antipsychotic medications that may affect cortical thickness (Tremblay, Abbasi, Zeighami, et al., 2020). The Parkinson's Progression Markers Initiative (PPMI) study dataset, using deformation-based morphometry (DBM) methodology and diffusion imaging analysis, revealed that males had greater atrophy and greater disruption in white matter network efficiency than females, despite having similar clinical disease severity and duration (Tremblay et al., 2020). Specifically, males with PD had significantly decreased local network efficiency compared to females in 111 regions, including frontal (including bilateral inferior and middle frontal gyrus), temporal, cingulate, insular and parietal cortices, in addition to subcortical regions including thalamus, basal ganglia (putamen, caudate and nucleus accumbens), hippocampus and amygdala (Tremblay et al., 2020). The findings of this study lead the authors to recommend future routine stratification of PD data by sex, and analysis of sex interactions, as has been similarly recommended for future AD studies (Tremblay et al., 2020).

Polysomnography (PSG) measures eye and muscle movements during sleep, identifying RBD and allowing for a high-diagnostic specificity of 98% for Lewy-related pathology (PD, DLB, and MSA) (McKeith et al., 2017; Outeiro et al., 2019). RBD is useful for diagnosing prodromal DLB because of how effective it is at predicting future Lewy body disorder development (Fujishiro, Ota, Yamagata, et al., 2021). Idiopathic RBD is widely accepted as being male-predominant, and because males commonly have complaints more compatible with RBD, such as a higher occurrence of sleep-related injury, some studies discuss a possible male referral bias for PSG (Bjørnarå, Dietrichs, & Toft, 2013; Bugalho & Salavisa, 2019). RBD does present differently across the sexes; males have higher muscular phasic activity than females and a higher prevalence of myoclonic and trunk events, whereas females have a higher prevalence of segmental motor events (Bugalho & Salavisa, 2019). These differences could translate into referral bias for males and thus diagnostic delays in females. There are very few studies on the clinical expression of RBD in females with LBD, one study on a small cohort of female PD patients found that dream enactment was typically less violent, however, females did have vivid action-filled dreams with vocalization and fewer mean hours of sleep each night possibly contributing to increased depression and anxiety (Mahale, Yadav, & Pal PKr., 2016).

Sympathetic cardiac metaiodobenzylguanidine (MIBG) scintigraphy is used to quantify postganglionic sympathetic cardiac innervation and can differentiate PD and DLB (decreased binding) from MSA and PSP (normal binding) (Bloem, Okun, & Klein, 2021; McKeith et al., 2017). MIBG uptake is determined by calculating the heart to mediastinum-uptake (H/M) ratio (Slaets, van Acker, Versijpt, et al., 2015). A study evaluating the use of MIBG cardiac scintigraphy to diagnose DLB in clinical practice found the H/M ratio was significantly lower in males and recommended sex and age-corrected H/M ratios to increase the diagnostic accuracy of MIBG cardiac scintigraphy (Slaets et al., 2015).

There are currently no established biomarkers of alpha-synuclein, or any CSF, blood, or peripheral tissue biomarkers clinically useful as diagnostic tests for PD, PDD or DLB (McKeith et al., 2017; Poewe et al., 2017). A cross-sectional study comparing total plasma alpha-synuclein found that in advanced PD stages, levels decreased only in males, and not in females (Caranci, Piscopo, Rivabene, et al., 2013). Additionally, lower plasma alpha-synuclein concentrations were significantly associated with cognitive impairment in males but not in females (Caranci et al., 2013).

Uric acid is another potential biomarker under investigation, which shows strong sex specificity, with serum urate appearing to be an inverse risk factor for PD in males only (Meoni et al., 2020). Urate levels correlate negatively with disease severity in males with PD but not in females with PD (McFarland, Burdett, Desjardins, Frosch, & Schwarzschild, 2013). Reduced levels of urate were also only found in brain tissue of male patients with PD, not female PD patients, compared to healthy controls (McFarland et al., 2013).

3.3 Risk factors

Risk factors associated with cognitive impairment or dementia in the context of LBD are the presence of hallucinations, older age, the overall severity of motor symptoms, presence of speech impairment, axial impairment such as postural instability or gait difficulty, low level of education, presence of depression, and male sex (Aarsland, Batzu, Halliday, et al., 2021). The Sydney Multicenter study followed patients with PD for 20 years and found that patients developed dementia at around 70 years of age, regardless of their age of PD onset, and that the cognitive deficits were similar between individuals within this sample (Reid, Hely, Morris, Loy, & Halliday, 2011). This study does not report information on the sex ratio of the cohort or investigate any sex differences. Another longitudinal study reported that the evolution of dementia in PD is associated with age, non-tremor-dominant motor phenotype and deficits with a posterior cortical basis, such as semantic fluency and the ability to copy an intersecting pentagons figure (Williams-Gray, Foltynie, Brayne, Robbins, & Barker, 2007).

3.4 Treatment

Levodopa is the gold-standard therapy for PD; however, the pharmacokinetics and pharmacodynamics of this medication differ between sexes according to retrospective and prospective studies designed to measure drug outcomes (Kumagai et al., 2014). Females show higher levodopa plasma concentrations related to lower body weight and reduced rates of drug clearance, resulting in greater levodopa bioavailability (Nyholm, Karlsson, Lundberg, & Askmark, 2010). This finding could be linked to levodopa-related complications in females with PD, such as higher rates of dyskinesia and greater severity of motor fluctuations (Warren Olanow, Kieburtz, Rascol, et al., 2013). Levodopa is also used to treat symptoms of parkinsonism in DLB, however, the doses are typically limited because Levodopa can cause agitation and worsening of visual hallucinations (Hershey & Coleman-Jackson, 2019; Panza, Lozupone, Watling, & Imbimbo, 2021).

Deep-brain stimulation (DBS) of the subthalamic nucleus is another well-established treatment reserved for advanced PD (Meoni et al., 2020). Females tend to show greater health-related quality of life improvements following DBS, yet some studies have found that roughly twice as many males undergo surgery compared to females (Hariz, Limousin, Zrinzo, et al., 2013).

Cholinesterase inhibitors (CHEIs), such as rivastigmine and donepezil, are used to manage cognitive symptoms in LBD (Aarsland et al., 2021; McKeith et al., 2017). Meta-analysis of Class I clinical trials supports the use of CHEIs for improving cognition, global function and activities of daily living; even if patients do not improve while on medication, they are less likely to deteriorate further (McKeith et al., 2017). CHEIs are also useful for treating neuropsychiatric symptoms and may reduce apathy and improve hallucinations and delusions (McKeith et al., 2017). The benefits of memantine, a N-methyl-D-aspartate (NMDA) receptor antagonist, for DLB and PDD are less clear; however, it is well-tolerated (McKeith et al., 2017). The cholinergic system is sexually dimorphic, it is anatomically larger in males with higher cell density, and it is more stable with aging than in females (Rhodes & Rubin, 1999). Despite this, surprisingly few randomized clinical trials of cholinesterase inhibitors have investigated sex differences in safety and efficacy (Ferretti et al., 2018). Whether sex plays a role in drug responses or adverse events remains to be studied (Ferretti et al., 2018).

3.5 Conclusion

Despite Parkinson's disease being the most studied movement disorder, literature investigating sex differences is scarce and mainly consists of cross-sectional studies (Georgiev, Hamberg, Hariz, Forsgren, & Hariz, 2017). Much remains to be understood on how sex and gender differences relate to LBDs with respect to phenotype, diagnostics, risk factors and treatment.

4. Sex and gender differences in frontotemporal dementia

4.1 Overview

Frontotemporal dementia describes a group of clinical syndromes united by underlying frontotemporal lobar degeneration (FTLD) pathology. The clinical syndromes associated with FTLD are heterogeneous and are based on whether the patients present with behavioral, language and/or motor impairments. The FTLD-related syndromes determined by the regional distribution of pathology are: (i) behavioral variant FTD (bvFTD), characterized by changes in personality and behavior, and neurodegeneration primarily in the frontal and temporal lobes; (ii) Semantic variant primary progressive aphasia (svPPA), characterized by gradual loss of semantic knowledge and is associated with left asymmetric anterior and inferior temporal lobe degeneration; (iii) non-fluent variant primary progressive aphasia (nfvPPA), characterized by effortful, non-fluent speech with atrophy predominantly in the left inferior frontal cortex; (iv) FTD-motor neuron disease (FTD-MND or FTD-amyotrophic lateral sclerosis (FTD-ALS)), is concomitant motor neuron loss resulting in muscle atrophy and weakness with behavioral and/or cognitive deficits; (v) Progressive supranuclear palsy

(PSP), which in its typical forms displays vertical gaze palsy, parkinsonian symptoms and falls; and (vi) Corticobasal syndrome (CBS) that displays progressive, asymmetric rigidity, limb apraxia and cognitive impairment. The two main pathological substrates of FTLD are tau and TAR DNA-binding protein 43 (TDP-43) with approximately 5% due to fused in Sarcoma (FUS).

Sex and gender differences have been increasingly investigated across FTLD-related syndromes. Given that little literature exists that specifically examines differences in gender rather than sex in FTLD patients, here, the language used in the original study will be maintained (i.e., male/female or men/women used in the original studies).

4.2 Frequency

FTLD has an estimated prevalence of 15–22 cases per 100,000 people (Onyike & Diehl-Schmid, 2013), accounting for approximately 2.7% of all dementia cases in patients greater than 65, and 10.2% in patients younger than 65 (Hogan, Jetté, Fiest, et al., 2016). Studies investigating sex and gender differences in frequency of FTLD-related syndromes have shown heterogeneity in their findings. Some studies have reported no sex differences in frequency while others have reported significant differences (Gustafson, 1993; Hogan, Jetté, et al., 2016; Johnson, Diehl, Mendez, et al., 2005; Llorca-bof, Ill, Gonz, et al., 2019; Stevens, van Duijn, Kamphorst, et al., 1998). Furthermore, when examining sex differences in relative frequency of genetic FTD cases, some sex differences have been observed, both with respect to frequency, as well as patient characteristics (discussed below).

One early post-mortem study of 30 non-AD related frontal lobe degeneration patients reported equal sex distribution (Gustafson, 1993), whereas another early study of FTLD patients (in the Dutch population) reported slight differences in male:female ratios in sporadic FTLD in comparison to healthy controls in their sample (M:F ratio being 3:5 in FTLD vs 7:10 in healthy controls) (Stevens et al., 1998). In a systematic review of 49 sporadic FTD case reports in China, no sex or gender differences were observed in patients with bvFTD, nfvPPA or svPPA (Ren, Huang, Xu, et al., 2012). When examining the prevalence of FTD in early-onset cases (age of onset < 65 years), a four times higher prevalence in males in comparison to females was observed in one study (Ratnavalli, Brayne, Dawson, & Hodges, 2002). A large study of 353 FTLD patients reported a significantly higher proportion of men diagnosed with bvFTD and svPPA in comparison to nfvPPA, where there was an overrepresentation of women (Johnson et al., 2005). The authors of this study argue that the discrepancy in reports on sex differences in the literature prior to this study is likely due to small sample sizes in previous studies. Furthermore, a large study of 531 FTD patients conducted in Italy has reported a higher frequency (percentage of bvFTD vs svPPA) of bvFTD in males (74%), and a higher frequency of svPPA in females (57%) (Pengo, Alberici, Libri, et al., 2022).

Finally, two studies that included FTD phenocopies (FTD cases with no or slow decline over at least 3 years after symptom onset) suggested that there is a higher prevalence of males with phenocopy FTD as Garcin, Lillo, Hornberger, et al. (2009) reported that men made up the vast majority (92%) of phenocopy FTD cases. This finding has been corroborated in a more recent systematic review on phenocopy bvFTD (Valente, Caramelli, Gambogi, et al., 2019), which has shown that phenocopy bvFTD patients are predominantly males, with no significant cognitive deficits and preserved executive function and episodic memory.

4.3 Biofluid and genetic biomarkers

Diagnostic biomarkers for FTD are still lacking as to date diagnosis is still largely based on clinical criteria, with image supported diagnosis now possible with bvFTD, svPPA and nfvPPA (Gorno-Tempini, Hillis, Weintraub, et al., 2011; Rascovsky, Hodges, Knopman, et al., 2011). Though multiple studies have been conducted investigating sex or gender differences, there remains a gap in the literature regarding sex or gender differences in biofluid biomarkers for FTLD. One large systematic review investigating biofluid-based biomarkers in FTLD (del Campo, Zetterberg, Gandy, et al., 2022), has reported a lack of research investigating gender and sex differences in biofluid-based biomarkers for FTLD, including progranulin, tau and NFL.

One large meta-analysis published in 2017 by Curtis, Masellis, Hsiung, et al. (2017) examined sex differences in the prevalence of genetic mutations in FTD and ALS. No sex differences were observed in the prevalence of C9orf72 (C9 open reading frame 72) or MAPT (microtubule associated protein tau) mutation carriers in FTD. However, for GRN (granulin) mutation carriers, there was a significantly higher (33%) prevalence in female patients. These authors suggest that given that GRN expression during embryogenesis is involved in sex differentiation in the brain, it is possible that differences in sex hormones during embryogenesis or potentially menopause can alter the risk for developing GRN mutation-related diseases. A second study examining age at symptom onset and death in genetic FTD was published in 2020 by Moore et al. (Moore, Nicholas, Grossman, et al., 2020). No differences in sex distribution were observed in C9orf72 and MAPT mutation carriers, however there was a significant overrepresentation of women in the GRN mutation carriers, similar to the above study. With respect to both age of onset as well as age at death, women GRN and C9orf72-related mutation carriers were significantly older, and no differences were observed in MAPT-mutation carriers. This study reported no sex differences in disease duration regardless of genetic group.

When comparing genetic and sporadic FTLD cases in a sample of 910 patients, de Boer, Riedl, van der Lee, et al. (2021) observed differences in sex distribution between sporadic and genetic bvFTD, whereby sporadic bvFTD was associated with a higher proportion of male cases, and genetic bvFTD was associated with an equal sex distribution. The authors suggest that the increased prevalence of males with sporadic bvFTD could be associated with risk factors including head trauma. With respect to age of onset, genetic FTD cases showed a significantly later age of symptom onset in females (median difference of 2.5 years), which although only a trend, was also apparent in sporadic cases of bvFTD, nfvPPA and svPPA (median difference of 3 years). In contrast, a rare UBQLN2 Mutation (ChrXp11.21) that causes dominant X-linked ALS and ALS-Dementia group, females had an earlier age of disease onset (Couratier, Corcia, Lautrette, Nicol, & Marin, 2017).

MicroRNAs (miRNA) are single-stranded non-protein coding RNA molecules that are associated with the inhibition or degradation of coding mRNA molecules (Bagga, Bracht, Hunter, et al., 2005; He & Hannon, 2004), and have been demonstrated as dysregulated in numerous neurodegenerative diseases including Parkinson's Disease, Alzheimer's Disease and ALS. MicroRNAs have been proposed as a peripheral biomarker for neurodegenerative disease. A recent review paper by Piscopo, Bellenghi, Manzini, et al. (2021) found that a combined score of three identified miRNAs when applying a Bayesian approach showed a sensitivity of 100% and specificity of 87.5% in distinguishing all FTD patients from healthy controls. Furthermore, FTD patients showed sex differences in miR-206 expression between male and female FTD patients, such that males with FTD had significantly different miR-206 expression in comparison to controls while females showed miR-206 levels similar to controls.

4.4 Neuroimaging biomarkers

Multiple neuroimaging studies conducted in FTLD populations have examined differences in structural MRI-related metrics, including cortical thickness, atrophy as well as post-mortem pathological protein deposition.

A recent study (Illán-Gala, Casaletto, Borrego-Écija, et al., 2021), suggested that women with bvFTD may have a higher brain reserve in comparison to men with bvFTD. In this study, both men and women with bvFTD showed reduced cortical thickness in frontotemporal regions in comparison to healthy controls of the same sex, however a stronger effect size was seen in women. Furthermore, in comparison to men with bvFTD, women with bvFTD showed reduced cortical thickness in several frontotemporal regions of both hemispheres. A significant group (bvFTD or healthy control) by sex interaction was observed in frontotemporal regions, suggesting that women with bvFTD had greater reduction of frontotempral grey matter at diagnosis, when accounting for age, education, MMSE and brain size. These authors suggest that these findings are indirect evidence of a higher brain reserve in women with bvFTD, since for a given amount of atrophy, women with bvFTD showed higher than expected scores on executive function and lower than expected apathy, sleep, and appetite behavior changes. Furthermore, cortical asymmetry has been observed in bvFTD patients, where women show greater left frontal cortex atrophy, and men show greater right frontal and/or bilateral temporal cortex atrophy (Johnson et al., 2005). These authors suggest that these sex-specific cortical asymmetry may reflect different sex-specific vulnerabilities to neurodegeneration. Finally, with respect to pathological protein deposition, one study reported no sex differences in terms of protein deposition as measured by post-mortem MRI or disease severity in FTLD syndromes (Whitwell, Jack, Senjem, et al., 2009).

4.5 Neuropsychological markers/clinical progression/disease onset

One recent study of 531 FTD patients conducted in Italy found sex differences in specific cognitive deficits. Here, females showed significantly greater deficits in language tasks, as well as executive, visual attention and visuospatial tasks (Pengo et al., 2022). Males exhibited more severe personality and behavioral symptoms, including apathy, irritability, poor judgement, aggressivity and hypersexuality. These authors suggest that the sex difference in specific deficits as well as distribution of diagnosis (as discussed above) may be indicative of sex-specific vulnerabilities to FTD.

Multiple studies have investigated sex differences in the misdiagnosis of FTLD as a psychiatric disorder. A systematic retrospective chart review (Woolley, Khan, Murthy, Miller, & Rankin, 2011) of sporadic FTLD and AD revealed a 2:1 ratio of men:women with a higher incidence of psychiatric misdiagnosis in women with bvFTD compared to men with bvFTD. These findings were corroborated by a large-scale study conducted in Barcelona that reported a higher proportion of men diagnosed with bvFTD in comparison to motor syndromes (PSP, CBS or FTD-MND) and language variants of FTD (svPPA, nfvPPA) (Llorca-bof et al., 2019). In this study, female patients also had a lower MMSE score at diagnosis and a higher incidence of depression compared to men in the same group. These authors suggest that women with bvFTD get diagnosed at more advanced stages because of misdiagnosis of a psychiatric illness. With respect to misdiagnosis of FTLD as a psychiatric disorder, one small retrospective chart review study of 28 patients (Zapata-Restrepo, Rivas, Miranda, et al., 2021) in a Columbian sample of patients diagnosed with probable bvFTD had reported not only a high rate of misdiagnosis at initial consultation (71%), but also differences in misdiagnoses by gender. Here, men were most likely to be diagnosed with anxiety or a psychotic disorder (33% and 22% respectively), whereas women were more likely to be diagnosed with depression or bipolar disorder (26% and 26%). Finally, with respect to disease progression, several studies (Brodaty, Seeher, & Gibson, 2012; Garcin et al., 2009) and one meta-analysis (Kansal, Mareddy, Sloane, et al., 2016) have shown no sex differences in time to death or years of life lost in FTD patients.

4.6 Risk factors (focus on both biological and SES)

With respect to sex-specific risk factors for FTD, one study (Dye, Miller, Singer, & Levine, 2012) has suggested previous hormone replacement therapy (HRT) as a risk factor for FTD. In this sample, 70% of women diagnosed with FTD had previously taken HRT, in comparison to approximately 24% in the surrounding population. These authors suggest the link between HRT and FTD risk, may be related to upregulation of tau in response to estradiol (E2) administration, which has been shown in previous research (Ferreira & Caceres, 1991). Finally, and in contrast, an estrogenic model of delayed onset for FTLD has been suggested (Flaherty, Kraft, Brothers, et al., 2017), where the inhibitory activity of estrogens on microglia-associated neuroinflammation may be associated with the delayed onset and progression of early onset neurodegenerative diseases such as FTLD, with replacement of hormones with HRT potentially offsetting menopause-associated declines.

In summary, there appear to be many unanswered questions with respect to the effect of sex and gender on FTLD and thus ongoing research into demographics, pathogenesis, genetics, biomarkers and co-morbidity factors should factor in a sex and gender analysis.

Key take home messages

- Sex differences have been described in frequency, incidence, prevalence of the most common causes of dementia, including sporadic and genetic forms.
- Sex differences also exist in biomarkers and clinical symptoms, calling for precision medicine approaches.
- Risk factors to dementia are modulated by both sex and gender; tailored preventative campaigns might be needed to account for them.
- It is important to account for and report sex stratified data, to avoid publication bias and better understand the individual needs of people along their patient journey.

References

2019 Alzheimer's Disease Facts and Figures. (2019). *Includes a special report on Alzheimer's detection in the primary care setting: Connecting patients and physicians.*

Aarsland, D., Batzu, L., Halliday, G. M., et al. (2021). Parkinson disease-associated cognitive impairment. *Nature Reviews Disease Primers*, 7(1), 47. https://doi.org/10.1038/s41572-021-00280-3.

Ahangar, A. A., Saadat, P., Heidari, B., Taheri, S. T., & Alijanpour, S. (2018). Sex difference in types and distribution of risk factors in ischemic and hemorrhagic stroke. *International Journal of Stroke*, *13*(1), 83–86. https://doi.org/10.1177/1747493017724626.

Akhter, F., Persaud, A., Zaokari, Y., Zhao, Z., & Zhu, D. (2021). Vascular dementia and underlying sex differences. *Frontiers in Aging Neuroscience*, *13*, 583. https://doi.org/10.3389/FNAGI.2021.720715/BIBTEX.

Alqarni, A., Jiang, J., Crawford, J. D., et al. (2021). Sex differences in risk factors for white matter hyperintensities in non-demented older individuals. *Neurobiology of Aging*, *98*, 197–204. https://doi.org/10.1016/J.NEUROBIOLAGING.2020.11.001.

Altmann, A., Tian, L., Henderson, V. W., & Greicius, M. D. (2014). Sex modifies the APOE-related risk of developing Alzheimer disease. *Annals of Neurology*, *75*(4), 563–573. https://doi.org/10.1002/ANA.24135.

Andersen, K., Launer, L. J., Dewey, M. E., et al. (1999). Gender differences in the incidence of AD and vascular dementia: The EURODEM Studies. EURODEM Incidence Research Group. *Neurology, 53*(9), 1992–1997. https://doi.org/10.1212/WNL.53.9.1992.

Apostolova, L. G., Dinov, I. D., Dutton, R. A., et al. (2006). 3D comparison of hippocampal atrophy in amnestic mild cognitive impairment and Alzheimer's disease. *Brain, 129*(11), 2867–2873. https://doi.org/10.1093/BRAIN/AWL274.

Appelros, P., Stegmayr, B., & Terent, A. (2009). Sex differences in stroke epidemiology: A systematic review. *Stroke, 40*(4), 1082–1090. https://doi.org/10.1161/STROKEAHA.108.540781.

Aukes, A. M., de Groot, J. C., Aarnoudse, J. G., & Zeeman, G. G. (2009). Brain lesions several years after eclampsia. *American Journal of Obstetrics & Gynecology, 200*(5), 504.e1–504.e5. https://doi.org/10.1016/J.AJOG.2008.12.033.

Azad, N. A., & al Bugami M, Loy-English I. (2007). Gender differences in dementia risk factors. *Gender Medicine, 4*(2), 120–129. https://doi.org/10.1016/S1550-8579(07)80026-X.

Bachman, D. L., Wolf, P. A., Linn, R. T., et al. (1993). Incidence of dementia and probable Alzheimer's disease in a general population: The Framingham Study. *Neurology, 43*(3 Pt. 1), 515–519. https://doi.org/10.1212/WNL.43.3_PART_1.515.

Bae, J. B., Lipnicki, D. M., Han, J. W., et al. (2020). Parity and the risk of incident dementia: A COSMIC study. *Epidemiology and Psychiatric Sciences, 29*, e176. https://doi.org/10.1017/S2045796020000876.

Bagga, S., Bracht, J., Hunter, S., et al. (2005). Regulation by let-7 and lin-4 miRNAs results in target mRNA degradation. *Cell, 122*(4), 553–563. https://doi.org/10.1016/j.cell.2005.07.031.

Barnes, L. L., Wilson, R. S., Bienias, J. L., Schneider, J. A., Evans, D. A., & Bennett, D. A. (2005). Sex differences in the clinical manifestations of Alzheimer disease pathology. *Archives of General Psychiatry, 62*(6), 685–691. https://doi.org/10.1001/ARCHPSYC.62.6.685.

Bayram, E., Coughlin, D. G., Banks, S. J., & Litvan, I. (2021). Sex differences for phenotype in pathologically defined dementia with Lewy bodies. *Journal of Neurology, Neurosurgery & Psychiatry, 92*(7), 745–750. https://doi.org/10.1136/jnnp-2020-325668.

Behl, P., Bocti, C., Swartz, R. H., et al. (2007). Strategic subcortical hyperintensities in cholinergic pathways and executive function decline in treated Alzheimer patients. *Archives in Neurology, 64*(2), 266–272. https://doi.org/10.1001/ARCHNEUR.64.2.266.

Bejanin, A., Iulita, M. F., Vilaplana, E., et al. (2021). Association of apolipoprotein E ε4 allele with clinical and multimodal biomarker changes of Alzheimer disease in adults with Down syndrome. *JAMA Neurology, 78*(8), 937–947. https://doi.org/10.1001/JAMANEUROL.2021.1893.

Benjamin, E. J., Virani, S. S., Callaway, C. W., et al. (2018). Heart disease and stroke statistics-2018 update: A Report From the American Heart Association. *Circulation, 137*(12), E67–E492. https://doi.org/10.1161/CIR.0000000000000558.

Benke, T., Delazer, M., Sanin, G., et al. (2013). Cognition, gender, and functional abilities in Alzheimer's disease: How are they related? *Journal of Alzheimer's Disease, 35*(2), 247–252. https://doi.org/10.3233/JAD-122383.

Besson, F. L., Joie, R.l., Doeuvre, L., et al. (2015). Neurobiology of disease cognitive and brain profiles associated with current neuroimaging biomarkers of preclinical alzheimer's disease. 10402. *The Journal of Neuroscience, 35*(29), 10402–10411. https://doi.org/10.1523/JNEUROSCI.0150-15.2015.

Bjørnarå, K. A., Dietrichs, E., & Toft, M. (2013). REM sleep behavior disorder in Parkinson's disease—Is there a gender difference? *Parkinsonism & Related Disorders, 19*(1), 120–122. https://doi.org/10.1016/j.parkreldis.2012.05.027.

Black, S. E., DeCarli, C., Greenberg, S. M., et al. (2011). *AHA/ASA Scientific Statement Vascular Contributions to Cognitive Impairment and Dementia A Statement for Healthcare Professionals From the American Heart Association/American Stroke Association Council on Epidemiology and Prevention, Council on Cardiovascular Nursing, Council on Cardiovascular Radiology and Intervention, and Council on Cardiovascular Surgery and Anesthesia*. https://doi.org/10.1161/STR.0b013e3182299496.

Black, S., Gao, F., & Bilbao, J. (2009). Understanding white matter disease: Imaging-pathological correlations in vascular cognitive impairment. *Stroke*, *40*(3 Suppl). https://doi.org/10.1161/STROKEAHA.108.537704.

Blair, L. J., Frauen, H. D., Zhang, B., et al. (2015). Tau depletion prevents progressive blood-brain barrier damage in a mouse model of tauopathy. *Acta Neuropathologica Communications*, *3*, 8. https://doi.org/10.1186/S40478-015-0186-2.

Bloem, B. R., Okun, M. S., & Klein, C. (2021). Parkinson's disease. *The Lancet*, *397*(10291), 2284–2303. https://doi.org/10.1016/S0140-6736(21)00218-X.

Bocti, C., Black, S., & Frank, C. (2007). Management of dementia with a cerebrovascular component. *Alzheimer's & Dementia*, *3*(4), 398–403. https://doi.org/10.1016/J.JALZ.2007.07.009.

Bocti, C., Swartz, R. H., Gao, F. Q., Sahlas, D. J., Behl, P., & Black, S. E. (2005). A new visual rating scale to assess strategic white matter hyperintensities within cholinergic pathways in dementia. *Stroke*, *36*(10), 2126–2131. https://doi.org/10.1161/01.STR.0000183615.07936.B6.

Bouter, C., Vogelgsang, J., & Wiltfang, J. (2019). Comparison between amyloid-PET and CSF amyloid-β biomarkers in a clinical cohort with memory deficits. *Clinica Chimica Acta: International Journal of Clinical Chemistry*, *492*, 62–68. https://doi.org/10.1016/J.CCA.2019.02.005.

Brier, M. R., Gordon, B., Friedrichsen, K., et al. (2016). Tau and Ab imaging, CSF measures, and cognition in Alzheimer's disease. *Science Translational Medicine*, *8*(338). https://doi.org/10.1126/SCITRANSLMED.AAF2362/SUPPL_FILE/8-338RA66_TABLE_S3.ZIP.

Brodaty, H., Seeher, K., & Gibson, L. (2012). Dementia time to death: A systematic literature review on survival time and years of life lost in people with dementia. *International Psychogeriatrics*, *24*(7), 1034–1045. https://doi.org/10.1017/S1041610211002924.

Buchhave, P., Minthon, L., Zetterberg, H., Wallin, Å. K., Blennow, K., & Hansson, O. (2012). Cerebrospinal fluid levels of β-amyloid 1-42, but not of tau, are fully changed already 5 to 10 years before the onset of Alzheimer dementia. *Archives of General Psychiatry*, *69*(1), 98–106. https://doi.org/10.1001/ARCHGENPSYCHIATRY.2011.155.

Buckley, R. F., Mormino, E. C., Chhatwal, J., et al. (2019). Associations between baseline amyloid, sex, and APOE on subsequent tau accumulation in cerebrospinal fluid. *Neurobiology Aging*, *78*, 178–185. https://doi.org/10.1016/J.NEUROBIOLAGING.2019.02.019.

Buckley, R. F., Mormino, E. C., Rabin, J. S., et al. (2019). Sex differences in the association of global amyloid and regional tau deposition measured by positron emission tomography in clinically normal older adults. *JAMA Neurology*, *76*(5), 542–551. https://doi.org/10.1001/JAMANEUROL.2018.4693.

Buerger, K., Ewers, M., Pirttilä, T., et al. (2006). CSF phosphorylated tau protein correlates with neocortical neurofibrillary pathology in Alzheimer's disease. *Brain*, *129*(11), 3035–3041. https://doi.org/10.1093/brain/awl269.

Bugalho, P., & Salavisa, M. (2019). Factors influencing the presentation of REM sleep behavior disorder: The relative importance of sex, associated neurological disorder, and context of referral to polysomnography. *Journal of Clinical Sleep Medicine*, *15*(12), 1789–1798. https://doi.org/10.5664/JCSM.8086.

Bushnell, C. D., Chaturvedi, S., Gage, K. R., et al. (2018). Sex differences in stroke: Challenges and opportunities. *Journal of Cerebral Blood Flow & Metabolism, 38*(12), 2179. https://doi.org/10.1177/0271678X18793324.

Cacioppo, J. T., Hawkley, L. C., & Thisted, R. A. (2010). Perceived social isolation makes me sad: 5-year cross-lagged analyses of loneliness and depressive symptomatology in the Chicago Health, Aging, and Social Relations Study. *Psychology and Aging, 25*(2), 453–463. https://doi.org/10.1037/A0017216.

Canevelli, M., Quarata, F., Remiddi, F., et al. (2017). Sex and gender differences in the treatment of Alzheimer's disease: A systematic review of randomized controlled trials. *Pharmacological Research, 115*, 218–223. https://doi.org/10.1016/j.phrs.2016.11.035.

Caranci, G., Piscopo, P., Rivabene, R., et al. (2013). Gender differences in Parkinson's disease: Focus on plasma alpha-synuclein. *Journal of Neural Transmission, 120*(8), 1209–1215. https://doi.org/10.1007/s00702-013-0972-6.

Cereda, E., Cilia, R., Klersy, C., et al. (2016). Dementia in Parkinson's disease: Is male gender a risk factor? *Parkinsonism & Related Disorders, 26*, 67–72. https://doi.org/10.1016/j.parkreldis.2016.02.024.

Chatterjee, S., Peters, S. A. E., Woodward, M., et al. (2016). Type 2 diabetes as a risk factor for dementia in women compared with men: A pooled analysis of 2.3 million people comprising more than 100,000 cases of dementia. *Diabetes Care, 39*(2), 300–307. https://doi.org/10.2337/DC15-1588.

Chêne, G., Beiser, A., Au, R., et al. (2015). Gender and incidence of dementia in the Framingham Heart Study from mid-adult life. *Alzheimer's & Dementia, 11*(3), 310–320. https://doi.org/10.1016/J.JALZ.2013.10.005.

Cholerton, B., Johnson, C. O., Fish, B., et al. (2018). Sex differences in progression to mild cognitive impairment and dementia in Parkinson's disease. *Parkinsonism & Related Disorders, 50*, 29–36. https://doi.org/10.1016/j.parkreldis.2018.02.007.

Chui, H. C., & Ramirez-Gomez, L. (2015). Clinical and imaging features of mixed Alzheimer and vascular pathologies. *Alzheimer's Research & Therapy*, 7(1). https://doi.org/10.1186/S13195-015-0104-7.

Cignarella, A., Kratz, M., & Bolego, C. (2010). Emerging role of estrogen in the control of cardiometabolic disease. *Trends Pharmacological Science, 31*(4), 183–189. https://doi.org/10.1016/J.TIPS.2010.01.001.

Cohen, D., Eisdorfer, C., Gorelick, P., et al. (1993). Sex differences in the psychiatric manifestations of Alzheimer's disease. *Journal of the American Geriatric Society, 41*(3), 229–232. https://doi.org/10.1111/J.1532-5415.1993.TB06697.X.

Cook, I. A., Morgan, M. L., Dunkin, J. J., et al. (2002). Estrogen replacement therapy is associated with less progression of subclinical structural brain disease in normal elderly women: A pilot study. *International Journal of Geriatric Psychiatry*, 17(7), 610–618. https://doi.org/10.1002/GPS.644.

Copeland, J. R. M., Mccracken, C. F. M., Dewey, M. E., et al. (1999). Undifferentiated dementia, Alzheimer's disease and vascular dementia: Age- and gender-related incidence in Liverpool. The MRC-ALPHA Study. *British Journal of Psychiatry, 175*(Nov), 433–438. https://doi.org/10.1192/BJP.175.5.433.

Coppus, A. M. W., Evenhuis, H. M., Verberne, G. J., et al. (2010). Early age at menopause is associated with increased risk of dementia and mortality in women with Down syndrome. *Journal of Alzheimer's Disease, 19*(2), 545–550. https://doi.org/10.3233/JAD-2010-1247.

Corder, E. H., Saunders, A. M., Strittmatter, W. J., et al. (1993). Gene dose of apolipoprotein E type 4 allele and the risk of Alzheimer's disease in late onset families. *Science, 261*(5123), 921–923. https://doi.org/10.1126/SCIENCE.8346443.

Corrigan, J. D., Wolfe, M., Mysiw, W. J., Jackson, R. D., & Bogner, J. A. (2003). Early identification of mild traumatic brain injury in female victims of domestic violence. *American Journal of Obstetrics and Gynecology, 188*(5), S71–S76. https://doi.org/10.1067/MOB.2003.404.

Cosgrave, M. P., Tyrrell, J., McCarron, M., Gill, M., & Lawlor, B. A. (1999). Age at onset of dementia and age of menopause in women with Down's syndrome. *Journal of Intellectual Disability Research, 43*(Pt. 6), 461–465. https://doi.org/10.1046/J.1365-2788.1999.00192.X.

Couratier, P., Corcia, P., Lautrette, G., Nicol, M., & Marin, B. (2017). ALS and frontotemporal dementia belong to a common disease spectrum. *Revue Neurologique, 173*(5), 273–279. https://doi.org/10.1016/j.neurol.2017.04.001.

Cras, T. Y., Bos, D., Arfan Ikram, M., et al. (2020). Determinants of the presence and size of intracranial aneurysms in the general population: The Rotterdam Study. *Stroke, 51*(7), 2103–2110. https://doi.org/10.1161/STROKEAHA.120.029296.

Curtis, A. F., Masellis, M., Hsiung, G. Y. R., et al. (2017). Sex differences in the prevalence of genetic mutations in FTD and ALS. *Neurology, 89*(15), 1633–1642. https://doi.org/10.1212/WNL.0000000000004494.

de Boer, S. C. M., Riedl, L., van der Lee, S. J., et al. (2021). Differences in sex distribution between genetic and sporadic frontotemporal dementia. *Journal of Alzheimer's Disease, 84*(3), 1153–1161. https://doi.org/10.3233/JAD-210688.

de Leeuw, F. E., de Groot, J. C., Achten, E., et al. (2001). Prevalence of cerebral white matter lesions in elderly people: a population based magnetic resonance imaging study. The Rotterdam Scan Study. *Journal of Neurology, Neurosurgery, and Psychiatry, 70*(1), 9–14. https://doi.org/10.1136/JNNP.70.1.9.

DeCarli, C., Massaro, J., Harvey, D., et al. (2005). Measures of brain morphology and infarction in the framingham heart study: Establishing what is normal. *Neurobiology Aging, 26*(4), 491–510. https://doi.org/10.1016/J.NEUROBIOLAGING.2004.05.004.

del Campo, M., Zetterberg, H., Gandy, S., et al. (2022). New developments of biofluid-based biomarkers for routine diagnosis and disease trajectories in frontotemporal dementia. *Alzheimer's & Dementia, 1–16.* https://doi.org/10.1002/alz.12643.

di Carlo, A., Baldereschi, M., Amaducci, L., et al. (2002). Incidence of dementia, Alzheimer's disease, and vascular dementia in Italy. The ILSA Study. *Journal of American Geriatric Society, 50*(1), 41–48. https://doi.org/10.1046/J.1532-5415.2002.50006.X.

di Cesare, M., Bentham, J., Stevens, G. A., et al. (2016). Trends in adult body-mass index in 200 countries from 1975 to 2014: A pooled analysis of 1698 population-based measurement studies with 19·2 million participants. *The Lancet, 387*(10026), 1377–1396. https://doi.org/10.1016/S0140-6736(16)30054-X.

Dickerson, B. C., Bakkour, A., Salat, D. H., et al. (2009). The cortical signature of Alzheimer's disease: Regionally specific cortical thinning relates to symptom severity in very mild to mild AD dementia and is detectable in asymptomatic amyloid-positive individuals. *Cerebral Cortex, 19*(3), 497–510. https://doi.org/10.1093/CERCOR/BHN113.

Dong, M. J., Peng, B., Lin, X. T., Zhao, J., Zhou, Y. R., & Wang, R. H. (2007). The prevalence of dementia in the People's Republic of China: A systematic analysis of 1980-2004 studies. *Age and Ageing, 36*(6), 619–624. https://doi.org/10.1093/ageing/afm128.

Donovan, N. J., Amariglio, R. E., Zoller, A. S., et al. (2014). Subjective cognitive concerns and neuropsychiatric predictors of progression to the early clinical stages of Alzheimer disease. *The American Journal of Geriatric Psychiatry, 22*(12), 1642–1651. https://doi.org/10.1016/J.JAGP.2014.02.007.

Duara, R., Loewenstein, D. A., Greig, M. T., et al. (2011). Pre-MCI and MCI: Neuropsychological, clinical, and imaging features and progression rates. *The American Journal of Geriatric Psychiatry, 19*(11), 951–960. https://doi.org/10.1097/JGP.0B013E3182107C69.

Dubois, B., Feldman, H. H., Jacova, C., et al. (2007). Research criteria for the diagnosis of Alzheimer's disease: Revising the NINCDS-ADRDA criteria. *The Lancet. Neurology, 6*(8), 734–746. https://doi.org/10.1016/S1474-4422(07)70178-3.

Dubois, B., Feldman, H. H., Jacova, C., et al. (2014). *Position paper advancing research diagnostic criteria for Alzheimer's disease: The IWG-2 criteria. Vol. 13*. www.thelancet.com/neurology.

Dulicek, P., Ivanova, E., Kostal, M., et al. (2018). Analysis of risk factors of stroke and venous thromboembolism in females with oral contraceptives use. *Clinical and Applied Thrombosis/Hemostasis, 24*(5), 797–802. https://doi.org/10.1177/1076029617727857.

Dye, R. V., Miller, K. J., Singer, E. J., & Levine, A. J. (2012). Hormone replacement therapy and risk for neurodegenerative diseases. *International Journal of Alzheimer's Disease, 2012*. https://doi.org/10.1155/2012/258454.

Edland, S. D., Rocca, W. A., Petersen, R. C., Cha, R. H., & Kokmen, E. (2002). Dementia and Alzheimer disease incidence rates do not vary by sex in Rochester, Minn. *Archives of Neurology, 59*(10), 1589–1593. https://doi.org/10.1001/ARCHNEUR.59.10.1589.

Edwards, L., la Joie, R., Iaccarino, L., et al. (2021). Multimodal neuroimaging of sex differences in cognitively impaired patients on the Alzheimer's continuum: Greater tau-PET retention in females. *Neurobiology Aging, 105*, 86–98. https://doi.org/10.1016/J.NEUROBIOLAGING.2021.04.003.

Eikelboom, W. S., Pan, M., Ossenkoppele, R., et al. (2022). Sex differences in neuropsychiatric symptoms in Alzheimer's disease dementia: A meta-analysis. *Alzheimer's Research & Therapy, 14*(1), 48. https://doi.org/10.1186/S13195-022-00991-Z.

Emre, M., Aarsland, D., Brown, R., et al. (2007). Clinical diagnostic criteria for dementia associated with Parkinson's disease. *Movement Disorders, 22*(12), 1689–1707. https://doi.org/10.1002/mds.21507.

Erkinjuntti, T., Román, G., & Gauthier, S. (2004). Treatment of vascular dementia—Evidence from clinical trials with cholinesterase inhibitors. *Journal of Neurological Science, 226*(1-2), 63–66. https://doi.org/10.1016/J.JNS.2004.09.018.

Erol, A., & Karpyak, V. M. (2015). Sex and gender-related differences in alcohol use and its consequences: Contemporary knowledge and future research considerations. *Drug and Alcohol Dependence, 156*, 1–13. https://doi.org/10.1016/J.DRUGALCDEP.2015.08.023.

Fagan, A. M., Henson, R. L., Li, Y., et al. (2021). Comparison of CSF biomarkers in Down syndrome and autosomal dominant Alzheimer's disease: A cross-sectional study. *The Lancet Neurology, 20*(8), 615–626. https://doi.org/10.1016/S1474-4422(21)00139-3.

Fagan, A. M., Roe, C. M., Xiong, C., Mintun, M. A., Morris, J. C., & Holtzman, D. M. (2007). Cerebrospinal fluid tau/β-amyloid42 ratio as a prediction of cognitive decline in nondemented older adults. *Archives of Neurology, 64*(3), 343–349. https://doi.org/10.1001/ARCHNEUR.64.3.NOC60123.

Fan, C. C., Banks, S. J., Thompson, W. K., et al. (2020). Sex-dependent autosomal effects on clinical progression of Alzheimer's disease. *Brain, 143*(7), 2272–2280. https://doi.org/10.1093/BRAIN/AWAA164.

Fatemi, F., Kantarci, K., Graff-Radford, J., et al. (2018). Sex differences in cerebrovascular pathologies on FLAIR in cognitively unimpaired elderly. *Neurology, 90*(6), e466–e473. https://doi.org/10.1212/WNL.0000000000004913.

Ferreira, A., & Caceres, A. (1991). Estrogen-enhanced neurite growth: Evidence for a selective induction of tau and stable microtubules. *Journal of Neuroscience, 11*(2), 392–400. https://doi.org/10.1523/jneurosci.11-02-00392.1991.

Ferretti, M. T., Iulita, M. F., Cavedo, E., et al. (2018). Sex differences in Alzheimer disease—The gateway to precision medicine. *Nature Reviews Neurology, 14*(8), 457–469. https://doi.org/10.1038/s41582-018-0032-9.

Flaherty, C., Kraft, J., Brothers, A., et al. (2017). The relationship between oestrogen and executive functioning in ALS females with emerging frontotemporal lobar degeneration

(FTLD) supports a neuroendocrine model of FTLD attenuation. *Amyotrophic Lateral Sclerosis and Frontotemporal Degeneration*, *18*(1-2), 74–85. https://doi.org/10.1080/21678421.2016.1249487.

Fortea, J., Vilaplana, E., Carmona-Iragui, M., et al. (2020). Clinical and biomarker changes of Alzheimer's disease in adults with Down syndrome: A cross-sectional study. *Lancet*, *395*(10242), 1988–1997. https://doi.org/10.1016/S0140-6736(20)30689-9.

Fortea, J., Zaman, S. H., Hartley, S., Rafii, M. S., Head, E., & Carmona-Iragui, M. (2021). Alzheimer's disease associated with Down syndrome: A genetic form of dementia. *Lancet Neurology*, *20*(11), 930–942. https://doi.org/10.1016/S1474-4422(21)00245-3.

Fox, N. C., Crum, W. R., Scahill, R. I., Stevens, J. M., Janssen, J. C., & Rossor, M. N. (2001). Imaging of onset and progression of Alzheimer's disease with voxel-compression mapping of serial magnetic resonance images. *The Lancet*, *358*(9277), 201–205. https://doi.org/10.1016/S0140-6736(01)05408-3.

Frisoni, G. B., Testa, C., Sabattoli, F., Beltramello, A., Soininen, H., & Laakso, M. P. (2005). Structural correlates of early and late onset Alzheimer's disease: voxel based morphometric study. *Journal of Neurology, Neurosurgery, and Psychiatry*, *76*(1), 112–114. https://doi.org/10.1136/JNNP.2003.029876.

Fujishima, M., & Kiyohara, Y. (2002). Incidence and risk factors of dementia in a defined elderly Japanese population: The Hisayama study. *Annals of the New York Academy of Sciences*, *977*, 1–8. https://doi.org/10.1111/J.1749-6632.2002.TB04793.X.

Fujishiro, H., Ota, K., Yamagata, M., et al. (2021). Early diagnosis of prodromal dementia with Lewy bodies using clinical history of probable REM sleep behaviour disorder and cardiac 123 I-MIBG scintigraphy in memory clinics. *Psychogeriatrics*, *21*(3), 288–295. https://doi.org/10.1111/PSYG.12662.

Gale, S. D., Baxter, L., & Thompson, J. (2016). Greater memory impairment in dementing females than males relative to sex matched healthy controls. *Journal of Clinical and Experimental Neuropsychology*, *38*(5), 527–533. https://doi.org/10.1080/13803395.2015.1132298.

Gamberger, D., Lavrač, N., Srivatsa, S., Tanzi, R. E., & Doraiswamy, P. M. (2017). Identification of clusters of rapid and slow decliners among subjects at risk for Alzheimer's disease. *Scientific Reports*, 7(1), 1–12. https://doi.org/10.1038/s41598-017-06624-y.

Gan, J., Liu, S., Wang, X., et al. (2021). Clinical characteristics of Lewy body dementia in Chinese memory clinics. *BMC Neurology*, *21*(1), 144. https://doi.org/10.1186/s12883-021-02169-w.

Ganguli, M., Dodge, H. H., Chen, P., Belle, S., & DeKosky, S. T. (2000). Ten-year incidence of dementia in a rural elderly US community population: The MoVIES Project. *Neurology*, *54*(5), 1109–1116. https://doi.org/10.1212/WNL.54.5.1109.

Gannon, O. J., Robison, L. S., Custozzo, A. J., & Zuloaga, K. L. (2019). Sex differences in risk factors for vascular contributions to cognitive impairment & dementia. *Neurochemistry International*, *127*, 38–55. https://doi.org/10.1016/J.NEUINT.2018.11.014.

Garcin, B., Lillo, P., Hornberger, M., et al. (2009). Determinants of survival in behavioral variant frontotemporal dementia. *Neurology*, *73*(20), 1656–1661. https://doi.org/10.1212/WNL.0b013e3181c1dee7.

Gemmill, A., & Weiss, J. (2022). The relationship between fertility history and incident dementia in the U.S. Health and Retirement Study. *The Journal of Gerontology. Series B Psychological Sciences and Social Sciences*, 77(6), 1118–1131. https://doi.org/10.1093/GERONB/GBAB183.

Georgiev, D., Hamberg, K., Hariz, M., Forsgren, L., & Hariz, G. M. (2017). Gender differences in Parkinson's disease: A clinical perspective. *Acta Neurologica Scandinavica*, *136*(6), 570–584. https://doi.org/10.1111/ane.12796.

Gerstorf, D., Ram, N., Hoppmann, C., Willis, S. L., & Schaie, K. W. (2011). Cohort differences in cognitive aging and terminal decline in the Seattle longitudinal study. *Developmental Psychology*, *47*(4), 1026. https://doi.org/10.1037/A0023426.

Gillis, E. E., & Sullivan, J. C. (2016). Sex differences in hypertension: Recent advances. *Hypertension*, *68*(6), 1322–1327. https://doi.org/10.1161/HYPERTENSIONAHA.116.06602.

Gilsanz, P., Mayeda, E. R., Glymour, M. M., et al. (2017). Female sex, early-onset hypertension, and risk of dementia. *Neurology*, *89*(18), 1886–1893. https://doi.org/10.1212/WNL.0000000000004602.

Girouard, H., Lessard, A., Capone, C., Milner, T. A., & Iadecola, C. (2008). The neurovascular dysfunction induced by angiotensin II in the mouse neocortex is sexually dimorphic. *American Journal of Physiology. Heart and Circulatory Physiology*, *294*(1). https://doi.org/10.1152/AJPHEART.01137.2007.

Gorelick, P. B., Scuteri, A., Black, S. E., et al. (2011). Vascular contributions to cognitive impairment and dementia. *Stroke*, *42*(9), 2672–2713. https://doi.org/10.1161/STR.0B013E3182299496.

Gorno-Tempini, M. L., Hillis, A. E., Weintraub, S., et al. (2011). Classification of primary progressive aphasia and its variants. *Neurology*, *76*(11), 1006–1014. https://doi.org/10.1212/WNL.0b013e31821103e6.

Gottesman, R. F., Schneider, A. L. C., Zhou, Y., et al. (2016). The ARIC-PET amyloid imaging study: Brain amyloid differences by age, race, sex, and APOE. *Neurology*, *87*(5), 473. https://doi.org/10.1212/WNL.0000000000002914.

Greenberg, S. M., Tennis, M. K., Brown, L. B., et al. (2000). Donepezil therapy in clinical practice: A randomized crossover study. *Archives of Neurology*, *57*(1), 94–99. https://doi.org/10.1001/ARCHNEUR.57.1.94.

Gumus, M., Multani, N., Mack, M. L., & Tartaglia, M. C. (2021). Progression of neuropsychiatric symptoms in young-onset versus late-onset Alzheimer's disease. *Geroscience*, *43*(1), 213. https://doi.org/10.1007/S11357-020-00304-Y.

Gustafson, L. (1993). Clinical picture of frontal lobe degeneration of non-Alzheimer type. *Dementia*, *4*(3-4), 143–148. https://doi.org/10.1159/000107313.

Gustavsson, A., Norton, N., Fast, T., et al. (2022). Global estimates on the number of persons across the Alzheimer's disease continuum. *Alzheimer's & Dementia*, *2*. https://doi.org/10.1002/ALZ.12694.

Hachinski, V. (2015). Stroke and potentially preventable dementias proclamation: Updated world stroke day proclamation. *Stroke*, *46*(11), 3039–3040. https://doi.org/10.1161/STROKEAHA.115.011237.

Hachinski, V. C., Bowler, J. V., & Loeb, C. (1993). Vascular dementia. *Neurology*, *43*(10), 2159–2160. https://doi.org/10.1212/WNL.43.10.2159-A.

Hachinski, V., Iadecola, C., Petersen, R. C., et al. (2006). National Institute of Neurological Disorders and Stroke–Canadian Stroke Network Vascular Cognitive Impairment Harmonization Standards. *Stroke*, *37*(9), 2220–2241. https://doi.org/10.1161/01.STR.0000237236.88823.47.

Hachinski, V. C., Potter, P., & Merskey, H. (1987). Leuko-araiosis. *Archives of Neurology*, *44*(1), 21–23. https://doi.org/10.1001/archneur.1987.00520130013009.

Hachinski, V. C., Potter, P., & Merskey, H. (2022). *Leuko-Araiosis*. https://jamanetwork.com/.

Hariz, G. M., Limousin, P., Zrinzo, L., et al. (2013). Gender differences in quality of life following subthalamic stimulation for Parkinson's disease. *Acta Neurologica Scandinavica*, *128*(4), 281–285. https://doi.org/10.1111/ane.12127.

Hayden, K. M., Zandi, P. P., Lyketsos, C. G., et al. (2006). Vascular risk factors for incident Alzheimer disease and vascular dementia: The Cache County study. *Alzheimer Disease and Associated Disorders*, *20*(2), 93–100. https://doi.org/10.1097/01.WAD.0000213814.43047.86.

He, L., & Hannon, G. J. (2004). MicroRNAs: Small RNAs with a big role in gene regulation. *Nature Reviews Genetics*, *5*(7), 522–531. https://doi.org/10.1038/nrg1379.
Hebert, L. E., Scherr, P. A., McCann, J. J., Beckett, L. A., & Evans, D. A. (2001). Is the risk of developing Alzheimer's disease greater for women than for men? *American Journal of Epidemiology*, *153*(2), 132–136. https://doi.org/10.1093/AJE/153.2.132.
Hely, M. A., Reid, W. G. J., Adena, M. A., Halliday, G. M., & Morris, J. G. L. (2008). The Sydney multicenter study of Parkinson's disease: The inevitability of dementia at 20 years. *Movement Disorders*, *23*(6), 837–844. https://doi.org/10.1002/MDS.21956.
Hershey, L. A., & Coleman-Jackson, R. (2019). Pharmacological management of dementia with Lewy bodies. *Drugs & Aging*, *36*(4), 309–319. https://doi.org/10.1007/s40266-018-00636-7.
Hogan, D. B., Fiest, K. M., Roberts, J. I., et al. (2016). The prevalence and Incidence of dementia with Lewy bodies: A systematic review. *Canadian Journal of Neurological Sciences*, *43*(Suppl. 1 (S1)), S83–S95. https://doi.org/10.1017/CJN.2016.2.
Hogan, D. B., Jetté, N., Fiest, K. M., et al. (2016). The prevalence and incidence of frontotemporal dementia: A systematic review. *Canadian Journal of Neurological Sciences*, *43*(S1), S96–S109. https://doi.org/10.1017/cjn.2016.25.
Hohman, T. J., Dumitrescu, L., Barnes, L. L., et al. (2018). Sex-specific association of apolipoprotein E with cerebrospinal fluid levels of tau. *JAMA Neurology*, *75*(8), 989–998. https://doi.org/10.1001/JAMANEUROL.2018.0821.
Holland, D., Desikan, R. S., Dale, A. M., & Mcevoy, L. K. (2013). Higher rates of decline for women and apolipoprotein E 4 carriers. *American Journal of Neuroradiology*. https://doi.org/10.3174/ajnr.A3601.
Hörder, H., Johansson, L., Guo, X., et al. (2018). Midlife cardiovascular fitness and dementia: A 44-year longitudinal population study in women. *Neurology*, *90*(15), e1298–e1305. https://doi.org/10.1212/WNL.0000000000005290.
Huang, W. K., Liu, C. H., Pang, S. T., et al. (2020). Type of androgen deprivation therapy and risk of dementia among patients with prostate cancer in Taiwan. *JAMA Network Open*, *3*(8), e2015189. https://doi.org/10.1001/JAMANETWORKOPEN.2020.15189.
Huo, N., Vemuri, P., Graff-Radford, J., et al. (2022). Sex differences in the association between midlife cardiovascular conditions or risk factors with midlife cognitive decline. *Neurology*, *98*(6), e623–e632. https://doi.org/10.1212/WNL.0000000000013174.
Iadecola, C. (2013). The pathobiology of vascular dementia. *Neuron*, *80*(4), 844–866. https://doi.org/10.1016/J.NEURON.2013.10.008.
Illán-Gala, I., Casaletto, K. B., Borrego-Écija, S., et al. (2021). Sex differences in the behavioral variant of frontotemporal dementia: A new window to executive and behavioral reserve. *Alzheimer's & Dementia*, *17*(8), 1329–1341. https://doi.org/10.1002/alz.12299.
Imfeld, P., Brauchli Pernus, Y. B., Jick, S. S., & Meier, C. R. (2013). Epidemiology, co-morbidities, and medication use of patients with Alzheimer's disease or vascular dementia in the UK. *Journal of Alzheimer's Disease*, *35*(3), 565–573. https://doi.org/10.3233/JAD-121819.
Iulita, M. F., & Girouard, H. (2017). Treating hypertension to prevent cognitive decline and dementia: Re-opening the debate. *Advances in Experimental Medicine and Biology*, *956*, 447–473. https://doi.org/10.1007/5584_2016_98.
Jack, C. R., Bennett, D. A., Blennow, K., et al. (2018). NIA-AA research framework: Toward a biological definition of Alzheimer's disease. *Alzheimer's & Dementia*, *14*(4), 535–562. https://doi.org/10.1016/j.jalz.2018.02.018.
Jack, C. R., Knopman, D. S., Jagust, W. J., et al. (2013). Tracking pathophysiological processes in Alzheimer's disease: An updated hypothetical model of dynamic biomarkers. *The Lancet Neurology*, *12*(2), 207–216. https://doi.org/10.1016/S1474-4422(12)70291-0.

Jack, C. R., Wiste, H. J., Weigand, S. D., et al. (2015). Age, sex, and APOE ε4 effects on memory, brain structure, and β-amyloid across the adult life span. *JAMA Neurology, 72*(5), 511–519. https://doi.org/10.1001/JAMANEUROL.2014.4821.

Jacobs, D., Sano, M., Marder, K., et al. (1994). Age at onset of Alzheimer's disease: Relation to pattern of cognitive dysfunction and rate of decline. *Neurology, 44*(7), 1215–1220. https://doi.org/10.1212/WNL.44.7.1215.

Jansen, W. J., Ossenkoppele, R., Knol, D. L., et al. (2015). Prevalence of cerebral amyloid pathology in persons without dementia: A meta-analysis. *JAMA - Journal of the American Medical Association, 313*(19), 1924–1938. https://doi.org/10.1001/jama.2015.4668.

Jellinger, K. A. (2018). Dementia with Lewy bodies and Parkinson's disease-dementia: Current concepts and controversies. *Journal of Neural Transmission (Vienna), 125*(4), 615–650. https://doi.org/10.1007/S00702-017-1821-9.

Johnson, J. K., Diehl, J., Mendez, M. F., et al. (2005). Frontotemporal lobar degeneration. *Archives of Neurology, 62*(6), 87–114. https://doi.org/10.1001/archneur.62.6.925.

Kaiser, N. C., Liang, L. J., Melrose, R. J., Wilkins, S. S., Sultzer, D. L., & Mendez, M. F. (2014). Differences in anxiety among patients with early- versus late-onset Alzheimer's disease. *The Journal of Neuropsychiatry and Clinical Neurosciences, 26*(1), 73–80. https://doi.org/10.1176/APPI.NEUROPSYCH.12100240.

Kansal, K., Mareddy, M., Sloane, K. L., et al. (2016). Survival in frontotemporal dementia phenotypes: A meta-analysis. *Dementia and Geriatric Cognitive Disorders, 41*(1-2), 109–122. https://doi.org/10.1159/000443205.

Kautzky-Willer, A., Harreiter, J., & Pacini, G. (2016). Sex and gender differences in risk, pathophysiology and complications of type 2 diabetes mellitus. *Endocrine Reviews, 37*(3), 278–316. https://doi.org/10.1210/ER.2015-1137.

Kawas, C., & Zonderman, A. (2000). Age-specific incidence rates of Alzheimer's disease: The Baltimore longitudinal study of aging. *Neurology, 54*(11), 2072–2077. https://doi.org/10.1212/WNL.54.11.2072. Published online.

Kim, S., Kim, M. J., Kim, S., et al. (2015). Gender differences in risk factors for transition from mild cognitive impairment to Alzheimer's disease: A CREDOS study. *Comprehensive Psychiatry, 62*, 114–122. https://doi.org/10.1016/J.COMPPSYCH.2015.07.002.

Koran, M. E. I., Wagener, M., & Hohman, T. J. (2017). Sex differences in the association between AD biomarkers and cognitive decline. *Brain Imaging Behavior, 11*(1), 205–213. https://doi.org/10.1007/S11682-016-9523-8.

Korczyn, A. D., Vakhapova, V., & Grinberg, L. T. (2012). Vascular dementia. *Journal of Neurological Science, 322*(1–2), 2–10. https://doi.org/10.1016/J.JNS.2012.03.027.

Koss, E., Edland, S., Fillenbaum, G., et al. (1996). Clinical and neuropsychological differences between patients with earlier and later onset of Alzheimer's disease: A CERAD analysis, Part XII. *Neurology, 46*(1), 136–141. https://doi.org/10.1212/WNL.46.1.136.

Kukull, W. A., Higdon, R., Bowen, J. D., et al. (2002). Dementia and Alzheimer disease incidence: A prospective cohort study. *Archives of Neurology, 59*(11), 1737–1746. https://doi.org/10.1001/ARCHNEUR.59.11.1737.

Kumagai, T., Nagayama, H., Ota, T., Nishiyama, Y., Mishina, M., & Ueda, M. (2014). Sex differences in the pharmacokinetics of levodopa in elderly patients with Parkinson disease. *Clinical Neuropharmacology, 37*(6), 173–176. https://doi.org/10.1097/WNF.0000000000000051.

Lai, F., Kammann, E., Rebeck, G. W., Anderson, A., Chen, Y., & Nixon, R. A. (1999). APOE genotype and gender effects on Alzheimer disease in 100 adults with Down syndrome. *Neurology, 53*(2), 331–336. https://doi.org/10.1212/WNL.53.2.331.

Lai, F., Mhatre, P. G., Yang, Y., Wang, M. C., Schupf, N., & Rosas, H. D. (2020). Sex differences in risk of Alzheimer's disease in adults with Down syndrome. *Alzheimer's Dementia (Amsterdam, Netherlands), 12*(1). https://doi.org/10.1002/DAD2.12084.

Lambert, M. A., Bickel, H., Prince, M., et al. (2014). Estimating the burden of early onset dementia; systematic review of disease prevalence. *European Journal of Neurology*, *21*(4), 563–569. https://doi.org/10.1111/ENE.12325.

Lambert, J. C., Ibrahim-Verbaas, C. A., Harold, D., et al. (2013). Meta-analysis of 74,046 individuals identifies 11 new susceptibility loci for Alzheimer's disease. *Nature Genetics*, *45*(12), 1452. https://doi.org/10.1038/NG.2802.

Lang, B., Kindy, M. S., Kozel, F. A., Schultz, S. K., & Taheri, S. (2018). Multi-parametric classification of vascular cognitive impairment and dementia: The impact of diverse cerebrovascular injury biomarkers. *Journal of Alzheimer's Disease*, *62*, 39–60. https://doi.org/10.3233/JAD-170733.

Laws, K. R., Irvine, K., & Gale, T. M. (2016). Sex differences in cognitive impairment in Alzheimer's disease. *World Journal of Psychiatry*, *6*(1), 54. https://doi.org/10.5498/WJP.V6.I1.54.

Leening, M. J. G., Ferket, B. S., Steyerberg, E. W., et al. (2014). Sex differences in lifetime risk and first manifestation of cardiovascular disease: Prospective population based cohort study. *British Medical Journal*, *349*. https://doi.org/10.1136/BMJ.G5992.

Lehrner, J., Hagmann, M., Marksteiner, J., et al. (2015). Gender-specific differences in cognitive profiles of patients with Alzheimer's disease: Results of the Prospective Dementia Registry Austria (PRODEM-Austria). *Journal of Alzheimer's Disease*. https://doi.org/10.3233/JAD-150188.

Leszek, J., Mikhaylenko, E. V., Belousov, D. M., et al. (2021). The links between cardiovascular diseases and Alzheimer's disease. *Current Neuropharmacology*, *19*(2), 152–169. https://doi.org/10.2174/1570159X18666200729093724.

Levine, D. A., Gross, A. L., Briceño, E. M., et al. (2021). Sex differences in cognitive decline among US adults. *JAMA Network Open*, *4*(2). https://doi.org/10.1001/JAMANETWORKOPEN.2021.0169.

Li, G., Shofer, J. B., Petrie, E. C., et al. (2017). Cerebrospinal fluid biomarkers for Alzheimer's and vascular disease vary by age, gender, and APOE genotype in cognitively normal adults. *Alzheimer's Research & Therapy*, *9*(1). https://doi.org/10.1186/S13195-017-0271-9.

Li, R., & Singh, M. (2014). Sex differences in cognitive impairment and Alzheimer's disease. *Frontiers in Neuroendocrinology*, *35*(3), 385–403. https://doi.org/10.1016/J.YFRNE.2014.01.002.

Liesinger, A. M., Graff-Radford, N. R., Duara, R., et al. (2018). Sex and age interact to determine clinicopathologic differences in Alzheimer's disease. *Acta Neuropathology*, *136*(6), 873–885. https://doi.org/10.1007/S00401-018-1908-X.

Lima, R., Wofford, M., & Reckelhoff, J. F. (2012). Hypertension in postmenopausal women. *Current Hypertension Reports*, *14*(3), 254–260. https://doi.org/10.1007/S11906-012-0260-0.

Lin, K. A., Choudhury, K. R., Rathakrishnan, B. G., Marks, D. M., Petrella, J. R., & Doraiswamy, P. M. (2015). Marked gender differences in progression of mild cognitive impairment over 8 years. *Alzheimer's & Dementia: Translational Research & Clinical Interventions*, *1*(2), 103–110. https://doi.org/10.1016/J.TRCI.2015.07.001.

Litvan, I., Goldman, J. G., Tröster, A. I., et al. (2012). Diagnostic criteria for mild cognitive impairment in Parkinson's disease: Movement Disorder Society Task Force guidelines. *Movement Disorders*, *27*(3), 349–356. https://doi.org/10.1002/MDS.24893.

Liu, Y. Y., Hu, L., Ji, C., et al. (2009). Effects of hormone replacement therapy on magnetic resonance imaging of brain parenchyma hyperintensities in postmenopausal women. *Acta Pharmacologica Sinica*, *30*(7), 1065–1070. https://doi.org/10.1038/APS.2009.81.

Liu, C. K., Lai, C. L., Tai, C. T., Lin, R. T., Yen, Y. Y., & Howng, S. L. (1998). Incidence and subtypes of dementia in southern Taiwan: Impact of socio-demographic factors. *Neurology*, *50*(6), 1572–1579. https://doi.org/10.1212/WNL.50.6.1572.

Liu, G., Sun, B., Yu, L., et al. (2020). The gender-based differences in vulnerability to ambient air pollution and cerebrovascular disease mortality: Evidences based on 26781 deaths. *Global Heart*, *15*(1). https://doi.org/10.5334/GH.849.

Livingston, G., Huntley, J., Sommerlad, A., et al. (2020). The Lancet Commissions Dementia prevention, intervention, and care: 2020 report of the Lancet Commission The Lancet Commissions. *The Lancet*, *396*, 413–446. https://doi.org/10.1016/S0140-6736(20)30367-6.

Llorca-bof, V., Ill, I., Gonz, R. B., et al. (2019). Sex-related differences in the clinical diagnosis of frontotemporal dementia. *medRxiv*. Published online.

Lyketsos, C. G., Lopez, O., Jones, B., Fitzpatrick, A. L., Breitner, J., & Dekosky, S. (2002). Prevalence of neuropsychiatric symptoms in dementia and mild cognitive impairment: Results from the cardiovascular health study. *Journal of American Medical Association*, *288*(12), 1475–1483. https://doi.org/10.1001/JAMA.288.12.1475.

Maas, A. H. E. M., & Appelman, Y. E. A. (2010). Gender differences in coronary heart disease. *Netherlands Heart Journal*, *18*(12), 598–603. https://doi.org/10.1007/S12471-010-0841-Y.

Madsen, T. E., Khoury, J., Alwell, K., et al. (2017). Sex-specific stroke incidence over time in the Greater Cincinnati/Northern Kentucky Stroke Study. *Neurology*, *89*(10), 990–996. https://doi.org/10.1212/WNL.0000000000004325.

Maggioli, E., McArthur, S., Mauro, C., et al. (2016). Estrogen protects the blood-brain barrier from inflammation-induced disruption and increased lymphocyte trafficking. *Brain, Behavior, and Immunity*, *51*, 212–222. https://doi.org/10.1016/J.BBI.2015.08.020.

Mahale, R. R., Yadav, R., & Pal PKr. (2016). Rapid eye movement sleep behaviour disorder in women with Parkinson's disease is an underdiagnosed entity. *Journal of Clinical Neuroscience*, *28*, 43–46. https://doi.org/10.1016/j.jocn.2015.08.046.

MANN DMA. (1988). Alzheimer's disease and Down's syndrome. *Histopathology*, *13*(2), 125–137. https://doi.org/10.1111/J.1365-2559.1988.TB02018.X.

Martinkova, J., Quevenco, F. C., Karcher, H., et al. (2021). Proportion of women and reporting of outcomes by sex in clinical trials for Alzheimer disease: A systematic review and meta-analysis. *JAMA Network Open*, *4*(9), e2124124. https://doi.org/10.1001/JAMANETWORKOPEN.2021.24124.

Mattsson, N., Lönneborg, A., Boccardi, M., Blennow, K., & Hansson, O. (2017). Clinical validity of cerebrospinal fluid Aβ42, tau, and phospho-tau as biomarkers for Alzheimer's disease in the context of a structured 5-phase development framework. *Neurobiology of Aging*, *52*, 196–213. https://doi.org/10.1016/J.NEUROBIOLAGING.2016.02.034.

Mattsson, N., Zetterberg, H., Hansson, O., et al. (2009). CSF biomarkers and incipient Alzheimer disease in patients with mild cognitive impairment. *Journal of American Medical Association*, *302*(4), 385–393. https://doi.org/10.1001/JAMA.2009.1064.

McCarrey, A. C., An, Y., Kitner-Triolo, M. H., Ferrucci, L., & Resnick, S. M. (2016). Sex differences in cognitive trajectories in clinically normal older adults. *Psychology Aging*, *31*(2), 166. https://doi.org/10.1037/PAG0000070.

McFarland, N. R., Burdett, T., Desjardins, C. A., Frosch, M. P., & Schwarzschild, M. A. (2013). Postmortem brain levels of urate and precursors in Parkinson's disease and related disorders. *Neurodegenerative Diseases*, *12*(4), 189–198. https://doi.org/10.1159/000346370.

McKeith, I. G., Boeve, B. F., Dickson, D. W., et al. (2017). Diagnosis and management of dementia with Lewy bodies: Fourth consensus report of the DLB Consortium. *Neurology*, *89*(1), 88–100. https://doi.org/10.1212/WNL.0000000000004058.

McKhann, G., Drachman, D., Folstein, M., Katzman, R., Price, D., & Stadlan, E. M. (1984). Clinical diagnosis of Alzheimer's disease. *Neurology*, *34*(7), 939. https://doi.org/10.1212/WNL.34.7.939.

Mehta, N., Rodrigues, C., Lamba, M., et al. (2017). Systematic review of sex-specific reporting of data: Cholinesterase inhibitor example. *Journal of the American Geriatric Society*, *65*(10), 2213–2219. https://doi.org/10.1111/JGS.15020.

Meoni, S., Macerollo, A., & Moro, E. (2020). Sex differences in movement disorders. *Nature Reviews. Neurology*, *16*(2), 84–96. https://doi.org/10.1038/s41582-019-0294-x.

Mhatre, P. G., Lee, J. H., Pang, D., et al. (2021). The association between sex and risk of Alzheimer's disease in adults with Down syndrome. *Journal of Clinical Medicine*, *10*(13). https://doi.org/10.3390/JCM10132966.

Mielke, M. M., Teresa Ferretti, M., Florencia Iulita, M., Hayden, K., & Khachaturian, A. S. (2018). Sex and gender in Alzheimer's disease—Does it matter? *Alzheimer's & Dementia*. https://doi.org/10.1016/j.jalz.2018.08.003. Published online.

Mielke, M. M., Vemuri, P., & Rocca, W. A. (2014). Clinical epidemiology of Alzheimer's disease: Assessing sex and gender differences. *Clinical Epidemiology*, *6*(1), 37–48. https://doi.org/10.2147/CLEP.S37929.

Miller, V. M., Jayachandran, M., Barnes, J. N., Mielke, M. M., Kantarci, K., & Rocca, W. A. (2020). Risk factors of neurovascular ageing in women. *Journal of Neuroendocrinology*, *32*(1). https://doi.org/10.1111/JNE.12777.

Miller, K. B., Miller, V. M., & Barnes, J. N. (2019). Pregnancy history, hypertension, and cognitive impairment in postmenopausal women. *Current Hypertension Reports*, *21*(12). https://doi.org/10.1007/S11906-019-0997-9.

Minoshima, S., Giordani, B., Berent, S., Frey, K. A., Foster, N. L., & Kuhl, D. E. (1997). Metabolic reduction in the posterior cingulate cortex in very early Alzheimer's disease. *Annals of Neurology*, *42*(1), 85–94. https://doi.org/10.1002/ANA.410420114.

Montagne, A., Zhao, Z., Zlokovic, B., & v. (2017). Alzheimer's disease: A matter of blood-brain barrier dysfunction? *The Journal of Experimental Medicine*, *214*(11), 3151–3169. https://doi.org/10.1084/JEM.20171406.

Moore, K. M., Nicholas, J., Grossman, M., et al. (2020). Age at symptom onset and death and disease duration in genetic frontotemporal dementia: An international retrospective cohort study. *The Lancet Neurology*, *19*(2), 145–156. https://doi.org/10.1016/S1474-4422(19)30394-1.

Morris, J. C., Roe, C. M., Xiong, C., et al. (2010). APOE predicts amyloid-beta but not tau Alzheimer pathology in cognitively normal aging. *Annals of Neurology*, *67*(1), 122–131. https://doi.org/10.1002/ANA.21843.

Muehlschlegel, S. (2018). Subarachnoid hemorrhage. *Continuum (Minneapolis Minn).*, *24*(6), 1623–1657. https://doi.org/10.1212/CON.0000000000000679.

Nagata, T., Nakajima, S., Shinagawa, S., et al. (2017). Psychosocial or clinico-demographic factors related to neuropsychiatric symptoms in patients with Alzheimer's disease needing interventional treatment: Analysis of the CATIE-AD study. *International Journal of Geriatric Psychiatry*, *32*(12), 1264–1271. https://doi.org/10.1002/GPS.4607.

Najar, J., Östling, S., Waern, M., et al. (2020). Reproductive period and dementia: A 44-year longitudinal population study of Swedish women. *Alzheimer's & Dementia*, *16*(8), 1153–1163. https://doi.org/10.1002/ALZ.12118.

Neu, S. C., Pa, J., Kukull, W., et al. (2017). Apolipoprotein E genotype and sex risk factors for Alzheimer disease: A meta-analysis. *JAMA Neurology*, *74*(10), 1178–1189. https://doi.org/10.1001/jamaneurol.2017.2188.

Nguyen, R., Fiest, K. M., McChesney, J., et al. (2016). the international incidence of traumatic brain injury: A systematic review and meta-analysis. *The Canadian Journal of Neurological Sciences*, *43*(6), 774–785. https://doi.org/10.1017/CJN.2016.290.

Nichols, E., Szoeke, C. E. I., Vollset, S. E., et al. (2019). Global, regional, and national burden of Alzheimer's disease and other dementias, 1990–2016: A systematic analysis for the Global Burden of Disease Study 2016. *The Lancet Neurology*, *18*(1), 88–106. https://doi.org/10.1016/S1474-4422(18)30403-4.

NIMH. *Major depression*. Accessed March 14, 2022 https://www.nimh.nih.gov/health/statistics/major-depression.

Niu, H., Álvarez-Álvarez, I., Guillén-Grima, F., & Aguinaga-Ontoso, I. (2017). Prevalence and incidence of Alzheimer's disease in Europe: A meta-analysis. *Neurología (English Edition)*, *32*(8), 523–532. https://doi.org/10.1016/J.NRLENG.2016.02.009.

Nyholm, D., Karlsson, E., Lundberg, M., & Askmark, H. (2010). Large differences in levodopa dose requirement in Parkinson's disease: Men use higher doses than women. *European Journal of Neurology*, *17*(2), 260–266. https://doi.org/10.1111/J.1468-1331.2009.02866.X.

O'brien, C., Taylor, J. T., Thomas, J. P., et al. (2022). Northumbria Research Link. *NIHR Health Technology Assessment Programme*. https://doi.org/10.3310/pgfar09070.

O'Brien, J. T., & Thomas, A. (2015). Vascular dementia. *The Lancet*, *386*(10004), 1698–1706. https://doi.org/10.1016/S0140-6736(15)00463-8.

O'Meara, J. G., Kardia, S. L. R., Armon, J. J., Brown, C. A., Boerwinkle, E., & Turner, S. T. (2004). Ethnic and sex differences in the prevalence, treatment, and control of dyslipidemia among hypertensive adults in the GENOA study. *Archives of Internal Medicine*, *164*(12), 1313–1318. https://doi.org/10.1001/ARCHINTE.164.12.1313.

Onyike, C. U., & Diehl-Schmid, J. (2013). The epidemiology of frontotemporal dementia. *International Review of Psychiatry*, *25*(2), 130–137. https://doi.org/10.3109/09540261.2013.776523.

Outeiro, T. F., Koss, D. J., Erskine, D., et al. (2019). Dementia with Lewy bodies: An update and outlook. *Molecular Neurodegeneration*, *14*(1), 5. https://doi.org/10.1186/s13024-019-0306-8.

Oveisgharan, S., Arvanitakis, Z., Yu, L., Farfel, J., Schneider, J. A., & Bennett, D. A. (2018). Sex differences in Alzheimer's disease and common neuropathologies of aging. *Acta Neuropathology*, *136*(6), 887–900. https://doi.org/10.1007/S00401-018-1920-1.

Panegyres, P. K., & Chen, H. Y. (2014). Early-onset Alzheimer's disease: A global cross-sectional analysis. *European Journal of Neurology*, *21*(9). https://doi.org/10.1111/ENE.12453.

Pankratz, V. S., Roberts, R. O., Mielke, M. M., et al. (2015). Predicting the risk of mild cognitive impairment in the mayo clinic study of aging. *Neurology*, *84*(14), 1433–1442. https://doi.org/10.1212/WNL.0000000000001437.

Panza, F., Lozupone, M., Watling, M., & Imbimbo, B. P. (2021). Pharmacological management of dementia with Lewy bodies with a focus on zonisamide for treating parkinsonism. *Expert Opinion on Pharmacotherapy*, *22*(3), 325–337. https://doi.org/10.1080/14656566.2020.1828350.

Pearson, J. D., Morrell, C. H., & Gordon-Salant, S. (1995). Gender differences in a longitudinal study of age-associated hearing loss. *The Journal of the Acoustical Society of America*, *97*, 1196. https://doi.org/10.1121/1.412231.

Pendlebury, S. T., & Rothwell, P. M. (2009). Prevalence, incidence, and factors associated with pre-stroke and post-stroke dementia: A systematic review and meta-analysis. *The Lancet Neurology*, *8*(11), 1006–1018. https://doi.org/10.1016/S1474-4422(09)70236-4.

Pengo, M., Alberici, A., Libri, I., et al. (2022). Sex influences clinical phenotype in frontotemporal dementia. *Neurological Sciences*, *2022*(1), 1–7. https://doi.org/10.1007/S10072-022-06185-7.

Pepine, C. J., & Nichols, W. W. (2006). Estrogen and different aspects of vascular disease in women and men. *Circulation Research*, *99*(5), 459–461. https://doi.org/10.1161/01.RES.0000241056.84659.59.

Petrea, R. E., Beiser, A. S., Seshadri, S., Kelly-Hayes, M., Kase, C. S., & Wolf, P. A. (2009). Gender differences in stroke incidence and poststroke disability in the Framingham heart study. *Stroke*, *40*(4), 1032–1037. https://doi.org/10.1161/STROKEAHA.108.542894.

Piscopo, P., Bellenghi, M., Manzini, V., et al. (2021). A sex perspective in neurodegenerative diseases: Micrornas as possible peripheral biomarkers. *International Journal of Molecular Sciences*, *22*(9). https://doi.org/10.3390/ijms22094423.

Podcasy, J. L., & Epperson, C. N. (2016). Considering sex and gender in Alzheimer disease and other dementias. *Dialogues in Clinical Neuroscience*, *18*(4), 437–446. https://doi.org/10.31887/DCNS.2016.18.4/CEPPERSON.

Poewe, W., Seppi, K., Tanner, C. M., et al. (2017). Parkinson disease. *Nature Reviews Disease Primers*, *3*(1), 17013. https://doi.org/10.1038/nrdp.2017.13.

Portegies, M. L. P., Koudstaal, P. J., & Ikram, M. A. (2016). Cerebrovascular disease. *Handbook of Clinical Neurology.*, *138*, 239–261. https://doi.org/10.1016/B978-0-12-802973-2.00014-8.

Porter, V. R., Buxton, W. G., Fairbanks, L. A., et al. (2003). Frequency and characteristics of anxiety among patients with Alzheimer's disease and related dementias. *Journal of Neuropsychiatry and Clinical Neurosciences*, *15*(2), 180–186. https://doi.org/10.1176/JNP.15.2.180.

Potter, R. R., III, Long, A. P., & Lichtenstein, M. L. (2020). Population prevalence of autosomal dominant Alzheimer's disease: A systematic review. *Alzheimer's & Dementia*, *16*(S10), e037129. https://doi.org/10.1002/ALZ.037129.

Pradier, C., Sakarovitch, C., le Duff, F., et al. (2014). The mini mental state examination at the time of Alzheimer's disease and related disorders diagnosis, according to age, education, gender and place of residence: A cross-sectional study among the French National Alzheimer database. *PLoS One*, *9*(8). https://doi.org/10.1371/JOURNAL.PONE.0103630.

Quinn, N. P., & Marsden, C. D. (1986). Menstrual-related fluctuations in Parkinson's disease. *Movement Disorders*, *1*(1), 85–87. https://doi.org/10.1002/MDS.870010112.

Rascovsky, K., Hodges, J. R., Knopman, D., et al. (2011). Sensitivity of revised diagnostic criteria for the behavioural variant of frontotemporal dementia. *Brain*, *134*(9), 2456–2477. https://doi.org/10.1093/brain/awr179.

Ratnavalli, E., Brayne, C., Dawson, K., & Hodges, J. R. (2002). The prevalence of frontotemporal dementia. *Neurology*, *58*(11), 1615–1621. https://doi.org/10.1212/WNL.58.11.1615.

Reekes, T. H., Higginson, C. I., Ledbetter, C. R., Sathivadivel, N., Zweig, R. M., & Disbrow, E. A. (2020). Sex specific cognitive differences in Parkinson disease. *NPJ Parkinsons Disease*, *6*(1). https://doi.org/10.1038/S41531-020-0109-1.

Regnault, V., Thomas, F., Safar, M. E., et al. (2012). Sex difference in cardiovascular risk: Role of pulse pressure amplification. *Journal of American College of Cardiology*, *59*(20), 1771–1777. https://doi.org/10.1016/J.JACC.2012.01.044.

Reid, W. G. J., Hely, M. A., Morris, J. G. L., Loy, C., & Halliday, G. M. (2011). Dementia in Parkinson's disease: A 20-year neuropsychological study (Sydney Multicentre Study). *Journal of Neurology, Neurosurgery & Psychiatry*, *82*(9), 1033–1037. https://doi.org/10.1136/jnnp.2010.232678.

Reiman, E. M., Caselli, R. J., Yun, L. S., et al. (1996). Preclinical evidence of Alzheimer's disease in persons homozygous for the epsilon 4 allele for apolipoprotein E. *New England Journal of Medicine*, *334*(12), 752–758. https://doi.org/10.1056/NEJM199603213341202.

Ren, R. J., Huang, Y., Xu, G., et al. (2012). History, present, and progress of frontotemporal dementia in China: A systematic review. *International Journal of Alzheimer's Disease*. https://doi.org/10.1155/2012/587215.

Rhodes, M. E., & Rubin, R. T. (1999). Functional sex differences ('sexual diergism') of central nervous system cholinergic systems, vasopressin, and hypothalamic-pituitary-adrenal axis activity in mammals: A selective review. *Brain Research. Brain Research Reviews*, *30*(2), 135–152. https://doi.org/10.1016/S0165-0173(99)00011-9.

Rhodin, J. A., Thomas, T. N., Clark, L., Garces, A., & Bryant, M. (2003). In vivo cerebrovascular actions of amyloid beta-peptides and the protective effect of conjugated estrogens. *Journal of Alzheimer's Disease, 5*(4), 275–286. https://doi.org/10.3233/JAD-2003-5403.

Roberts, R. O., Knopman, D. S., Geda, Y. E., et al. (2014). Association of diabetes with amnestic and nonamnestic mild cognitive impairment. *Alzheimer's & Dementia, 10*(1), 18–26. https://doi.org/10.1016/J.JALZ.2013.01.001.

Robison, L. S., Gannon, O. J., Salinero, A. E., & Zuloaga, K. L. (2019). Contributions of sex to cerebrovascular function and pathology. *Brain Research, 1710*, 43–60. https://doi.org/10.1016/J.BRAINRES.2018.12.030.

Rocca, W. A., Cha, R. H., Waring, S. C., & Kokmen, E. (1998). Incidence of dementia and Alzheimer's disease: A reanalysis of data from Rochester, Minnesota, 1975–1984. *American Journal of Epidemiology, 148*(1), 51–62. https://doi.org/10.1093/OXFORDJOURNALS.AJE.A009560.

Rocca, W. A., Lohse, C. M., Smith, C. Y., Fields, J. A., Machulda, M. M., & Mielke, M. M. (2021). Association of premenopausal bilateral oophorectomy with cognitive performance and risk of mild cognitive impairment. *JAMA Network Open, 4*(11), e2131448. https://doi.org/10.1001/JAMANETWORKOPEN.2021.31448.

Román, G. (2003). Vascular dementia: A historical background. *International Psychogeriatrics, 15*(Suppl. 1), 11–13. https://doi.org/10.1017/S1041610203008901.

Román, G. (2022). Vascular dementia: A historical background. *International Psychogeriatrics, IS*, 11–13. https://doi.org/10.1017/S1041610203008901.

Rosenberg, G. A. (2017). Extracellular matrix inflammation in vascular cognitive impairment and dementia. *Clinical Science (London), 131*(6), 425–437. https://doi.org/10.1042/CS20160604.

Rosenfeld, C. S. (2017). Sex-dependent differences in voluntary physical activity. *Journal of Neuroscience Research, 95*(1-2), 279. https://doi.org/10.1002/JNR.23896.

Ruitenberg, A., Ott, A., van Swieten, J. C., Hofman, A., & Breteler, M. M. B. (2001). Incidence of dementia: Does gender make a difference? *Neurobiology Aging, 22*(4), 575–580. https://doi.org/10.1016/S0197-4580(01)00231-7.

Ryan, C. L., & Bauman, K. (2015). *Educational Attainment in the United States.* Population Characteristics Demographic Portrait of Educational Attainment. Published online 2016. Accessed February 27, 2022 http://census.gov/library/publications/2011/dec/c2010br-02.html.

Sabra, D., Intzandt, B., Desjardins-Crepeau, L., et al. (2021). Sex moderations in the relationship between aortic stiffness, cognition, and cerebrovascular reactivity in healthy older adults. *PLoS One, 16*(9). https://doi.org/10.1371/JOURNAL.PONE.0257815.

Sachdev, P. S., Parslow, R., Wen, W., Anstey, K. J., & Easteal, S. (2009). Sex differences in the causes and consequences of white matter hyperintensities. *Neurobiology Aging, 30*(6), 946–956. https://doi.org/10.1016/J.NEUROBIOLAGING.2007.08.023.

Salminen, L. E., Tubi, M. A., Bright, J., Thomopoulos, S. I., Wieand, A., & Thompson, P. M. (2022). Sex is a defining feature of neuroimaging phenotypes in major brain disorders. *Human Brain Mapping, 43*(1), 500–542. https://doi.org/10.1002/HBM.25438.

Savica, R., Grossardt, B. R., Bower, J. H., Boeve, B. F., Ahlskog, J. E., & Rocca, W. A. (2013). Incidence of dementia with Lewy bodies and Parkinson disease dementia. *JAMA Neurology, 70*(11), 1396. https://doi.org/10.1001/jamaneurol.2013.3579.

Schaie, K. W., & Willis, S. L. (2010). The Seattle Longitudinal Study of Adult Cognitive Development. *ISSBD Bulletin, 57*(1), 24. Accessed February 27, 2022. /pmc/articles/PMC3607395/.

Scheinin, N. M., Wikman, K., Jula, A., et al. (2014). Cortical 11C-PIB uptake is associated with age, APOE genotype, and gender in "healthy aging". *Journal of Alzheimer's Disease, 41*(1), 193–202. https://doi.org/10.3233/JAD-132783.

Schmitz-Steinkrüger, H., Lange, C., Apostolova, I., et al. (2021). Impact of age and sex correction on the diagnostic performance of dopamine transporter SPECT. *European Journal of Nuclear Medicine and Molecular Imaging*, *48*(5), 1445–1459. https://doi.org/10.1007/S00259-020-05085-2.

Schoonenboom, N. S. M., Pijnenburg, Y. A. L., Mulder, C., et al. (2004). Amyloid beta(1-42) and phosphorylated tau in CSF as markers for early-onset Alzheimer disease. *Neurology*, *62*(9), 1580–1584. https://doi.org/10.1212/01.WNL.0000123249.58898.E0.

Schupf, N., Kapell, D., Nightingale, B., Rodriguez, A., Tycko, B., & Mayeux, R. (1998). Earlier onset of Alzheimer's disease in men with Down syndrome. *Neurology*, *50*(4), 991–995. https://doi.org/10.1212/WNL.50.4.991.

Schupf, N., Pang, D., Patel, B. N., et al. (2003). Onset of dementia is associated with age at menopause in women with Down's syndrome. *Annals of Neurology*, *54*(4), 433–438. https://doi.org/10.1002/ANA.10677.

Schupf, N., Winsten, S., Patel, B., et al. (2006). Bioavailable estradiol and age at onset of Alzheimer's disease in postmenopausal women with Down syndrome. *Neuroscience Letters*, *406*(3), 298–302. https://doi.org/10.1016/J.NEULET.2006.07.062.

Seab, J. P., Jagust, W. J., Wong, S. T. S., Roos, M. S., Reed, B. R., & Budinger, T. F. (1988). Quantitative NMR measurements of hippocampal atrophy in Alzheimer's disease. *Magnetic Resonance in Medicine*, *8*(2), 200–208. https://doi.org/10.1002/MRM.1910080210.

Seier, M., & Hiller, A. (2017). Parkinson's disease and pregnancy: An updated review. *Parkinsonism & Related Disorders*, *40*, 11–17. https://doi.org/10.1016/J.PARKRELDIS.2017.05.007.

Sepehrband, F., Barisano, G., Sheikh-Bahaei, N., et al. (2021). Volumetric distribution of perivascular space in relation to mild cognitive impairment. *Neurobiological Aging*, *99*, 28–43. https://doi.org/10.1016/J.NEUROBIOLAGING.2020.12.010.

Shaw, L. J., Bairey Merz, C. N., Pepine, C. J., et al. (2006). Insights from the NHLBI-Sponsored Women's Ischemia Syndrome Evaluation (WISE) Study: Part I: Gender differences in traditional and novel risk factors, symptom evaluation, and gender-optimized diagnostic strategies. *Journal of the American College of Cardiology*, *47*(3 Suppl), S4–S20. https://doi.org/10.1016/J.JACC.2005.01.072.

Shimizu, Y., Sawada, N., Iwasaki, M., et al. (2019). Reproductive history and risk of cognitive impairment in Japanese women. *Maturitas*, *128*, 22–28. https://doi.org/10.1016/J.MATURITAS.2019.06.012.

Sinai, A., Mokrysz, C., Bernal, J., et al. (2018). Predictors of age of diagnosis and survival of Alzheimer's disease in down syndrome. *Journal of Alzheimer's Disease*, *61*(2), 717–728. https://doi.org/10.3233/JAD-170624.

Slaets, S., van Acker, F., Versijpt, J., et al. (2015). Diagnostic value of MIBG cardiac scintigraphy for differential dementia diagnosis. *International Journal of Geriatric Psychiatry*, *30*(8), 864–869. https://doi.org/10.1002/gps.4229.

Smith, R., Strandberg, O., Mattsson-Carlgren, N., et al. (2020). The accumulation rate of tau aggregates is higher in females and younger amyloid-positive subjects. *Brain*, *143*(12), 3805. https://doi.org/10.1093/BRAIN/AWAA327.

Song, X., Wu, J., Zhou, Y., et al. (2020). Reproductive and hormonal factors and risk of cognitive impairment among Singapore Chinese women. *American Journal of Obstetrics and Gynecology*, *223*(3), 410.e1–410.e23. https://doi.org/10.1016/J.AJOG.2020.02.032.

Sperling, R. A., Aisen, P. S., Beckett, L. A., et al. (2011). Toward defining the preclinical stages of Alzheimer's disease: Recommendations from the National Institute on Aging-Alzheimer's Association workgroups on diagnostic guidelines for Alzheimer's disease. *Alzheimer's & Dementia*, 7(3), 280–292. https://doi.org/10.1016/j.jalz.2011.03.003.

Spina, S., la Joie, R., Petersen, C., et al. (2021). Comorbid neuropathological diagnoses in early versus late-onset Alzheimer's disease. *Brain*, *144*(7), 2186. https://doi.org/10.1093/BRAIN/AWAB099.

Stern, Y. (2009). Cognitive reserve. *Neuropsychologia*, *47*(10), 2015–2028. https://doi.org/10.1016/J.NEUROPSYCHOLOGIA.2009.03.004.

Stern, Y., Gurland, B., Tatemichi, T. K., Tang, M. X., Wilder, D., & Mayeux, R. (1994). Influence of education and occupation on the incidence of Alzheimer's disease. *Journal of Medical Association*, *271*(13), 1004–1010. https://doi.org/10.1001/JAMA.1994.03510370056032.

Stern, Y., Zarahn, E., Hilton, H. J., Flynn, J., DeLaPaz, R., & Rakitin, B. (2003). Exploring the neural basis of cognitive reserve. *Journal of Clinical and Experimental Neuropsychology*, *25*(5), 691–701. https://doi.org/10.1076/JCEN.25.5.691.14573.

Stevens, M., van Duijn, C. M., Kamphorst, W., et al. (1998). Familial aggregation in frontotemporal dementia. *Neurology*, *50*(6), 1541–1545. https://doi.org/10.1212/WNL.50.6.1541.

Stocker, H., Möllers, T., Perna, L., & Brenner, H. (2018). The genetic risk of Alzheimer's disease beyond APOE ε4: Systematic review of Alzheimer's genetic risk scores. *Translational Psychiatry*, *8*(1), 1–9. https://doi.org/10.1038/s41398-018-0221-8.

Strittmatter, W. J., Saunders, A. M., Schmechel, D., et al. (1993). Apolipoprotein E: High-avidity binding to beta-amyloid and increased frequency of type 4 allele in late-onset familial Alzheimer disease. *Proceedings of the National Academy of Sciences of the United States of America*, *90*(5), 1977–1981. https://doi.org/10.1073/PNAS.90.5.1977.

Subramanian, I., Mathur, S., Oosterbaan, A., Flanagan, R., Keener, A. M., & Moro, E. (2022). Unmet needs of women living with Parkinson's disease: Gaps and controversies. *Movement Disorders*, *37*(3), 444–455. https://doi.org/10.1002/MDS.28921.

Sundermann, E. E., Biegon, A., Rubin, L. H., Lipton, R. B., Landau, S., & Maki, P. M. (2017). Does the female advantage in verbal memory contribute to underestimating Alzheimer's disease pathology in women versus men? *Journal of Alzheimer's Disease*, *56*(3), 947–957. https://doi.org/10.3233/JAD-160716.

Sundermann, E. E., Biegon, A., Rubin, L. H., et al. (2016). Better verbal memory in women than men in MCI despite similar levels of hippocampal atrophy. *Neurology*, *86*(15), 1368–1376. https://doi.org/10.1212/WNL.0000000000002570.

Sundermann, E. E., Maki, P., Biegon, A., et al. (2019). Sex-specific norms for verbal memory tests may improve diagnostic accuracy of amnestic MCI. *Neurology*, *93*(20), e1881. https://doi.org/10.1212/WNL.0000000000008467.

Sundermann, E. E., Maki, P. M., Rubin, L. H., Lipton, R. B., Landau, S., & Biegon, A. (2016). Female advantage in verbal memory: Evidence of sex-specific cognitive reserve. *Neurology*, *87*(18), 1916. https://doi.org/10.1212/WNL.0000000000003288.

Tao, Y., Peters, M. E., Drye, L. T., et al. (2018). Sex differences in the neuropsychiatric symptoms of patients with Alzheimer's disease. *American Journal of Alzheimer's Disease and Other Dementias*, *33*(7), 450–457. https://doi.org/10.1177/1533317518783278.

Tifratene, K., Robert, P., Metelkina, A., Pradier, C., & Dartigues, J. F. (2015). Progression of mild cognitive impairment to dementia due to AD in clinical settings. *Neurology*, *85*(4), 331–338.

Tiraboschi, P., Hansen, L. A., Thal, L. J., & Corey-Bloom, J. (2004). The importance of neuritic plaques and tangles to the development and evolution of AD. *Neurology*, *62*(11), 1984–1989. https://doi.org/10.1212/01.WNL.0000129697.01779.0A.

Towfighi, A., Zheng, L., & Ovbiagele, B. (2009). Sex-specific trends in midlife coronary heart disease risk and prevalence. *Archives of Internal Medicine*, *169*(19), 1762–1766. https://doi.org/10.1001/ARCHINTERNMED.2009.318.

Tremblay, C., Abbasi, N., Zeighami, Y., et al. (2020). Sex effects on brain structure in de novo Parkinson's disease: A multimodal neuroimaging study. *Brain*, *143*(10), 3052–3066. https://doi.org/10.1093/BRAIN/AWAA234.

Turan, N., Heider, R. A. J., Zaharieva, D., Ahmad, F. U., Barrow, D. L., & Pradilla, G. (2016). Sex differences in the formation of intracranial aneurysms and incidence and outcome of subarachnoid hemorrhage: Review of experimental and human studies. *Translational Stroke Research*, 7(1), 12–19. https://doi.org/10.1007/S12975-015-0434-6.

Turcano, P., & Savica, R. (2020). Sex differences in movement disorders. In *Vol. 175. Handbook of clinical neurology* (pp. 275–282). Elsevier https://linkinghub.elsevier.com/retrieve/pii/B9780444641236000199.

Valente, E. S., Caramelli, P., Gambogi, L. B., et al. (2019). Phenocopy syndrome of behavioral variant frontotemporal dementia: A systematic review. *Alzheimer's Research and Therapy*, *11*(1), 1–15. https://doi.org/10.1186/s13195-019-0483-2.

van den Heuvel, D. M. J., Admiraal-Behloul, F., ten Dam, V. H., et al. (2004). Different progression rates for deep white matter hyperintensities in elderly men and women. *Neurology*, *63*(9), 1699–1701. https://doi.org/10.1212/01.WNL.0000143058.40388.44.

Varrone, A., Dickson, J. C., Tossici-Bolt, L., et al. (2013). European multicentre database of healthy controls for [123I]FP- CIT SPECT (ENC-DAT): Age-related effects, gender differences and evaluation of different methods of analysis. *European Journal of Nuclear Medicine and Molecular Imaging*, *40*(2), 213–227. https://doi.org/10.1007/S00259-012-2276-8/TABLES/7.

Vasconcellos, L. F. R., & Pereira, J. S. (2015). Parkinson's disease dementia: Diagnostic criteria and risk factor review. *Journal of Clinical and Experimental Neuropsychology*, *37*(9), 988–993. https://doi.org/10.1080/13803395.2015.1073227.

Vemuri, P., Knopman, D. S., Lesnick, T. G., et al. (2017). Evaluation of amyloid protective factors and Alzheimer disease neurodegeneration protective factors in elderly individuals. *JAMA Neurology*, *74*(6), 718–726. https://doi.org/10.1001/jamaneurol.2017.0244.

Vijayan, M., & Reddy, P. H. (2016). Stroke, Vascular Dementia, and Alzheimer's disease: Molecular links. *Journal of Alzheimer's Disease*, *54*(2), 427–443. https://doi.org/10.3233/JAD-160527.

Villavisanis, D. F., Berson, E. R., Lauer, A. M., Cosetti, M. K., & Schrode, K. M. (2020). Sex-based differences in hearing loss: Perspectives from non-clinical research to clinical outcomes. *Otology & Neurotology*, *41*(3), 290. https://doi.org/10.1097/MAO.0000000000002507.

Visser, P. J., Verhey, F., Knol, D. L., et al. (2009). Prevalence and prognostic value of CSF markers of Alzheimer's disease pathology in patients with subjective cognitive impairment or mild cognitive impairment in the DESCRIPA study: A prospective cohort study. *The Lancet. Neurology*, *8*(7), 619–627. https://doi.org/10.1016/S1474-4422(09)70139-5.

Walker, Z., Possin, K. L., Boeve, B. F., & Aarsland, D. (2015). Lewy body dementias. *The Lancet*, *386*(10004), 1683–1697. https://doi.org/10.1016/S0140-6736(15)00462-6.

Ward, A., Crean, S., Mercaldi, C. J., et al. (2012). Prevalence of apolipoprotein E4 genotype and homozygotes (APOE e4/4) among patients diagnosed with Alzheimer's disease: A systematic review and meta-analysis. *Neuroepidemiology*, *38*(1), 1–17. https://doi.org/10.1159/000334607.

Warren Olanow, C., Kieburtz, K., Rascol, O., et al. (2013). Factors predictive of the development of Levodopa-induced dyskinesia and wearing-off in Parkinson's disease. *Movement Disorders*, *28*(8), 1064–1071. https://doi.org/10.1002/MDS.25364.

Wattmo, C., & Wallin, Å. K. (2017). Early- versus late-onset Alzheimer's disease in clinical practice: Cognitive and global outcomes over 3 years. *Alzheimer's Research and Therapy*, *9*(1). https://doi.org/10.1186/s13195-017-0294-2.

Weber, C. M., & Clyne, A. M. (2021). Sex differences in the blood–brain barrier and neurodegenerative diseases. *APL Bioengineering*, *5*(1), 11509. https://doi.org/10.1063/5.0035610.

Whitwell, J. L., Jack, C. R., Senjem, M. L., et al. (2009). MRI correlates of protein deposition and disease severity in postmortem frontotemporal lobar degeneration. *Neurodegenerative Diseases*, *6*(3), 106–117. https://doi.org/10.1159/000209507.

Wildman, R. P., Colvin, A. B., Powell, L. H., et al. (2008). Associations of endogenous sex hormones with the vasculature in menopausal women: The Study of Women's Health Across the Nation (SWAN). *Menopause*, *15*(3), 414–421. https://doi.org/10.1097/GME.0B013E318154B6F5.

Williams-Gray, C. H., Foltynie, T., Brayne, C. E. G., Robbins, T. W., & Barker, R. A. (2007). Evolution of cognitive dysfunction in an incident Parkinson's disease cohort. *Brain*, *130*(7), 1787–1798. https://doi.org/10.1093/brain/awm111.

Winblad, B., Engedal, K., Soininen, H., et al. (2001). A 1-year, randomized, placebo-controlled study of donepezil in patients with mild to moderate AD. *Neurology*, *57*(3), 489–495. https://doi.org/10.1212/WNL.57.3.489.

Wingo, T. S., Lah, J. J., Levey, A. I., & Cutler, D. J. (2012). Autosomal recessive causes likely in early-onset Alzheimer disease. *Archives of Neurology*, *69*(1), 59–64. https://doi.org/10.1001/archneurol.2011.221.

Woolley, J. D., Khan, B. K., Murthy, N. K., Miller, B. L., & Rankin, K. P. (2011). The diagnostic challenge of psychiatric symptoms in neurodegenerative disease. *The Journal of Clinical Psychiatry*, *72*(02), 126–133. https://doi.org/10.4088/jcp.10m06382oli.

World Health Organization. (2011). *Gender mainstreaming for health managers: A practical approach*. Accessed June 19, 2022 https://www.who.int/publications/i/item/9789241501057.

Wu, C. H., Yeh, H. T., Hsieh, C. S., et al. (2021). Evolutionary trajectories and genomic divergence in localized breast cancers after ipsilateral breast tumor recurrence. *Cancers (Basel)*, *13*(8). https://doi.org/10.3390/CANCERS13081821.

Yang, H. D., Kim, D. H., Lee, S. B., & Young, L. D. (2016). History of Alzheimer's disease. *Dementia and Neurocognitive Disorders*, *15*(4), 115–121. https://doi.org/10.12779/DND.2016.15.4.115.

Yoon, S. S., Gu, Q., Nwankwo, T., Wright, J. D., Hong, Y., & Burt, V. (2015). Trends in blood pressure among adults with hypertension: United States, 2003 to 2012. *Hypertension*, *65*(1), 54–61. https://doi.org/10.1161/HYPERTENSIONAHA.114.04012.

Yoshitake, T., Kiyohara, Y., Kato, I., et al. (1995). Incidence and risk factors of vascular dementia and Alzheimer's disease in a defined elderly Japanese population: The Hisayama Study. *Neurology*, *45*(6), 1161–1168. https://doi.org/10.1212/WNL.45.6.1161.

Zapata-Restrepo, L., Rivas, J., Miranda, C., et al. (2021). The psychiatric misdiagnosis of behavioral variant frontotemporal dementia in a Colombian sample. *Frontiers in Neurology*, *12*(November), 1–7. https://doi.org/10.3389/fneur.2021.729381.

CHAPTER SEVEN

Sex and gender differences in epilepsy

Lauren Hophing[a,†], Paulina Kyriakopoulos[b,†], and Esther Bui[a,*]
[a]Krembil Brain Institute, University Health Network, University of Toronto, Toronto, Canada
[b]Division of Neurology, Department of Medicine, The Ottawa Hospital, Ottawa, Canada
*Corresponding author. e-mail address: esther.bui@uhn.ca

Contents

† Equal contribution as first author.

International Review of Neurobiology, Volume 164
ISSN 0074-7742
https://doi.org/10.1016/bs.irn.2022.06.012

1. Introduction

Sex and gender differences in epilepsy are important influencing factors in epilepsy care. The term sex reflects the biological differences between males and females and refers to biological factors such as genes, hormones and reproductive organs. In epilepsy, the hormonal differences between the sexes are important as they impact specific treatment considerations for female patients at various life stages. In contrast, gender is a term that reflects societal influences that distinguish men, women and gender diverse people (Bartz et al., 2020; Mauvais-Jarvis et al., 2020) and can impact a patient's lived experience with epilepsy. An individual's gender identity can be different from the sex determined at time of conception. Although medical literature has explicitly reported on differences between sex and gender increasingly over the last decade, the terms sex and gender are still frequently used interchangeably. In this chapter, we will use the terms sex (male, female) and gender (men, women) based on the definitions noted above when possible.

Approximately 50 million people worldwide have epilepsy (GBD 2016 Neurology Collaborators, 2019) including about 150,000–450,000 transgender people (Johnson & Kaplan, 2017). In general, males and females have a near equal incidence and prevalence of epilepsy (Banerjee, Filippi, & Allen Hauser, 2009; Christensen et al., 2007; Fiest et al., 2017; Hauser, Annegers, & Kurland, 1993; Kotsopoulos, van Merode, Kessels, de Krom, & Knottnerus, 2002; McHugh & Delanty, 2008). More specifically, near equal occurrence between the sexes (Christensen, Kjeldsen, Andersen, Friis, & Sidenius, 2005; Manford, Hart, Sander, & Shorvon, 1992) has been mainly reported with focal epilepsies such as temporal lobe epilepsy (Christensen et al., 2005). No significant sex difference has been observed with hippocampal sclerosis, the most common pathology associated with temporal lobe epilepsy (Briellmann et al., 1999). When focal epilepsies were subdivided, however, a male-predominance was seen in those associated with a known structural abnormality, especially in the 30- to 59-year-old age group (Christensen et al., 2005). In this age group, higher risks of traumatic brain injuries (Bruns & Hauser, 2003) and post-traumatic brain injury epilepsy have been described in males, which may partly explain this observation (Karlander, Ljungqvist, & Zelano, 2021; Xu et al., 2017). Of the generalized epilepsies, genetic generalized epilepsy syndromes such as juvenile myoclonic epilepsy and juvenile absence epilepsy have a clear female-predominance. The biological factors that may lead to this difference have not yet been elucidated, though sex hormones are likely important factors (Christensen et al., 2005).

2. Sex hormones and epilepsy

We will present a simplified review of the proposed mechanisms and regulation of sex hormones to highlight the most relevant features for epilepsy. Endogenous steroid sex hormones, particularly estrogen and progesterone, play an important role in sex differences in epilepsy by variably affecting neuronal excitability through genomic and non-genomic mechanisms. Within minutes to days, these steroid hormones can regulate gene expression by acting as transcription factors after binding intracellular steroid receptors. Alternatively, certain neuroactive metabolites can have a very rapid effect through binding of membrane receptors (Reddy, 2014).

Animal studies looking at the effect of these hormones on seizures have used various epilepsy models. Seizures can be induced by chemoconvulsants including kainic acid, pilocarpine or pentylenetetrazol or with electrical stimulation such as in the kindling model, commonly used for temporal lobe epilepsy, where repeat electrical stimulation leads to a progressive increase in seizure intensity (Kandratavicius et al., 2014). Clinically, a neuroexcitatory response to estrogen has been noted as early as the 1950s. A study of 16 women with epilepsy that were given intravenous premarin, a conjugated estrogen, showed an increase in epileptiform activity in 11 of them and seizure occurrence in 4 (Logothetis, Harner, Morrell, & Torres, 1959). This has been mostly supported by animal studies where estrogen appears to facilitate the occurrence of seizures induced by either chemoconvulsants or electrical stimulation (Edwards, Burnham, Mendonca, Bowlby, & MacLusky, 1999; Hom & Buterbaugh, 1986; Woolley, 2000; Woolley & Timiras, 1962). Seemingly paradoxically, a delay in seizure onset and a decrease in seizure-associated mortality was noted in ovariectomized female rats given estradiol prior to kainic acid-induced seizures (Velísková, Velísek, Galanopoulou, & Sperber, 2000). This hints at a more complex relationship between estrogen and seizures (Velísková et al., 2000). Three main forms of estrogen are dominant at different times during a woman's life: estrone (E1) in the post-menopausal period, estradiol (E2) during the pre-menopausal years and estriol (E3) during pregnancy (Scharfman & MacLusky, 2006). Of the three, estradiol most effectively acts on estrogen receptors and is the subtype thought to have the more potent proconvulsant effect (Scharfman & MacLusky, 2006).

Progesterone has a mainly neuroinhibitory effect that has been supported by animal models showing a protective effect against seizures triggered by chemoconvulsants or electrical stimulation (Wu & Burnham, 2018).

Its primary action is through its neuroactive metabolite, allopregnanolone, which is synthesized in the brain where it can rapidly enhance chloride conductance after binding to GABA receptors (Reddy, 2014). A human study of seven female patients with focal epilepsy, who received an intravenous progesterone infusion near the beginning of their menstrual cycle when progesterone levels were presumed to be low, found that four had a significant reduction in epileptiform discharges (Bäckström, Zetterlund, Blom, & Romano, 1984).

Testosterone has variable effects on neuronal excitability in animal models (Frye, Rhodes, Walf, & Harney, 2001; Mejías-Aponte, Jiménez-Rivera, & Segarra, 2002; Pesce, Acevedo, Bustamante, Miranda, & Pinardi, 2000). Its effects are mainly through its metabolites including estradiol with its predominantly proconvulsant effect, as well as androsterone and 5α-androstane-3α, 17β-diol with mainly anticonvulsant effects (Harden & Pennell, 2013). Aromatase inhibition has been proposed as a possible adjunct treatment for men with epilepsy as it reduces the conversion of testosterone to estradiol, but additional research is needed to further understand this relationship (Harden & MacLusky, 2004, 2005).

3. Sex hormone changes and regulation during the lifespan

The synthesis of sex hormones is regulated by the hypothalamic-pituitary-gonadal axis. Pulsatile hypothalamic release of gonadotropin-releasing hormone (GnRH) stimulates the anterior pituitary to release luteinizing hormone (LH) and follicle-stimulating hormone (FSH). LH binds separate target cells within male and female gonads and leads to the cholesterol transfer needed to start steroidogenesis by increasing expression of a regulatory protein (Harden & Pennell, 2013; Maggi, 2020).

At puberty, with the onset of these pulsatile GnRH secretions, both females and males will have an increase in sex hormones, mainly estradiol and testosterone, respectively (Breehl & Caban, 2020). After this point, whereas sex hormones remain relatively stable in males with gradual decreases later in life (Decaroli & Rochira, 2017; Oettel & Mukhopadhyay, 2004), estrogen and progesterone continue to cyclically vary in females until menopause (Harden & Pennell, 2013; Maggi, 2020). The change in the estrogen-to-progesterone ratio throughout the cycle may affect seizure occurrence in some females and will be discussed in more detail in Section 4. In males, increased testosterone levels provide negative feedback to the pituitary and hypothalamus. In females, the feedback effect of progesterone and estrogen is similarly inhibitory throughout much of the menstrual cycle, but with a stimulatory estrogen effect leading to the predominant LH surge noted pre-ovulation (Harden & Pennell, 2013; Maggi, 2020).

During perimenopause, there is an overall trend of decrease in estradiol and progesterone. Smaller cyclical rises in progesterone and erratic fluctuations in estradiol can result in intermittent increases in the estrogen-to-progesterone ratio (Burger et al., 2007). In established menopause, the lower estradiol levels and relatively higher estrone levels result in an overall decrease in pro-convulsant effect (Burger, Dudley, Robertson, & Dennerstein, 2002; Burger, Hale, Dennerstein, & Robertson, 2008; Sveinsson & Tomson, 2014) (Figs. 1 and 2).

4. Catamenial epilepsy

An in vivo modeling of hormonal influences on seizures can be found in catamenial epilepsy. Catamenial epilepsy is a term used to describe the clustering of seizures experienced in about 40–45% of women with epilepsy (Herzog, 2015; Herzog et al., 2004) at various time points during the menstrual cycle, likely related to varying sex hormone levels. Although variable definitions were previously used to quantify this cyclical increase in seizure frequency, a practical guideline consists of noting at least a doubling of seizure frequency during a specific, consistent period of the menstrual cycle (Herzog, 2008a; Herzog, Klein, & Ransil, 1997).

Three distinct patterns of seizure clustering have been described that are illustrated in Fig. 3. The most commonly observed and treatment-responsive pattern is an increase during the perimenstrual period (denoted as catamenial type 1, C1) and typically occurs between days −3 to +3 (by counting backwards or forwards from day 1 or + 1, the first day of menses). This appears to be associated with an increased estradiol-to-progesterone ratio with a relative decrease in progesterone's protective anticonvulsant effect (Harden & Pennell, 2013; Herzog, 2008a; Herzog et al., 1997). Catamenial type 2 or C2 describes a periovulatory increase in seizure around days +10 to −13 during the rapid estradiol surge which may have a proconvulsant effect (Harden & Pennell, 2013; Herzog, 2008a; Herzog et al., 1997). The catamenial type 3 or C3 pattern is described during

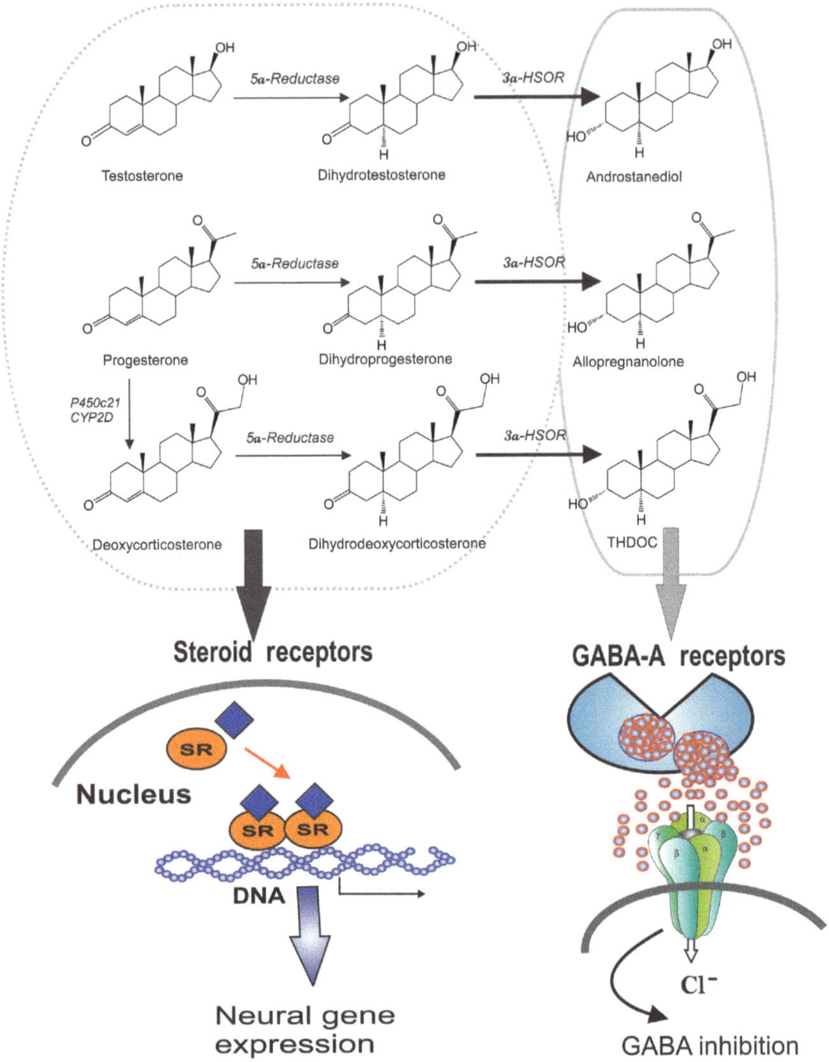

Fig. 1 Neuroendocrine mechanisms of sex differences in seizure susceptibility. The top panel of the figure shows the synthesis of steroid hormones. The bottom panel illustrates their genomic (via binding to intracellular receptors) and non-genomic mechanisms of action (such as enhancing chloride conductance by binding to membrane GABA-A receptors). THDOC is allotetrahydrodeoxycorticosterone. *Reproduced from Reddy D. S. (2020). Brain structural and neuroendocrine basis of sex differences in epilepsy. Handbook of Clinical Neurology, 175, 223–233. https://doi.org/10.1016/B978-0-444-64123-6.00016-3, with automatic permission from Elsevier.*

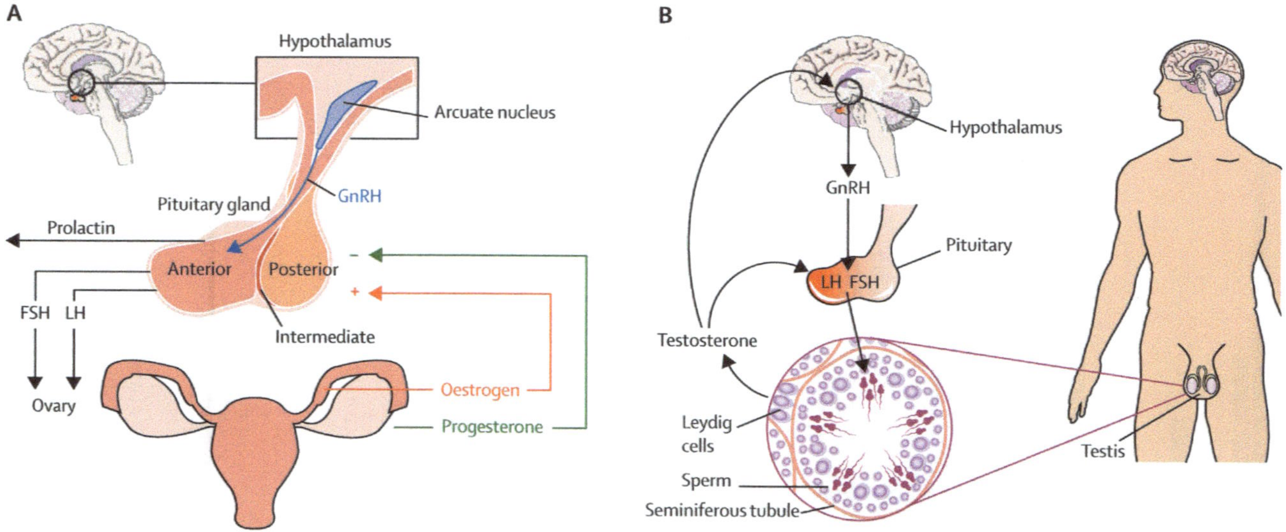

Fig. 2 Hypothalamic-pituitary-gonadal axis in females and males. This simplified illustration of the hypothalamic-pituitary-gonadal axis highlights the key hormonal players in epilepsy in females (A) and males (B). GnRH = gonadotropin-releasing hormone. FSH = follicle-stimulating hormone. LH = luteinizing hormone. *Reproduced from Harden, C. L., & Pennell, P. B. (2013). Neuroendocrine considerations in the treatment of men and women with epilepsy. The Lancet. Neurology, 12(1), 72–83. https://doi.org/10.1016/S1474-4422(12)70239-9, with automatic permission from Elsevier.*

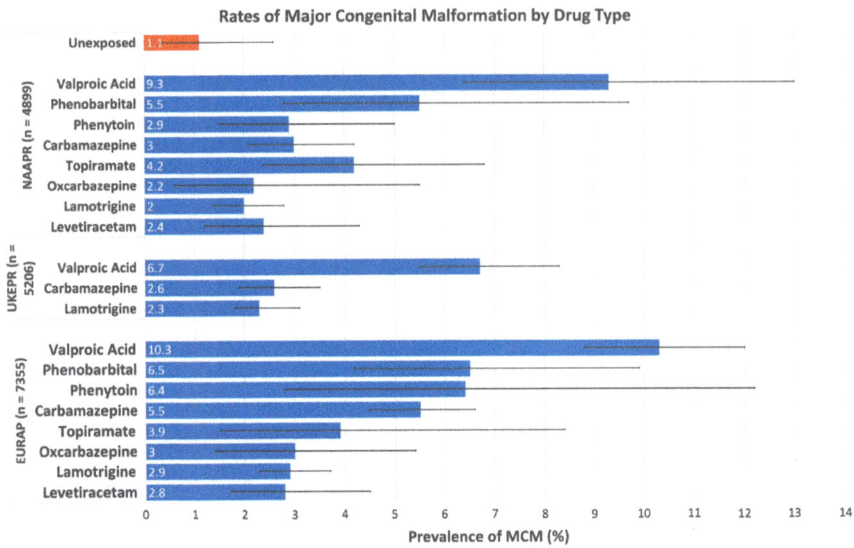

Fig. 3 Rates of major congenital malformations by drug type.

anovulatory cycles with low progesterone-producing luteal phases between days +10 and the next cycle's day +3 where there is again an increased estradiol-to-progesterone ratio with a proposed reduction in progesterone's protective effect (Harden & Pennell, 2013; Herzog, 2008a; Herzog et al., 1997).

A diagnosis of catamenial epilepsy is made by tracking seizure frequency in relation to the patient's menstrual cycle for a minimum of 3 months (Voinescu, 2019). Adjunct measures can include tracking basal temperature to monitor for an increase by at least 0.5 °C the day after ovulation or use of ovulation strips to track ovulation to detect a C2 pattern or the anovulatory cycles of a C3 pattern. Looking for a low progesterone level during the luteal phase may also be considered when a C3 pattern is suspected (Koppel & Harden, 2014).

Treatment of catamenial epilepsy begins with optimization of a patient's seizure control using appropriately chosen antiseizure medication for their seizure type and, if indicated, consideration of epilepsy surgery. Targeted treatments for catamenial epilepsy are not yet supported by robust data as highlighted in a recent Cochrane review (Maguire & Nevitt, 2021). Treatment options can be considered depending on (1) whether the patient's menstrual cycle is regular or not and (2) their specific pattern of catamenial epilepsy. Given the differences in underlying pathophysiology, it is also reasonable to suspect that the different subtypes may require different treatment approaches, but further work is needed to determine this.

If the patient's menstrual cycle is regular, the initial goal of intervention may be to provide increased anticonvulsant protection. Given the decrease in progesterone proposed as contributing to the C1 and C3 patterns, natural progesterone supplementation has been studied. However, the use of natural progesterone lozenges is supported only in the C1 pattern in a single randomized, double-blind, placebo-controlled trial in women with focal onset epilepsy. No significant decrease in seizure frequency (set as at least 50% reduction) was noted compared to placebo when all catamenial patterns were grouped together. However, a pre-specified secondary analysis showed significant seizure reduction in the subgroup of women with a C1 pattern that consisted of at least a threefold decrease in seizures from their baseline (37.8% in the treatment group compared to 11.1% in the placebo group with a P value of 0.037). These patients were given 200 mg oral natural progesterone lozenges twice daily between days 14 and 28 (Herzog et al., 2012). A gradual taper was done closer to the onset of menses with 100 mg bid on days 26–27 prior to discontinuation on day 28. In patients with shorter cycles (<26 days), the taper began with onset of menses (Herzog et al., 2012; Voinescu, 2019).

Several non-hormonal options have been used for years as first-line options for catamenial epilepsy although there is little supporting evidence for them including acetazolamide, benzodiazepines and menstrual suppression. A small retrospective questionnaire study of women with focal or generalized epilepsy and a C1 catamenial pattern showed a significant reduction in seizures that was not different between those taking acetazolamide continuously or intermittently at daily doses between 125 and 750 mg (Lim, Foldvary, Mascha, & Lee, 2001). Doses of acetazolamide at 250–500 mg daily from 3 to 7 days prior to menses have been subsequently proposed to treat C1 patterns (Harden & Pennell, 2013). A small double-blind cross-over study supported the use of clobazam at 20–30 mg daily for 10 days starting 2 days before the interval of perimenstrual seizure worsening (Feely, Calvert, & Gibson, 1982). There is also some suggestion that intermittent use of other antiseizure medications, such as levetiracetam 500–750 mg twice daily reported in a five patient case series (Liu et al., 2015), may be considered.

Alternatively, an ad hoc increase in a patient's regular antiseizure medications can be tried but with caution. Drugs such as phenytoin can rise exponentially with small incremental changes due to its zero-order kinetics (Harden & Pennell, 2013). Although there is no evidence to support these treatments in the C2 or C3 subtypes, some physicians might still try them (Voinescu, 2019). Menstrual suppression is an alternative approach which can be considered in patients with unpredictable cycles or lack of response to the treatments described above to minimize hormonal fluctuations. This may be tried in all forms of catamenial epilepsy (Voinescu, 2019). In a single small open-label study, medroxyprogesterone acetate (a synthetic progestin-only agent) was associated with near 40% reduction in seizures at 1-year follow-up (Mattson, Cramer, Caldwell, & Siconolfi, 1984). Some authors propose a 150 mg dose administered intramuscularly every 10–12 weeks (Harden & Pennell, 2013; Voinescu, 2019) with note of the risk of osteoporosis with long-term use. Continuous oral contraceptive pills are also used as an alternative to similarly target amenorrhea, but there is currently no supportive data and the possible interactions with antiseizure medications should be considered (Harden & Pennell, 2013; Voinescu, 2019) (Tables 1 and 2).

Table 1 Summary of catamenial epilepsy subtypes.

	Proposed timeline	**Hormone**	**Proposed proconvulsant mechanism**
Perimenstrual (Catamenial type 1, C1)	Days −3 to +3	Progesterone	A more rapid decrease in progesterone preceding the estrogen decrease leading to an increased estradiol-to-progesterone ratio
Periovulatory (Catamenial type 2, C2)	Days +10 to −13	Estradiol	A rapid estradiol increase
Inadequate luteal phase with anovulatory cycles (Catamenial type 3, C3)	Day +10 to the next cycle's day +3	Progesterone	Decreased progesterone secretion from an underdeveloped corpus luteum throughout the luteal phase with an associated increase in the estradiol-to-progesterone ratio

Table 2 Summary of treatment options for catamenial epilepsy.

Treatment goals	Treatment types	
Increase anticonvulsant treatment during the higher risk period of the cycle	Hormonal	– Progesterone lozenges 200 mg oral bid between days 14 and 28, 100 mg bid on days 26 and 27 then stopped on day 28. In patients with shorter cycles (less than 26 days), the taper can begin with onset of menses. * *The evidence supports its use if there is at least a three-fold increase in seizures during the C1 period.*
	Non-Hormonal	– Acetazolamide 250–500 mg daily from 3 to 7 days prior to menses – Clobazam 20–30 mg daily for 10 days starting 2 days before the period of seizure worsening – Small increase of baseline antiseizure drugs
Reduce cyclical hormone fluctuations	Hormonal	– Medroxyprogesterone acetate 150 mg every 10–12 weeks via intramuscular injection – Continuous oral contraceptive pills

5. Hormonal influences of antiseizure medications

Another important interaction involves antiseizure medications (ASMs) and hormones. ASMs can impact endogenous hormones such as estrogen, progesterone, testosterone, insulin, and thyroid hormones as well as exogenous hormones in the form of combined hormonal oral contraceptives (Svalheim, Sveberg, Mochol, & Tauboll, 2015). These relationships can be bidirectional and their important interactions are highlighted below.

5.1 Endocrine effects of antiseizure medications

One of the most well studied ASMs with respect to its endocrine effects is valproate. Valproate has been associated with higher risks of anovulatory menstrual cycles, hyperandrogenism, polycystic ovaries and polycystic ovarian syndrome (PCOS) (Taubøll, Isojärvi, & Herzog, 2021). A study by Rättyä et al. showed that female androgen levels can increase within 3 months of valproate exposure (Rättyä et al., 2001). The risk of developing PCOS and hyperandrogenism is highest when valproate is initiated before the age of 20 (Svalheim et al., 2015; Taubøll et al., 2021). Young women should be counseled on these risks if valproate is considered due to the subsequent effects on the menstrual cycle and fertility. These effects may be reversed after valproate is discontinued, especially if it is stopped before the exposed individual reaches 20 years of age (Isojärvi, Taubøll, & Herzog, 2005; Markoula et al., 2020).

The development of PCOS with valproate exposure is felt to be related to the adverse medication side effects of metabolic syndrome, weight gain, hyperinsulinemia, and low serum insulin-like growth factor binding protein-1 concentrations. Hyperinsulinemia and high levels of insulin-like growth factor are stimulants of ovarian androgen synthesis. This has been found to be a significant factor in the pathogenesis of PCOS in women who are obese, but not in women with low BMI (Isojärvi et al., 2005).

In addition to the above proposed mechanism, animal models have also shown that valproate has direct effects on ovarian steroidogenesis and reduces the conversion of testosterone to estrogen in the ovary through its enzyme-inhibiting properties (Isojärvi et al., 2005; Svalheim et al., 2015). These animal models demonstrate that there is a gonadal effect of valproate that contributes to the pathogenesis of PCOS through the resultant androgen-dominant ovarian environment (Isojärvi et al., 2005).

Hepatic enzyme inducing antiseizure medications (EIASMs) include phenobarbital, phenytoin, carbamazepine, topiramate and their derivatives. These medications induce cytochrome P450 hepatic enzyme activity, which can increase the metabolism of other drugs and hormones. EIASMs increase concentrations of sex hormone binding globulin (SHBG), which binds to circulating androgens and estrogens. Elevated SHBG concentrations consequently result in reduced levels of free, circulating androgen and estrogen (Isojärvi et al., 2005; Markoula et al., 2020; Svalheim et al., 2015; Taubøll et al., 2021). These ASM hormonal effects have not been found to correlate with menstrual disorders or reproductive endocrine disorders in this population; however, it has been postulated to contribute to reduced libido (Morrell et al., 2005). Serum dehydroepiandrosterone sulfate (DHAES) concentrations have also been shown to be lower with the use of EIASMs, although the clinical implications of this is not well understood (Markoula et al., 2020; Svalheim et al., 2015). No significant changes in functional levels of serum LH, FSH, or prolactin have been demonstrated with the use of these medications. Interestingly, lower serum levels of T4, free T4, and free T3 have been shown with EIASM use; however, when assessed clinically, there appears to be no clinical implications of these low levels in terms of functional hypothyroidism (Isojärvi et al., 2005; Svalheim et al., 2015). Thyroid hormone replacement is not generally recommended with the use of EIASMs in the absence of clinical signs of hypothyroidism however EIASMs may be less preferred in women with pre-existing difficult-to-control thyroid disorders.

The two most commonly used medications in women with epilepsy are lamotrigine and levetiracetam; neither have been shown to have significant effects on sex hormone levels or menstrual disorders (Markoula et al., 2020; Tauboll et al., 2021). Both lamotrigine and levetiracetam have been associated with patient reports of improved sexual function (Markoula et al., 2020). The endocrine effects of common ASMs are summarized in Table 3.

5.2 Antiseizure medications and hormonal contraception

Hormonal contraception and ASMs can interact with one another in a bidirectional manner. This carries implications for both contraceptive success and seizure control in women with epilepsy. Hormonal contraception refers to a variety of contraceptive methods, including the combined (estrogen and progestin containing) oral contraceptive pills, patch, or vaginal ring, as well as progestin only pills, implants, depot injections, IUDs, and emergency contraception (Reimers, Brodtkorb, & Sabers, 2015). Most emergency contraception contains high concentrations of levonorgestrel.

The efficacy of hormonal contraception can be reduced by interactions with EIASMs such as phenobarbital, phenytoin, topiramate, carbamazepine, oxcarbazepine, and eslicarbazepine. This is important to be aware of, because the rate of failure of the oral contraceptive pill in women with epilepsy is reported to be between 3% and 6%, which is 3–6 times the rate of failure in the general population (Reimers et al., 2015). Women with epilepsy should be counseled about the risk of contraceptive failure when hormonal contraception is used concomitantly with ASMs. Decisions around contraceptive methods and antiseizure medications should be made in an informed manner, understanding these interactions.

EIASMs can reduce the efficacy of hormonal contraception by decreasing circulating concentrations of estrogen and unbound progesterone. This is achieved in two ways: (1) by increasing circulating levels of SHBG; and

Table 3 Summary of endocrine effects of ASMs.

Antiseizure medication	SHBG	DHEAS	Estradiol	Free androgen index	LH	FSH	Prolactin	T4 and free T4	T3
Phenytoin	↑	↓	↔	↓	↔	↔	↔	↓	↓
Phenobarbital	↑	↓	↓	↓	↔	↔	↔	N/A	N/A
Carbamazepine	↑	↓	↓	↓	↔	↔	↔	↓	↓
Valproate	↔	N/A	↓	↑	↔	↔	N/A	N/A	N/A
Lamotrigine	N/A	N/A	N/A	↔	N/A	N/A	N/A	N/A	N/A

(2) by directly increasing the metabolism of the exogenous estrogen and progesterone that is contained in hormonal contraception (Reimers et al., 2015; Tauboll et al., 2021). Carbamazepine can reduce the circulating level of estradiol by as much as 40–50% (Reimers et al., 2015). Similar effects can be seen with topiramate, eslicarbazepine, and perampanel (Reimers et al., 2015). It is very important to counsel women about the risk of incomplete contraceptive efficacy to avoid contraceptive failure. Ethosuximide, valproate, gabapentin, levetiracetam, zonisamide, and lacosamide have not been found to reduce the efficacy or metabolism of hormonal contraception (Bangar, Shastri, El-Sayeh, & Cavanna, 2016; Reimers et al., 2015).

Additional counseling should also be directed toward the use of emergency contraception for women taking EIASMs. The efficacy of emergency hormonal contraception (the "Morning After Pill") can similarly be affected by the aforementioned EIASMs, and the desired effect of the emergency medication may not be achieved. For women taking EIASMs, insertion of a copper IUD is the recommended method of emergency contraception (Reimers et al., 2015). Contraceptive methods that are unaffected by ASMs include barrier methods, intrauterine devices, and tubal ligation.

The relationship between ASMs and hormonal contraceptives is bidirectional. The estrogen component of combined oral contraceptives decreases certain ASMs through induction of glucuronidation, an important metabolic pathway. This is most significantly seen with lamotrigine and to a lesser extent valproate and oxcarbazepine. Induction of glucuronidation can decrease serum drug levels by up to 50% (Tauboll et al., 2021). This effect can develop and resolve within a matter of days, and can have profound effects on seizure frequency. Due to the volatility of drug levels induced by exogenous hormones, combined hormonal oral contraceptives should be avoided, especially for women on lamotrigine therapy (Reimers et al., 2015; Tauboll et al., 2021).

6. Sexual and reproductive dysfunction in epilepsy

Reproductive and sexual dysfunction have been reported to be higher in both men and women with epilepsy compared to the general population (Henning et al., 2019; Herzog, 2008b; Koppel & Harden, 2014; Mameniškienė, Guk, & Jatužis, 2017). Sexual dysfunction can include erectile dysfunction in men, dyspareunia in women, anorgasmia in both, as well as low satisfaction with their sex lives (Henning et al., 2019). In a study by Henning et al. (Henning et al., 2019), ASMs including enzyme inducers were not associated with sexual dysfunction, despite earlier reports of decreased sexual function and low levels of testosterone in patients on EIASMs (Herzog et al., 2005).

In men with epilepsy, semen-related abnormalities (abnormal sperm motility, morphology, or low sperm counts) are frequently seen (Herzog, 2008b) and have been observed to be independently affected by ASM exposure including valproic acid or enzyme-inducers (Isojärvi et al., 2004; Ocek et al., 2018). Although lower birth rates have been reported in men with epilepsy, to what extent psychosocial factors influence these rates remains unclear (Herzog, 2008b).

Women with epilepsy have also been reported to have lower birth rates compared to the general population (Markoula et al., 2020). Studies of birth rate must be interpreted with caution as lower birth rates do not necessarily equate to infertility. Women with epilepsy also face psychosocial factors such as lower marriage rates, higher divorce rates, and fewer sexual experiences in population studies (Markoula et al., 2020). These factors impact the frequency of unprotected intercourse, which in turn affects the recorded birth rate.

Despite potential confounding factors on recorded birth rate, reproductive endocrine disorders have been shown to be more common in women with epilepsy compared to the general population. These disorders include menstrual disorders, hypothalamic amenorrhea, premature ovarian failure, and polycystic ovarian syndrome (Harden, Koppel, Herzog, Nikolov, & Hauser, 2003; Herzog, 2008b; Koppel & Harden, 2014). Seizures can lead to direct alterations of the hypothalamic-pituitary-gonadal axis, resulting in altered pulsatility of GnRH secretion. This in turn can disturb the secretion of LH and FSH, resulting in dysregulation of estrogen and progesterone levels. Variability in LH and FSH secretion, and higher levels of estradiol and DHEAS have been noted in women with temporal lobe epilepsy compared to the general population (Markoula et al., 2020). As well, generalized tonic-clonic seizures and focal seizures with impaired awareness can also be associated with a post-ictal rise in prolactin, which may affect the hypothalamic-pituitary-ovarian axis to result in decreased levels of GnRH (Markoula et al., 2020). Clinically, women with epilepsy appear to have a higher frequency of anovulatory menstrual cycles. This has been described in women with both generalized epilepsies and temporal lobe epilepsy (Cummings, Giudice, & Morrell, 1995; Morrell et al., 2002). Rates of premature ovarian failure are also higher in women with epilepsy than in healthy controls and women with poorly controlled epilepsy demonstrate lower levels of anti-Mullerian hormone, a measure of ovarian reserve (Herzog, 2008b; Markoula et al., 2020). PCOS is another important consideration when assessing fertility in women with epilepsy, as it is known to affect reproductive function. The relationship between PCOS and women with epilepsy has largely been associated with valproate use. The prevalence of PCOS in all women with epilepsy has been reported to be as high as two- to fourfold that of the general population (Herzog, 2006). Women with epilepsy who are not on any ASM, however, have generally not been shown to have an increased risk of PCOS (Markoula et al., 2020) with the exception of medically refractory left temporal lobe epilepsy, where alterations of the hypothalamic-pituitary-gonadal axis are significant enough to increase the risk of PCOS (Herzog, 1993; Kalinin & Zheleznova, 2007).

Women with epilepsy experiencing infertility and seeking fertility therapy should be reassured by a recent study by Larsen et al., (Larsen, Jølving, Fedder, & Nørgård, 2020) demonstrating no significant difference between live births per embryo in women with epilepsy using assisted reproductive technology (ART) compared to the general population. The use of ASM(s), polypharmacy, or EIASM during ART treatment was also not found to have a significant effect on the chance of live birth after embryo transfer (Larsen et al., 2020). Furthermore, a recent study by Pennell et al. in 2018 (Pennell et al., 2018) prospectively compared a cohort of women with epilepsy and healthy controls trying to conceive, and showed no difference between the two groups in time to pregnancy as well as live birth rates. A limitation in this study was the exclusion of women with pre-existing infertility or fertility issues such as PCOS therefore limiting its generalizability (Pennell et al., 2018). Overall, these findings highlight that, in the absence of pre-existing fertility issues, women with epilepsy have an equal chance of conceiving compared to women without epilepsy.

7. Special populations

7.1 Pediatric and adolescent concerns

The management of epilepsy in girls and young women requires unique considerations related to the important transitions that occur in their development at this age. The diagnosis of a chronic neurological condition can raise challenges with respect to psychosocial well-being, self-identity, sexuality, and menarche, which should be recognized by their health care provider.

Many individuals experience their first seizure during their teenage years, around the time of significant developmental changes, such as menarche. Among girls with epilepsy, an association has been shown between the time of first seizure onset and the average age of menarche. Girls who develop seizures before puberty tend to have an older mean age of menarche than girls who develop seizures after puberty. In addition, the age of menarche is also a significant predictor of age of seizure onset. Seizure onset in women with epilepsy most commonly occurs 2 years before, to 6 years after, the year of menarche. This suggests that seizure debut in this population may be associated with the processes of both adrenarche and gonadarche (which on average extends over a period of 4.5 years), as opposed to menarche alone (Herzog, Mandle, & MacEachern, 2019; Tauboll et al., 2021).

Receiving a diagnosis of a chronic condition such as epilepsy can be especially challenging for young girls, with many psychosocial implications. This population tends to report greater impact of the diagnosis on their quality of life, increased anxiety, and worse self-image when compared to their male counterparts (Zupanc & Haut, 2008). Young women are also less likely to acquire their driver's license at the age of 18, compared to young men with epilepsy (Zupanc & Haut, 2008). Studies of stigma and depression have not revealed gender differences in adolescents, but both stigma and comorbid mental health concerns are prevalent and important components of health-related quality of life in this age group (Zupanc & Haut, 2008).

Adolescence is an important time for the development of sexuality and self-identity. Chronic conditions such as epilepsy can have implications on the evolution of one's sexual identity. Young women with epilepsy have been shown to have unmet sexual and reproductive healthcare needs (Kirkpatrick et al., 2020). A Norwegian study found that youths with epilepsy were more likely to have higher risk sexual behaviors and were less likely to use contraception (Stefanidou & Montouris, 2020). It is important to discuss sexual and reproductive healthcare with young women with epilepsy due to the complex hormonal, reproductive, and sexual implications of epilepsy and its management.

7.2 Transgender populations

Transgender persons with epilepsy have unique needs that may intersect with their epilepsy care in terms of gender-affirming health care and cosmetic side effects. Gender-affirming treatments may target the suppression of puberty in young individuals, with the use of gonadotropin-releasing hormone analogs, or in older patients, the process of feminization or masculinization may be achieved with exogenous hormone therapies. It is important to note that during male to female transitions, the interactions between EIASMs and hormonal therapies can result in undesired effects. EIASMs can lower the effective dose of exogenous estrogen and progesterone, and estrogen therapies can have a negative effect on seizure control. Feminizing hormones and ASMs may therefore need to be measured and monitored to ensure adequate effects of respective treatments. Conversely, in female to male transitions, EIASMs can increase SHBG concentrations resulting in reduced levels of free testosterone. This can lead to erectile dysfunction and reduced libido in transitioned males (Johnson & Kaplan, 2017; Stefanidou & Montouris, 2020).

The cosmetic side effects of ASMs may also be of particular relevance to women with epilepsy and transgender individuals, with implications for body image. Women are more than twice as likely to report cosmetic side effects of ASMs than men, with resultant changes in medication (Chen et al., 2015). The ASM side effects of weight gain and hair loss have been reported as the most significantly intolerable cosmetic effects for women (Chen et al., 2015). Other notable side effects include acne, hirsutism, and gingival hyperplasia, which can have impacts on quality of life. These adverse effects are most commonly seen with the use of valproate and phenytoin (Chen et al., 2015). It is important to be aware of these adverse effects so specific medications can be avoided in patients who are at higher risk of developing cosmetic side effects or in those who express significant concerns about the undesired effects. An alternative medication such as levetiracetam could be considered to avoid these potential effects.

8. Pregnancy and epilepsy

Epilepsy is one of the most common neurological disorders seen in pregnancy, and is the leading neurological disorder which requires continuous treatment during pregnancy (Vélez-Ruiz & Pennell, 2016). Approximately 15 million women with epilepsy are of childbearing age (Tomson et al., 2019). Overall, the majority of women with epilepsy will have a normal pregnancy, delivery, and healthy child. However, it is important to be aware that a diagnosis of epilepsy does increase the risk of maternal, obstetrical, and neonatal complications in pregnancy, with a 10 times greater risk of maternal mortality than the general population, and increased rates of cesarean section

(Borthen, Eide, Daltveit, & Gilhus, 2011; Panelli et al., 2021; Tomson et al., 2019). The management of epilepsy in women, therefore, requires special considerations during childbearing years. Particular attention must be made to preconceptual counseling, teratogenicity of ASMs, changes in seizure control and drug metabolism during pregnancy, and recommendations regarding peripartum care and breastfeeding.

8.1 Preconception care

Unplanned pregnancies occur in ≥50% of the general population, with a similar incidence seen among women with epilepsy (Herzog, Mandle, Cahill, Fowler, & Hauser, 2017). Early and routine preconception counseling in this population is therefore particularly important, as these discussions can have implications for family planning and pregnancy outcomes. Conversations around pregnancy planning should be raised routinely during clinic visits with women of childbearing potential to ensure that patients who do not have active plans to start a family are using appropriate contraception, or those who do intend to conceive in the near future understand the associated risks and can be optimized appropriately.

For women with epilepsy who are planning to conceive, proactive preconception care has been shown to improve the outcomes of pregnancy. These outcomes include improved maternal seizure control, fewer major congenital malformations in the child, and an increased likelihood of the patient being on antiseizure drug monotherapy with a drug other than valproate at the time of conception (Abe et al., 2014; Betts & Fox, 1999). Initial preconception visits should review the woman's epilepsy history to confirm her diagnosis, epilepsy syndrome, and appropriateness of current management. The type and dose of ASMs must be evaluated, balancing the risks of seizures versus the risks of teratogenicity. These risks should be discussed during the visit through a process of joint decision-making. If ongoing ASM is indicated, the goals of care during this period are to consider if the patient can be on monotherapy with a medication other than valproate, targeting the lowest effective medication dose. If a change to ASM is considered, this should ideally be initiated as early as 1 year in advance, to allow sufficient time to assess the effectiveness of the new regimen before conception. The best predictor of seizure control *during* pregnancy is seizure control 9–12 months *prior* to pregnancy (Kinney & Morrow, 2016; Tomson et al., 2019).

8.2 Preconception folic acid

An integral component of preconception care involves the initiation of folic acid. All women of childbearing age who are taking ASMs should be advised to supplement with folic acid regardless of family planning status, due to the risks of unplanned pregnancies in women with epilepsy. In the general population, a dose of 0.4 mg of folic acid daily is recommended for all women of childbearing age to reduce the risk of congenital malformations, such as neural tube defects, in the child. It is suggested that higher doses of folic acid may be required in women with epilepsy taking ASMs that interfere with folate absorption such as phenytoin, carbamazepine, or barbiturates, or those with greater risks of neural tube defects, such as valproate. Robust data on the optimal periconceptional dose of folic acid in women with epilepsy is lacking, but the ILAE Task Force for the management of epilepsy in pregnancy recommends at least a dose of 0.4 mg daily, up to a dose of 5 mg daily. This should be initiated at least 3 months prior to pregnancy and continued throughout the pregnancy (Tomson et al., 2019).

Studies specifically looking at the effects of folic acid supplementation in women with epilepsy are inconclusive in terms of reducing the risk of major congenital malformations but positive associations with better neurodevelopmental and behavioral outcomes in children. The Neurodevelopmental Effects of Antiepileptic Drugs (NEAD) Study found improved IQ scores in 6-year-old children of women with epilepsy who took preconception folic acid compared to those who did not (Meador et al., 2013). The Norwegian moBa study also found an association between folic acid supplementation and reduced risks of autism spectrum disorder (Bjørk et al., 2018). Prospective epilepsy pregnancy registries (EURAP and UKEPR), however, have failed to demonstrate a lower risk of major congenital malformations in women with epilepsy who were taking folic acid supplementation (Morrow et al., 2009; Tomson et al., 2018); however, these studies were not designed to study the effects of folic acid supplementation.

8.3 Seizure management in pregnancy

The most important predictor of seizure burden during pregnancy is the frequency of seizures 9–12 months before pregnancy (Kinney & Morrow, 2016; Tomson et al., 2019). Historically, it has been understood that two-thirds of pregnant women will not experience any change to their seizure frequency from their pre-pregnancy baseline (Battino et al., 2013). More recently, the Maternal Outcomes and Neurodevelopmental Effects of Antiepileptic Drugs (MONEAD) Study Group found no change in seizure frequency in pregnant women throughout all trimesters, delivery, and peripartum phases, compared to nonpregnant controls, specifically when ASM dose adjustments were made to maintain the target drug levels prior to conception (Pennell et al., 2020). In women who experience seizures in the month prior to conception, a 15-fold increase in seizure risk during pregnancy has been shown, whereas in those who are seizure free 9–12 months prior to conception, more than 80% will remain seizure free during pregnancy (Tomson et al., 2019). In women with catamenial epilepsy, there tends to be an improvement in seizure frequency during pregnancy. Cagnetti et al. demonstrated at least a 50% reduction in seizure frequency in approximately 44% of women with catamenial epilepsy (Cagnetti, Lattanzi, Foschi, Provinciali, & Silvestrini, 2014).

When women with epilepsy experience an increase in seizure frequency, it is most often seen during the second and third trimesters, as opposed to the first trimester. With increasing recognition of changes in drug clearance during pregnancy, this may be attributed to falling levels of ASMs (especially lamotrigine, oxcarbazepine, and levetiracetam) during the second and third trimesters. The highest incidence of seizures is seen around the time of labor and delivery, although this only occurs in 1–2% of pregnancies in women with epilepsy (Stephen, Harden, Tomson, & Brodie, 2019; Tomson et al., 2019). Predictors of worsened seizure frequency during pregnancy include focal epilepsy syndromes, polytherapy of ASMs, decreased ASM concentrations compared to preconception baselines, medication nonadherence, sleep deprivation, and severe emesis resulting in poor oral medication retention (Reisinger, Newman, Loring, Pennell, & Meador, 2013; Tomson et al., 2019).

Optimal seizure control during pregnancy is important, as seizures during pregnancy can pose risks to the mother and fetus. Traumatic maternal injuries, which may occur during seizures, can cause trauma to the uterus and potentially threaten the fetus (Hiilesmaa & Teramo, 2013). During generalized tonic clonic seizures, the mother may enter a state of hypoxia with resultant lactic acidosis, which poses risks of fetal hypoxia and acidosis if there is transfer through the placenta (Hiilesmaa & Teramo, 2013). Generalized seizures during pregnancy have not been associated with increased risks of major congenital malformations in the child (Hiilesmaa & Teramo, 2013). Focal seizures without secondary generalization are unlikely to have significant fetal implications; however, there are case reports of brief fetal distress during focal seizures with impaired awareness (Sahoo & Klein, 2005).

A retrospective cohort study by Chen et al. (Chen, Chiou, Lin, & Lin, 2009) further demonstrated the implications of seizures during pregnancy with regards to pregnancy outcomes, showing a significantly increased odds of small for gestational age infants in women with epilepsy who had seizures during pregnancy versus women with epilepsy who did not have seizures during pregnancy. Conversely, a retrospective study by Borthen et al. (Borthen et al., 2011) did not demonstrate any increased risk of obstetrical complications (including pre-eclampsia, bleeding in pregnancy, induction, cesarean section and major congenital malformations) in women with active epilepsy. Importantly, in this study active epilepsy was defined as seizures within the 5 years preceding delivery of the newborn, and women with epilepsy who had seizures *during* pregnancy were not clearly delineated.

With regards to maternal risks, epilepsy and uncontrolled seizures confer up to a 10-fold increased risk of maternal mortality compared to the general population. Fortunately, maternal mortality is a rare outcome in the general population (MacDonald, Bateman, McElrath, & Hernández-Díaz, 2015). Almost 80% of deaths in women with epilepsy are related to sudden unexpected death in epilepsy (SUDEP) (Edey, Moran, & Nashef, 2014). Women with epilepsy furthermore have 3 times the risk of severe maternal morbidity (SMM), as defined by the Centers for Disease Control and Prevention, and 4 times the risk of non-transfusion SMM compared with women of the general population (Panelli et al., 2021). Status epilepticus is infrequent during pregnancy and in the EURAP population-based study, it was only reported in 0.6% of all pregnancies, with convulsive status in 0.3% (Battino et al., 2013). In this study, convulsive status epilepticus was associated with one case of perinatal death, and three children born with major congenital malformations (Battino et al., 2013; Sveberg, Svalheim, & Taubøll, 2015).

A diagnosis of epilepsy, regardless of seizure activity or ASMs exposure, also carries a small but significantly higher risk of obstetrical complications such as spontaneous miscarriage, hypertensive disorders, hemorrhage, induction of labor, cesarean section, preterm birth, and fetal growth restriction according to a 2015 meta-analysis (Viale et al., 2015). The risk of cesarean section in women with epilepsy is reported to be up to 1.5 times the risk in healthy controls (Stephen et al., 2019). It is important to note that epilepsy itself is not an indication for cesarean section, but seizures during labor can sometimes interfere with a woman's ability to labor successfully.

Women with epilepsy should be counseled on these maternal and fetal risks during their preconception visits. Optimization of seizure control should be strongly emphasized, and the physician should target treatment optimization ideally a year before conception.

8.4 Antiseizure medications in pregnancy

8.4.1 Teratogenic risks

Antiseizure medications can carry risks to the fetus including major congenital malformations (MCM), fetal growth restriction, and poor developmental and neurobehavioral outcomes (Tomson et al., 2019). Certain ASMs carry higher risks of teratogenicity, and it is important to understand these risks to inform the pharmacological management of women with epilepsy during pregnancy.

It is well understood that intrauterine exposure to ASMs can increase the risk of MCMs above that of the general population, with differential effects from exposures to individual ASMs. MCMs are defined as any structural abnormality of surgical, medical, or cosmetic importance discovered in a newborn. The most common malformations associated with ASM exposure include cleft lip, cleft palate, neural tube defects, skeletal malformations including hypoplasia of the mid face and fingers, hypospadias, and congenital heart malformations (especially septal defects) (Meador et al., 2006).

With ASM monotherapy, the greatest risk of MCM has been widely shown to be associated with exposure to valproate. Three large pregnancy registries: North American Anti-epileptic Drug Pregnancy Registry (NAAPR), International Registry of Antiepileptic Drugs and Pregnancy (EURAP), and UK Epilepsy and Pregnancy Register (UKEPR) have shown consistent findings through observational data, demonstrating the greatest risk of MCM with exposure to valproate and the comparatively low risk with lamotrigine and levetiracetam. Fig. 4 demonstrates the rates of MCMs associated with individual ASMs as seen in each of the respective pregnancy registries (Campbell et al., 2014; Hernández-Díaz et al., 2012; Tomson et al., 2018). Consistent findings have also been reported in meta-analyses by Veroniki et al. and Weston et al. (Veroniki et al., 2017; Weston et al., 2016). Veroniki et al. found no increased odds of MCM with the use of lamotrigine or levetiracetam in pregnant women with epilepsy. Weston et al. did not find any significantly increased risk of MCM with exposure to lamotrigine, gabapentin, levetiracetam, oxcarbazepine, primidone, or zonisamide (Veroniki et al., 2017; Weston et al., 2016).

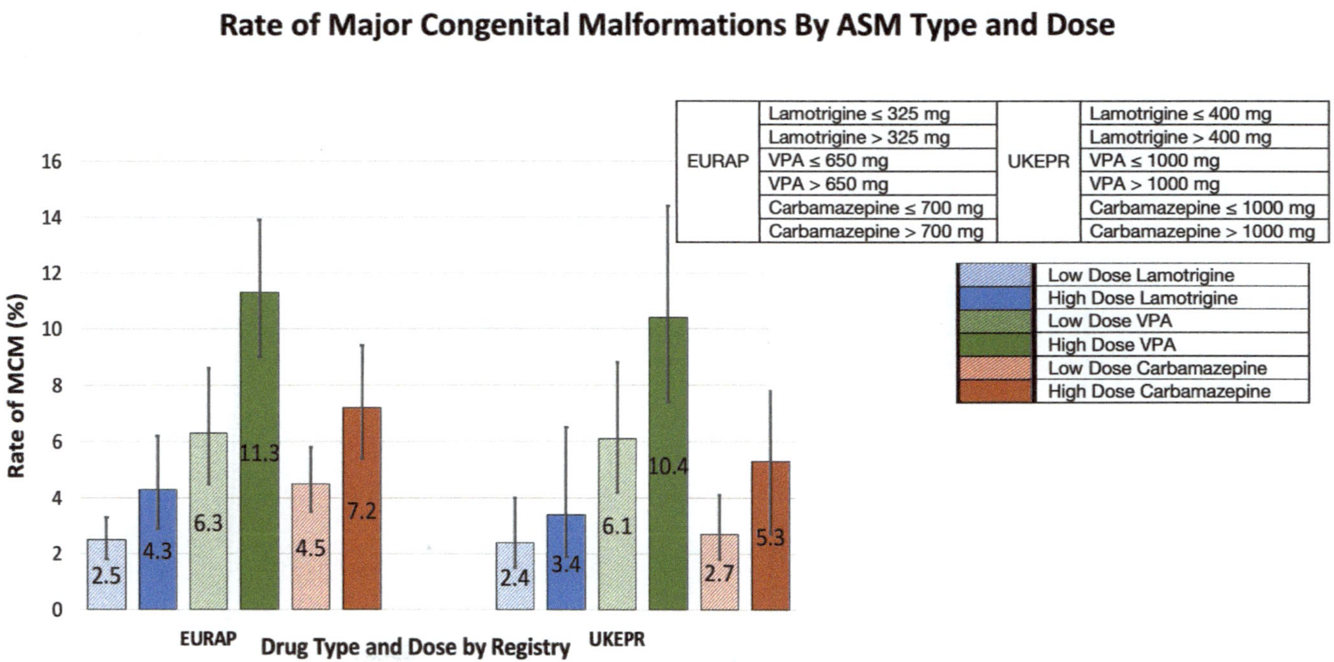

Fig. 4 Rates of major congenital malformations by drug dose.

Table 4 Most common major congenital malformations by drug.

Antiseizure drug	Major congenital malformations
Oxcarbazepine	**Cardiac**, cleft lip or palate, skeletal (polydactyly)
Topiramate	**Cardiac, cleft lip or** palate, neural tube defects
Carbamazepine	**Cardiac,** hypospadias, renal (hydronephrosis), neural tube defects, GI defect, skeletal (polydactyly), cleft lip or palate
Phenytoin	**Cardiac, neural tube defects, cleft lip or palate**
Phenobarbital	**Cardiac,** neural tube defects, skeletal (polydactyly), renal, hypospadias, cleft lip or palate
Valproate	**Hypospadias, Cardiac defect**, **Neural Tube Defects, Cleft lip or palate,** skeletal (polydactyly), renal, GI defect, neurodevelopmental
Lamotrigine	**Hypospadias**, **cardiac,** GI defect, cleft lip or palate, renal (hydronephrosis), neural tube defects
Levetiracetam	**Cardiac, cleft lip or palate, hypospadias**, GI defect, renal defect

MCMs reported for individual ASMs, with the most frequently reported MCMs bolded.

The specific MCMs that are frequently seen with exposure to individual ASMs is summarized in Table 4.

In addition to individual ASM effects, dose-dependent relationships have also been reported for valproate, lamotrigine, phenobarbital, and carbamazepine with regards to the associated risk of MCM (Campbell et al., 2014; Hernández-Díaz et al., 2012; Tomson et al., 2018). Based on registry data from EURAP, lamotrigine doses of $\leq$325 mg/day are associated with the lowest risk of MCM. The risks of MCM increase with higher doses of valproate, although an increased risk of MCM is still seen with valproate doses as low as 650 mg/day.

Fetal growth restriction has been most consistently and significantly reported with intrauterine exposure to topiramate. The odds of microcephaly or birth weight less than the 10th percentile is as high as 4.8 and 3.1, respectively, in children who are exposed to Topiramate compared to unexposed children born to healthy control women (Veiby, Daltveit, Engelsen, & Gilhus, 2014). An increased risk of neonatal small for gestational age (SGA) or small head circumference (although not necessarily meeting criteria for microcephaly) has also been associated with intrauterine exposures to primidone, phenobarbital, carbamazepine, and valproate (Tomson et al., 2019). The NEAD study demonstrated that the risk of SGA or microcephaly as a neonate in children with intrauterine exposures to carbamazepine or valproate normalized by an age of 36 months, when compared to children exposed to phenytoin or lamotrigine in-utero (Pennell et al., 2012). A greater odds of SGA has also been reported with the use of ASM polytherapy, with a reported odds of 2.3 times that of the general population (Veiby, Daltveit, Engelsen, & Gilhus, 2009). These outcomes are important to be aware of, as SGA is associated with higher risks of stillbirth, and long-term health outcomes, and microcephaly is associated with neurodevelopmental sequelae (Pennell et al., 2012).

ASM exposure during pregnancy has also been found to have effects on neurodevelopmental outcomes. The greatest risk has been demonstrated with intrauterine exposure to valproate. An increased risk of autism spectrum disorder, attention deficit hyperactivity disorder, learning disabilities, and reduced IQ by 7–11 points has been shown in these children in a dose-dependent fashion (Bromley et al., 2013; Christensen et al., 2019; Meador et al., 2009, 2013, 2011). No significant effect on IQ or early childhood development has been noted with fetal exposure to lamotrigine, levetiracetam, oxcarbazepine, or phenytoin during pregnancy (Tomson et al., 2019).

When considering pharmacologic management in pregnant women with epilepsy, selection of an ASM should avoid valproate and topiramate. Lamotrigine and levetiracetam are the most commonly used and recommended ASMs for pregnant women with epilepsy, given their favorable profile with respect to teratogenicity (Kinney & Morrow, 2016; Shih et al., 2017; Tomson et al., 2019). Whenever possible, monotherapy should be targeted with the lowest effective ASM dose. An increased risk of MCM has been reported with ASM polytherapy (Veroniki et al., 2017), however, the magnitude of the risk has been shown to be more greatly affected by the type and dose of the ASMs (i.e., the inclusion and dose of valproate) as opposed to the actual number of combined ASMs (Holmes, Mittendorf, Shen, Smith, & Hernandez-Diaz, 2011).

8.4.2 Therapeutic drug monitoring in pregnancy

During pregnancy, physiological changes alter ASM clearance and subsequently result in reduced serum drug levels as the pregnancy progresses. These changes can occur as a result of altered drug absorption, distribution, renal clearance, and hepatic metabolism. The extent of change in ASM clearance is most significantly associated with the drug-elimination pathway (Tomson et al., 2019; Voinescu et al., 2018).

Significant alterations in ASM clearance have been reported with the use of lamotrigine, levetiracetam, and oxcarbazepine in pregnancy. Topiramate, phenobarbital, phenytoin, and zonisamide have also been shown to decline in serum concentrations during pregnancy (Tomson et al., 2019). The greatest reductions in the serum concentrations of lamotrigine and levetiracetam occur during the first trimester due to significant increases in drug clearance. Studies show around a 1.5-fold increase in lamotrigine clearance by the end of the first trimester and up to a 2-fold increase in levetiracetam clearance (Karanam et al., 2018; Voinescu et al., 2018). Female fetuses and genetic polymorphisms of uridine diphosphate-glucuronosyltransferase (UGT) are associated with higher rates of lamotrigine clearance during pregnancy (Stephen et al., 2019).

It is important to be aware of these pharmacologic changes during pregnancy, as the risk of seizures increases when the concentration of ASM falls below 65% of the preconception target baseline concentration (Voinescu et al., 2018). The preconception target baseline ASM concentration should be determined during preconception visits by measuring the serum concentrations of ASMs once seizure control is optimized. Serum drug monitoring can then guide ASM dose adjustments during pregnancy, to ensure that drug levels do not fall more than 35% of the target concentration, thereby increasing the risk of breakthrough seizures. The ILAE Task Force on Women and Pregnancy recommends at least monthly drug monitoring for Lamotrigine, Levetiracetam, Phenobarbital, Phenytoin, Oxcarbazepine, Gabapentin, Zonisamide, and Topiramate during pregnancy (Tomson et al., 2019). Additional drug levels can be obtained in the context of breakthrough seizures or adverse medication effects.

If ASM serum levels are not obtainable and women are prescribed lamotrigine, levetiracetam, or oxcarbazepine, one can consider increasing the medication dose by 30–50% after the first trimester (Tomson et al., 2019). This is an empiric dose adjustment that can reduce the risk of breakthrough seizures, given the known increase in the clearance of these drugs that occurs during the first trimester.

Post-partum, there is a normalization of ASM clearance that occurs. A cautious empiric postpartum drug taper back to near preconception dose should be prescribed to prevent drug toxicity especially with lamotrigine. The rate of normalization of ASM clearance postpartum is not well studied, but generally the return to pre-pregnancy baseline occurs over the first 2–3 weeks postpartum, if not within the first 1–2 months. Drug levels can continue to be monitored during the postpartum phase to inform reductions in the ASM dose, but this is not necessary in most cases (Tomson et al., 2019).

8.5 Advanced epilepsy therapies and pregnancy

8.5.1 Neurostimulation

In addition to antiseizure medications, candidacy for epilepsy surgery or neurostimulation therapies should be contemplated early and may be influenced by timing of family planning. Epilepsy surgery is generally underutilized in the treatment of drug resistant focal epilepsy and surgical candidacy should be more frequently considered in the treatment of women with epilepsy. A recent study by Kandraju et al. demonstrated that women with epilepsy who are good surgical candidates have improved seizure control and reduced antiseizure drug burden during pregnancy, if they undergo surgical resection prior to pregnancy (Kandraju et al., 2020). An observational study by Fabris et al. (2016) looked at the effect of epilepsy surgery on fertility rates in women with epilepsy. They found no significant effect on the number of pregnancies or births in women with epilepsy following epilepsy surgery, but did suggest that there may be an indirect effect on birth rates through improved societal factors such as marriage rates and social integration (Fabris et al., 2016). Studies of sexuality have also found that surgical resection improves experiences of sexuality in women with epilepsy, as a result of subsequent improvements in seizure frequency (Markoula et al., 2020). These studies overall suggest that women with focal, drug resistant epilepsy should not delay surgical epilepsy evaluation and intervention.

Fewer studies have investigated the effects of neurostimulation on women with epilepsy. Vagus nerve stimulation (VNS) is a relatively safe and effective therapy for drug resistant epilepsy in those who are not candidates for resective surgery. Theoretical concerns have been raised regarding increased risks of obstetrical complications during pregnancy and labor in women with VNS. These concerns are derived from the role that the vagus nerve plays in mediating uterine contractions and regulation of uterine blood flow (Ding et al., 2021; Suller Marti et al., 2019). Animal models of VNS in pregnancy have not found any major congenital malformations or alterations in gonadotropin levels as a rest of neurostimulation (Suller Marti et al., 2019). Overall, studies suggest that VNS is reasonably safe in pregnancy, however, further research is needed in this field.

Even fewer studies have investigated the effects of responsive neurostimulation (RNS) on pregnancy outcomes. RNS is a neuromodulatory approach that delivers stimulation in response to abnormal intracranial EEG signals. Li et al. found no increased rates of obstetrical complications or major congenital malformations in women with epilepsy treated with RNS compared to those treated with ASMs (Li et al., 2021).

The effects of deep brain stimulation (DBS) on pregnancy outcomes has also been minimally studied in women with epilepsy. DBS has been more frequently used in the treatment of movement disorders and psychiatric diseases, where the majority of pregnancy and obstetrical outcomes have been investigated. In these studies, no significant adverse effects related to DBS have been found. In women with epilepsy, similar findings have been reported (Bóné et al., 2021; Scelzo et al., 2015).

8.5.2 Ketogenic diet

The safety of the ketogenic diet in pregnancy has not been well studied in humans. Only two case reports have been published that describe obstetrical and neonatal outcomes in women who were prescribed the ketogenic diet during pregnancy. In one of these cases, the child was born with bilateral ear deformities, and in the other case, there were no complications. Given the nature of this publication, causality with exposure to the ketogenic diet cannot be determined. Recent international recommendations based on expert consensus on ketogenic diet therapies consider pregnancy a contraindication to ketogenic diet therapy (Cervenka et al., 2021).

8.6 Postpartum care

8.6.1 Breastfeeding

For all children, the World Health Organization recommends exclusive breastfeeding for at least the first 6 months of life. Breastfeeding has been shown to have a multitude of benefits for both the infant and mother (Kramer & Kakuma, 2012). The same recommendations and benefits are applicable for infants born to mothers with epilepsy. Some mothers with epilepsy may express concerns regarding the risks of ASM exposure to their children through breast milk. All ASMs are lipophilic and therefore are present in breast milk; however, the concentration of ASMs that can be found in breast milk are low. Breastfeeding by mothers taking ASMs has not been shown to be associated with poor neurodevelopmental outcomes as per the NEAD and MoBa studies (Meador et al., 2010; Veiby, Engelsen, & Gilhus, 2013) and may have a cognitive benefit with an IQ of 4 points higher noted in breastfed infants of women taking ASMs (Meador et al., 2014). Women with epilepsy should be reassured and encouraged to breastfeed their infants.

8.6.2 *Postpartum mood disorders*

Pregnancy in particular infers a unique risk for psychiatric illness. An increased risk of postpartum depression and anxiety has been noted particularly in medically refractory cases (Bjørk, Veiby, Engelsen, & Gilhus, 2015). The increased vulnerability may be partly explained by the higher occurrence of pre-existing mood disorder in people with epilepsy that may be heightened by pregnancy (Verreault et al., 2014). Specific concerns have been described include fear of congenital malformations (Turner et al., 2008) or child care concerns such as how to safely care for the baby when outside the home or bathe them (Bagshaw, Crawford, & Chappell, 2008).

9. Menopause and bone health

Menopause is considered to have occurred when a woman has gone a full year without menses and is typically noted between the ages of 48 and 52 years. In contrast, perimenopause can last 4.8–5.5 years and includes the first year after cessation of menses as well as the menopausal transition when the menstrual cycle changes begin (Burger et al., 2007). Premature ovarian failure, characterized by undergoing menopause before 40 years (Burger et al., 2007), has been described in women with epilepsy with a higher incidence seen in women with a history of catamenial epilepsy (Klein, Serje, & Pezzullo, 2001). A correlation between higher seizure frequency and earlier age at onset of menopause has also been described (Harden et al., 2003). This is thought to be related to dysfunction of the hypothalamic-pituitary axis described in the previous section leading to poor maturation and early loss of follicles (Harden & Pennell, 2013).

The initial perimenopausal period is characterized by unpredictable fluctuations in estradiol levels from cycle-to-cycle with less prominent luteal progesterone rises, leading to periods with a relatively high estrogen-to-progesterone ratio. This is followed by an eventual decrease in estradiol levels. As a woman moves from the early to late menopausal transition, the duration of the menstrual cycle increases, followed by skipped cycles before eventual sustained amenorrhea. Changes in the amount of bleeding per cycle (which can be lighter or heavier) are also noted. Mood changes, bloating and breast tenderness may be observed during periods of high estradiol. However, it is the vasomotor symptoms, particularly hot flashes thought to be related to decreasing estradiol levels in the later phase of the menopausal transition, which can be most distressing to some women and prompt medical treatment which can be hormonal or non-hormonal

(Burger et al., 2007). The post-menopausal period is characterized by relatively stable, lower levels of hormones (Burger et al., 2002, 2008; Sveinsson & Tomson, 2014).

Seizure patterns may change during this period of transition. Although no prospective research has been completed, three retrospective, cross-sectional studies have been done based on patients' self-reporting of seizure frequency (Abbasi, Krumholz, Kittner, & Langenberg, 1999; Harden, Pulver, Ravdin, & Jacobs, 1999; McAuley, Koshy, Moore, Peebles, & Reeves, 2000). Two of these (Abbasi et al., 1999; McAuley et al., 2000) did not distinguish between the perimenopausal and menopausal period and no significant changes in seizure pattern were reported. The third study showed that women with a pre-existing susceptibility to changes in seizure frequency with hormonal fluctuations, as suspected from a history of catamenial epilepsy, were more likely to experience increases in seizures during the perimenopausal period and decreases thereafter (Harden et al., 1999).

Although known to be the most effective treatment for vasomotor symptoms (Burger et al., 2007), estrogen-containing hormone replacement therapy (HRT) is generally discouraged in women with epilepsy because of concern for increased seizure frequency. This is supported by a single double-blinded randomized controlled trial that looked at the relationship between seizure frequency and hormone replacement therapy. A combination of conjugated equine estrogen (CEE) and medroxyprogesterone acetate (MPA) was used daily for 3 months at a dose of 0.625 and 2.5 mg or at double-dose. A dose-dependent increase in seizures was noted as well as a decrease in lamotrigine levels of up to 30% (Harden et al., 2006). Of note, MPA is not metabolized to anticonvulsant neurosteroid allopregnanolone (Harden et al., 2006). In these patients, non-hormonal treatments can be considered including paroxetine, venlafaxine, and gabapentin which have the best supportive evidence (Burger et al., 2007). However, vasomotor symptoms may require the more effective estrogenic treatments if they are quite disabling. Of particular relevance to patients with epilepsy would be an associated sleep impairment from vasomotor symptoms, which may be a seizure trigger. In such cases, it has been proposed that a natural progesterone be used along with a simplified estrogen compound such as 17-beta-estradiol (Erel & Guralp, 2011; Harden, 2008; Peebles, McAuley, Moore, Malone, & Reeves, 2000).

Osteoporosis is two to six times more prevalent in people with epilepsy compared to the general population (Fernandez, Cooke, & Patel, 2019).

The elevated risk is multifactorial in nature, contributed to by ASMs as well as socioeconomic consequences of epilepsy itself (e.g., poor diet, reduced physical activity) (Diemar, Sejling, Eiken, Andersen, & Jørgensen, 2019). Bone health is a particularly relevant issue for women with epilepsy as female gender and the post-menopausal decline in estrogen are two independent risk factors for osteoporosis (Compston, McClung, & Leslie, 2019). In this population, there is an increased risk of fractures with associated morbidity and mortality. This risk is elevated in women with epilepsy, even when seizure-related injuries are excluded (Diemar et al., 2019).

ASMs can be associated with bone loss and reduced bone mineral density in a phenomenon called anticonvulsant osteomalacia (Diemar et al., 2019). Overall, the most profound negative impacts on bone health have been described with phenytoin, phenobarbital, and primidone (Pack, 2011). Pack et al. demonstrated significant femoral neck bone loss in young women after just 1 year of exposure to phenytoin, suggesting that bone mineral density screening should be initiated as early as after 1 year of exposure to phenytoin (Pack, Morrell, Randall, McMahon, & Shane, 2008). Carbamazepine and valproate have mixed data, but both have also been associated with reduced bone mineral density and increased fracture risk (Pack, 2011). Animal studies of levetiracetam suggest that reductions in bone strength and bone formation can occur (Siniscalchi et al., 2020; Svalheim, Røste, Nakken, & Taubøll, 2011). Lamotrigine has been found to have a limited, if any, effect on bone health (Siniscalchi et al., 2020; Svalheim et al., 2011).

The mechanisms by which ASMs contribute to osteoporosis is complex. EIASMs increase the hydroxylation of vitamin D, leading to hypocalcemia, and bone loss. Several ASMs such as phenytoin, carbamazepine, phenobarbital, and valproate, have a direct effect on bone cells which results in increased bone mineral turnover (Diemar et al., 2019). Furthermore, chronic hyponatremia, as can be seen with carbamazepine, oxcarbazepine, eslicarbazepine, levetiracetam, and valproate, is also associated with an increased risk of osteoporosis and fractures (Diemar et al., 2019).

Consensus guidelines on the management of bone disease in patients with epilepsy are lacking. Generally, dual energy X-ray absorptiometry (DXA) is recommended every 2–5 years for adults taking ASMs that are associated with negative effects on bone health (Siniscalchi et al., 2020; Svalheim et al., 2011). Daily supplementation of vitamin D and calcium is recommended for female postmenopausal women. The recommended daily dose of vitamin D is 800 IU when taken with calcium, and 1800–4000 IU when taken alone (Fernandez, Mohammed, & Patel, 2018).

10. Summary

Consideration of sex and gender issues is fundamental in understanding and managing epilepsy. The influences of sex and gender are seen throughout an individual's life cycle and include the developing brain, childhood and adolescence, as well as pregnancy and aging. Sex and gender factors influence not only diagnostic considerations but also critical therapeutic choices. Epilepsy treatment is most impacted by the teratogenicity of antiseizure medications, especially valproate and topiramate. Patients and their health care professionals need to be aware of the sex and gender issues in order to make informed choices for their health and their family's health.

References

Abbasi, F., Krumholz, A., Kittner, S. J., & Langenberg, P. (1999). Effects of menopause on seizures in women with epilepsy. *Epilepsia*, *40*(2), 205–210. https://doi.org/10.1111/j.1528-1157.1999.tb02076.x.

Abe, K., Hamada, H., Yamada, T., Obata-Yasuoka, M., Minakami, H., & Yoshikawa, H. (2014). Impact of planning of pregnancy in women with epilepsy on seizure control during pregnancy and on maternal and neonatal outcomes. *Seizure*, *23*(2), 112–116. https://doi.org/10.1016/j.seizure.2013.10.003.

Bäckström, T., Zetterlund, B., Blom, S., & Romano, M. (1984). Effects of intravenous progesterone infusions on the epileptic discharge frequency in women with partial epilepsy. *Acta Neurologica Scandinavica*, *69*(4), 240–248. https://doi.org/10.1111/j.1600-0404.1984.tb07807.x.

Bagshaw, J., Crawford, P., & Chappell, B. (2008). Problems that mothers' with epilepsy experience when caring for their children. *Seizure*, *17*(1), 42–48. https://doi.org/10.1016/j.seizure.2007.06.004.

Banerjee, P. N., Filippi, D., & Allen Hauser, W. (2009). The descriptive epidemiology of epilepsy-a review. *Epilepsy Research*, *85*(1), 31–45. https://doi.org/10.1016/j.eplepsyres.2009.03.003.

Bangar, S., Shastri, A., El-Sayeh, H., & Cavanna, A. E. (2016). Women with epilepsy: Clinically relevant issues. *Functional Neurology*, *31*(3), 127.

Bartz, D., Chitnis, T., Kaiser, U. B., Rich-Edwards, J. W., Rexrode, K. M., Pennell, P. B., et al. (2020). Clinical advances in sex-and gender-informed medicine to improve the health of all: A review. *JAMA Internal Medicine*, *180*(4), 574–583.

Battino, D., Tomson, T., Bonizzoni, E., Craig, J., Lindhout, D., Sabers, A., et al. (2013). Seizure control and treatment changes in pregnancy: Observations from the EURAP epilepsy pregnancy registry. *Epilepsia*, *54*(9), 1621–1627.

Betts, T., & Fox, C. (1999). Proactive pre-conception counselling for women with epilepsy-is it effective? *Seizure*, *8*(6), 322–327. https://doi.org/10.1053/seiz.1999.0325.

Bjørk, M., Riedel, B., Spigset, O., Veiby, G., Kolstad, E., Daltveit, A. K., et al. (2018). Association of folic acid supplementation during pregnancy with the risk of autistic traits in children exposed to antiepileptic drugs in utero. *JAMA Neurology*, *75*(2), 160–168.

Bjørk, M. H., Veiby, G., Engelsen, B. A., & Gilhus, N. E. (2015). Depression and anxiety during pregnancy and the postpartum period in women with epilepsy: A review of frequency, risks and recommendations for treatment. *Seizure*, *28*, 39–45.

Bóné, B., Kovács, N., Balás, I., Horváth, R. A., Dóczi, T., & Janszky, J. (2021). Pregnancy and deep brain stimulation therapy for epilepsy. *Epileptic Disorders, 23*(4), 633–638.

Borthen, I., Eide, M. G., Daltveit, A. K., & Gilhus, N. E. (2011). Obstetric outcome in women with epilepsy: A hospital-based, retrospective study. *BJOG: An International Journal of Obstetrics and Gynaecology, 118*(8), 956–965. https://doi.org/10.1111/j.1471-0528.2011.03004.x.

Breehl, L., & Caban, O. (2020). Physiology, puberty. *StatPearls.*

Briellmann, R. S., Jackson, G. D., Mitchell, L. A., Fitt, G. J., Kim, S. E., & Berkovic, S. F. (1999). Occurrence of hippocampal sclerosis: Is one hemisphere or gender more vulnerable? *Epilepsia, 40*(12), 1816–1820. https://doi.org/10.1111/j.1528-1157.1999.tb01604.x.

Bromley, R. L., Mawer, G. E., Briggs, M., Cheyne, C., Clayton-Smith, J., García-Fiñana, M., et al. (2013). The prevalence of neurodevelopmental disorders in children prenatally exposed to antiepileptic drugs. *Journal of Neurology, Neurosurgery & Psychiatry, 84*(6), 637–643.

Bruns, J., Jr., & Hauser, W. A. (2003). The epidemiology of traumatic brain injury: A review. *Epilepsia, 44*(s10), 2–10. https://doi.org/10.1046/j.1528-1157.44.s10.3.x.

Burger, H. G., Dudley, E. C., Robertson, D. M., & Dennerstein, L. (2002). Hormonal changes in the menopause transition. *Recent Progress in Hormone Research, 57*, 257–276.

Burger, H. G., Hale, G. E., Dennerstein, L., & Robertson, D. M. (2008). Cycle and hormone changes during perimenopause: The key role of ovarian function. *Menopause (New York, N.Y.), 15*(4 Pt 1), 603–612. https://doi.org/10.1097/gme.0b013e318174ea4d.

Burger, H., Woods, N. F., Dennerstein, L., Alexander, J. L., Kotz, K., & Richardson, G. (2007). Nomenclature and endocrinology of menopause and perimenopause. *Expert Review of Neurotherapeutics*, 7(sup1), S35–S43.

Cagnetti, C., Lattanzi, S., Foschi, N., Provinciali, L., & Silvestrini, M. (2014). Seizure course during pregnancy in catamenial epilepsy. *Neurology, 83*(4), 339–344.

Campbell, E., Kennedy, F., Russell, A., Smithson, W. H., Parsons, L., Morrison, P. J., et al. (2014). Malformation risks of antiepileptic drug monotherapies in pregnancy: Updated results from the UK and Ireland epilepsy and pregnancy registers. *Journal of Neurology, Neurosurgery & Psychiatry, 85*(9), 1029–1034.

Cervenka, M. C., Wood, S., Bagary, M., Balabanov, A., Bercovici, E., Brown, M. G., et al. (2021). International recommendations for the management of adults treated with ketogenic diet therapies. Neurology: Clinical. *Practice, 11*(5), 385–397.

Chen, Y. H., Chiou, H. Y., Lin, H. C., & Lin, H. L. (2009). Affect of seizures during gestation on pregnancy outcomes in women with epilepsy. *Archives of Neurology, 66*(8), 979–984.

Chen, B., Choi, H., Hirsch, L. J., Moeller, J., Javed, A., Kato, K., et al. (2015). Cosmetic side effects of antiepileptic drugs in adults with epilepsy. *Epilepsy & Behavior, 42*, 129–137.

Christensen, J., Kjeldsen, M. J., Andersen, H., Friis, M. L., & Sidenius, P. (2005). Gender differences in epilepsy. *Epilepsia, 46*(6), 956–960. https://doi.org/10.1111/j.1528-1167.2005.51204.x.

Christensen, J., Pedersen, L., Sun, Y., Dreier, J. W., Brikell, I., & Dalsgaard, S. (2019). Association of prenatal exposure to valproate and other antiepileptic drugs with risk for attention-deficit/hyperactivity disorder in offspring. *JAMA Network Open, 2*(1), e186606.

Christensen, J., Vestergaard, M., Pedersen, M. G., Pedersen, C. B., Olsen, J., & Sidenius, P. (2007). Incidence and prevalence of epilepsy in Denmark. *Epilepsy Research, 76*(1), 60–65. https://doi.org/10.1016/j.eplepsyres.2007.06.012.

Compston, J. E., McClung, M. R., & Leslie, W. D. (2019). Osteoporosis. *Lancet (London, England), 393*(10169), 364–376. https://doi.org/10.1016/S0140-6736(18)32112-3.

Cummings, L. N., Giudice, L., & Morrell, M. J. (1995). Ovulatory function in epilepsy. *Epilepsia, 36*(4), 355–359.

Decaroli, M. C., & Rochira, V. (2017). Aging and sex hormones in males. *Virulence, 8*(5), 545–570. https://doi.org/10.1080/21505594.2016.1259053.

Diemar, S. S., Sejling, A. S., Eiken, P., Andersen, N. B., & Jørgensen, N. R. (2019). An explorative literature review of the multifactorial causes of osteoporosis in epilepsy. *Epilepsy & Behavior, 100*, 106511.

Ding, J., Wang, L., Wang, C., Gao, C., Wang, F., & Tao, S. (2021). Is vagal-nerve stimulation safe during pregnancy? A mini review. *Epilepsy Research, 106671*.

Edey, S., Moran, N., & Nashef, L. (2014). SUDEP and epilepsy-related mortality in pregnancy. *Epilepsia, 55*(7), e72–e74.

Edwards, H. E., Burnham, W. M., Mendonca, A., Bowlby, D. A., & MacLusky, N. J. (1999). Steroid hormones affect limbic afterdischarge thresholds and kindling rates in adult female rats. *Brain Research, 838*(1–2), 136–150. https://doi.org/10.1016/s0006-8993(99)01619-4.

Erel, T., & Guralp, O. (2011). Epilepsy and menopause. *Archives of Gynecology and Obstetrics, 284*(3), 749–755. https://doi.org/10.1007/s00404-011-1936-4.

Fabris, R. R., Cascino, T. G., Mandrekar, J., Marsh, W. R., Meyer, F. B., & Cascino, G. D. (2016). Drug-resistant focal epilepsy in women of childbearing age: Reproduction and the effect of epilepsy surgery. *Epilepsy & Behavior, 60*, 17–20.

Feely, M., Calvert, R., & Gibson, J. (1982). Clobazam in catamenial epilepsy. A model for evaluating anticonvulsants. *Lancet (London, England), 2*(8289), 71–73. https://doi.org/10.1016/s0140-6736(82)91691-9.

Fernandez, H., Cooke, M., & Patel, T. (2019). Epilepsy and lifestyle behaviors related to bone health. *Epilepsia, 60*(11), 2306–2313. https://doi.org/10.1111/epi.16351.

Fernandez, H., Mohammed, H. T., & Patel, T. (2018). Vitamin D supplementation for bone health in adults with epilepsy: A systematic review. *Epilepsia, 59*(4), 885–896. https://doi.org/10.1111/epi.14015.

Fiest, K. M., Sauro, K. M., Wiebe, S., Patten, S. B., Kwon, C. S., Dykeman, J., et al. (2017). Prevalence and incidence of epilepsy: A systematic review and meta-analysis of international studies. *Neurology, 88*(3), 296–303.

Frye, C. A., Rhodes, M. E., Walf, A. A., & Harney, J. P. (2001). Testosterone reduces pentylenetetrazole-induced ictal activity of wildtype mice but not those deficient in type I 5alpha-reductase. *Brain Research, 918*(1–2), 182–186. https://doi.org/10.1016/s0006-8993(01)02967-5.

GBD 2016 Neurology Collaborators. (2019). Global, regional, and national burden of neurological disorders, 1990-2016: A systematic analysis for the global burden of disease study 2016. *The Lancet. Neurology, 18*(5), 459–480. https://doi.org/10.1016/S1474-4422(18)30499-X.

Harden, C. L. (2008). Hormone replacement therapy: Will it affect seizure control and AED levels? *Seizure, 17*(2), 176–180. https://doi.org/10.1016/j.seizure.2007.11.026.

Harden, C. L., Herzog, A. G., Nikolov, B. G., Koppel, B. S., Christos, P. J., Fowler, K., et al. (2006). Hormone replacement therapy in women with epilepsy: A randomized, double-blind, placebo-controlled study. *Epilepsia, 47*(9), 1447–1451.

Harden, C. L., Koppel, B. S., Herzog, A. G., Nikolov, B. G., & Hauser, W. A. (2003). Seizure frequency is associated with age at menopause in women with epilepsy. *Neurology, 61*(4), 451–455.

Harden, C., & MacLusky, N. J. (2004). Aromatase inhibition, testosterone, and seizures. *Epilepsy & Behavior, 5*(2), 260–263. https://doi.org/10.1016/j.yebeh.2003.12.001.

Harden, C., & MacLusky, N. J. (2005). Aromatase inhibitors as add-on treatment for men with epilepsy. *Expert Review of Neurotherapeutics, 5*(1), 123–127. https://doi.org/10.1586/14737175.5.1.123.

Harden, C. L., & Pennell, P. B. (2013). Neuroendocrine considerations in the treatment of men and women with epilepsy. *The Lancet. Neurology*, *12*(1), 72–83. https://doi.org/10.1016/S1474-4422(12)70239-9.

Harden, C. L., Pulver, M. C., Ravdin, L., & Jacobs, A. R. (1999). The effect of menopause and perimenopause on the course of epilepsy. *Epilepsia*, *40*(10), 1402–1407. https://doi.org/10.1111/j.1528-1157.1999.tb02012.x.

Hauser, W. A., Annegers, J. F., & Kurland, L. T. (1993). Incidence of epilepsy and unprovoked seizures in Rochester, Minnesota: 1935-1984. *Epilepsia*, *34*(3), 453–468. https://doi.org/10.1111/j.1528-1157.1993.tb02586.x.

Henning, O., Johannessen Landmark, C., Traeen, B., Svendsen, T., Farmen, A., Nakken, K. O., et al. (2019). Sexual function in people with epilepsy: Similarities and differences with the general population. *Epilepsia*, *60*(9), 1984–1992. https://doi.org/10.1111/epi.16311.

Hernández-Díaz, S., Smith, C. R., Shen, A., Mittendorf, R., Hauser, W. A., Yerby, M., et al. (2012). Comparative safety of antiepileptic drugs during pregnancy. *Neurology*, *78*(21), 1692–1699. https://doi.org/10.1212/WNL.0b013e3182574f39.

Herzog, A. G. (1993). A relationship between particular reproductive endocrine disorders and the laterality of epileptiform discharges in women with epilepsy. *Neurology*, *43*(10), 1907–1910. https://doi.org/10.1212/wnl.43.10.1907.

Herzog, A. G. (2006). Menstrual disorders in women with epilepsy. *Neurology*, *66*(6 Suppl 3), S23–S28. https://doi.org/10.1212/wnl.66.66_suppl_3.s23.

Herzog, A. G. (2008a). Catamenial epilepsy: Definition, prevalence pathophysiology and treatment. *Seizure*, *17*(2), 151–159. https://doi.org/10.1016/j.seizure.2007.11.014.

Herzog, A. G. (2008b). Disorders of reproduction in patients with epilepsy: Primary neurological mechanisms. *Seizure*, *17*(2), 101–110. https://doi.org/10.1016/j.seizure.2007.11.025.

Herzog, A. G. (2015). Catamenial epilepsy: Update on prevalence, pathophysiology and treatment from the findings of the NIH progesterone treatment trial. *Seizure*, *28*, 18–25. https://doi.org/10.1016/j.seizure.2015.02.024.

Herzog, A. G., Drislane, F. W., Schomer, D. L., Pennell, P. B., Bromfield, E. B., Dworetzky, B. A., et al. (2005). Differential effects of antiepileptic drugs on sexual function and hormones in men with epilepsy. *Neurology*, *65*(7), 1016–1020.

Herzog, A. G., Fowler, K. M., Smithson, S. D., Kalayjian, L. A., Heck, C. N., Sperling, M. R., et al. (2012). Progesterone vs placebo therapy for women with epilepsy: A randomized clinical trial. *Neurology*, *78*(24), 1959–1966.

Herzog, A. G., Harden, C. L., Liporace, J., Pennell, P., Schomer, D. L., Sperling, M., et al. (2004). Frequency of catamenial seizure exacerbation in women with localization-related epilepsy. *Annals of Neurology*, *56*(3), 431–434.

Herzog, A. G., Klein, P., & Ransil, B. J. (1997). Three patterns of catamenial epilepsy. *Epilepsia*, *38*(10), 1082–1088. https://doi.org/10.1111/j.1528-1157.1997.tb01197.x.

Herzog, A. G., Mandle, H. B., Cahill, K. E., Fowler, K. M., & Hauser, W. A. (2017). Predictors of unintended pregnancy in women with epilepsy. *Neurology*, *88*(8), 728–733.

Herzog, A. G., Mandle, H. B., & MacEachern, D. B. (2019). Does the age of seizure onset relate to menarche and does it matter? *Seizure*, *69*, 1–6.

Hiilesmaa, V. K., & Teramo, K. A. (2013). Fetal and maternal risks with seizures. *Epilepsy in Women*, 115–127.

Holmes, L. B., Mittendorf, R., Shen, A., Smith, C. R., & Hernandez-Diaz, S. (2011). Fetal effects of anticonvulsant polytherapies: Different risks from different drug combinations. *Archives of Neurology*, *68*(10), 1275–1281.

Hom, A. C., & Buterbaugh, G. G. (1986). Estrogen alters the acquisition of seizures kindled by repeated amygdala stimulation or pentylenetetrazol administration in ovariectomized female rats. *Epilepsia*, *27*(2), 103–108. https://doi.org/10.1111/j.1528-1157.1986.tb03510.x.

Isojärvi, J. I. T., Löfgren, E., Juntunen, K. S. T., Pakarinen, A. J., Päivänsalo, M., Rautakorpi, I., et al. (2004). Effect of epilepsy and antiepileptic drugs on male reproductive health. *Neurology*, *62*(2), 247–253.

Isojärvi, J. I., Taubøll, E., & Herzog, A. G. (2005). Effect of antiepileptic drugs on reproductive endocrine function in individuals with epilepsy. *CNS Drugs*, *19*(3), 207–223.

Johnson, E. L., & Kaplan, P. W. (2017). Caring for transgender patients with epilepsy. *Epilepsia*, *58*(10), 1667–1672.

Kalinin, V. V., & Zheleznova, E. V. (2007). Chronology and evolution of temporal lobe epilepsy and endocrine reproductive dysfunction in women: Relationships to side of focus and catameniality. *Epilepsy & Behavior*, *11*(2), 185–191. https://doi.org/10.1016/j.yebeh.2007.04.014.

Kandraju, S. S., Jose, M., Salini, R. A., Pavithran, V., Samuel Peedicail, J., Menon, R. N., et al. (2020). Women with drug-resistant epilepsy: Surgery or pregnancy first? *Epilepsia*, *61*(8), 1758–1763.

Kandratavicius, L., Balista, P. A., Lopes-Aguiar, C., Ruggiero, R. N., Umeoka, E. H., Garcia-Cairasco, N., et al. (2014). Animal models of epilepsy: Use and limitations. *Neuropsychiatric Disease and Treatment*.

Karanam, A., Pennell, P. B., French, J. A., Harden, C. L., Allien, S., Lau, C., et al. (2018). Lamotrigine clearance increases by 5 weeks gestational age: Relationship to estradiol concentrations and gestational age. *Annals of Neurology*, *84*(4), 556–563.

Karlander, M., Ljungqvist, J., & Zelano, J. (2021). Post-traumatic epilepsy in adults: A nationwide register-based study. *Journal of Neurology, Neurosurgery, and Psychiatry*, *92*(6), 617–621. Advance online publication. https://doi.org/10.1136/jnnp-2020-325382.

Kinney, M. O., & Morrow, J. (2016). Epilepsy in pregnancy. *BMJ (Clinical Research ed.)*, *353*, i2880. https://doi.org/10.1136/bmj.i2880.

Kirkpatrick, L., Collins, A., Sogawa, Y., Talabi, M. B., Harrison, E., & Kazmerski, T. M. (2020). Sexual and reproductive healthcare for adolescent and young adult women with epilepsy: A qualitative study of pediatric neurologists and epileptologists. *Epilepsy & Behavior*, *104*, 106911.

Klein, P., Serje, A., & Pezzullo, J. C. (2001). Premature ovarian failure in women with epilepsy. *Epilepsia*, *42*(12), 1584–1589.

Koppel, B. S., & Harden, C. L. (2014). Gender issues in the neurobiology of epilepsy: A clinical perspective. *Neurobiology of Disease*, *72*(Pt B), 193–197. https://doi.org/10.1016/j.nbd.2014.08.033.

Kotsopoulos, I. A., van Merode, T., Kessels, F. G., de Krom, M. C., & Knottnerus, J. A. (2002). Systematic review and meta-analysis of incidence studies of epilepsy and unprovoked seizures. *Epilepsia*, *43*(11), 1402–1409. https://doi.org/10.1046/j.1528-1157.2002.t01-1-26901.x.

Kramer, M. S., & Kakuma, R. (2012). Optimal duration of exclusive breastfeeding. *The Cochrane Database of Systematic Reviews*, *2012*(8), CD003517. https://doi.org/10.1002/14651858.CD003517.pub2.

Larsen, M. D., Jølving, L. R., Fedder, J., & Nørgård, B. M. (2020). The efficacy of assisted reproductive treatment in women with epilepsy. *Reproductive Biomedicine Online*, *41*(6), 1015–1022.

Li, Y., Eliashiv, D., LaHue, S. C., Rao, V. R., Martini, M. L., Panov, F., et al. (2021). Pregnancy outcomes of refractory epilepsy patients treated with brain-responsive neurostimulation. *Epilepsy Research*, *169*, 106532.

Lim, L. L., Foldvary, N., Mascha, E., & Lee, J. (2001). Acetazolamide in women with catamenial epilepsy. *Epilepsia*, *42*(6), 746–749. https://doi.org/10.1046/j.1528-1157.2001.33600.x.

Liu, J., Yu, N., Sun, H., Liang, Y., Xie, Y., & Chen, K. (2015). Intermittent levetiracetam treatment in five patients with catamenial epilepsy. *The Journal of the Pakistan Medical Association, 65*(7), 793–795.

Logothetis, J., Harner, R., Morrell, F., & Torres, F. (1959). The role of estrogens in catamenial exacerbation of epilepsy. *Neurology, 9*(5), 352.

MacDonald, S. C., Bateman, B. T., McElrath, T. F., & Hernández-Díaz, S. (2015). Mortality and morbidity during delivery hospitalization among pregnant women with epilepsy in the United States. *JAMA Neurology, 72*(9), 981–988.

Maggi, R. (2020). The hypothalamus-pituitary-ovary Axis. *Female Reproductive Dysfunction*, 3–21.

Maguire, M. J., & Nevitt, S. J. (2021). Treatments for seizures in catamenial (menstrual-related) epilepsy. *The Cochrane Database of Systematic Reviews, 9*(9). CD013225 https://doi.org/10.1002/14651858.CD013225.pub3.

Mameniškienė, R., Guk, J., & Jatužis, D. (2017). Family and sexual life in people with epilepsy. *Epilepsy & Behavior, 66*, 39–44. https://doi.org/10.1016/j.yebeh.2016.10.012.

Manford, M., Hart, Y. M., Sander, J. W., & Shorvon, S. D. (1992). National General Practice Study of epilepsy (NGPSE): Partial seizure patterns in a general population. *Neurology, 42*(10), 1911–1917. https://doi.org/10.1212/wnl.42.10.1911.

Markoula, S., Siarava, E., Keramida, A., Chatzistefanidis, D., Zikopoulos, A., Kyritsis, A. P., et al. (2020). Reproductive health in patients with epilepsy. *Epilepsy & Behavior, 113*, 107563.

Mattson, R. H., Cramer, J. A., Caldwell, B. V., & Siconolfi, B. C. (1984). Treatment of seizures with medroxyprogesterone acetate: Preliminary report. *Neurology, 34*(9), 1255–1258. https://doi.org/10.1212/wnl.34.9.1255.

Mauvais-Jarvis, F., Merz, N. B., Barnes, P. J., Brinton, R. D., Carrero, J. J., DeMeo, D. L., et al. (2020). Sex and gender: Modifiers of health, disease, and medicine. *The Lancet, 396*(10250), 565–582.

McAuley, J. W., Koshy, S. J., Moore, J. L., Peebles, C. T., & Reeves, A. L. (2000). Characterization and health risk assessment of postmenopausal women with epilepsy. *Epilepsy & Behavior, 1*(5), 353–355. https://doi.org/10.1006/ebeh.2000.0114.

McHugh, J. C., & Delanty, N. (2008). Epidemiology and classification of epilepsy: Gender comparisons. *International Review of Neurobiology, 83*, 11–26. https://doi.org/10.1016/S0074-7742(08)00002-0.

Meador, K. J., Baker, G. A., Browning, N., Clayton-Smith, J., Combs-Cantrell, D. T., Cohen, M., et al. (2009). Cognitive function at 3 years of age after fetal exposure to antiepileptic drugs. *New England Journal of Medicine, 360*(16), 1597–1605.

Meador, K. J., Baker, G. A., Browning, N., Clayton-Smith, J., Combs-Cantrell, D. T., Cohen, M., et al. (2010). Effects of breastfeeding in children of women taking antiepileptic drugs. *Neurology, 75*(22), 1954–1960.

Meador, K. J., Baker, G. A., Browning, N., Cohen, M. J., Bromley, R. L., Clayton-Smith, J., et al. (2013). Fetal antiepileptic drug exposure and cognitive outcomes at age 6 years (NEAD study): A prospective observational study. *The Lancet Neurology, 12*(3), 244–252.

Meador, K. J., Baker, G. A., Browning, N., Cohen, M. J., Bromley, R. L., Clayton-Smith, J., et al. (2014). Breastfeeding in children of women taking antiepileptic drugs: Cognitive outcomes at age 6 years. *JAMA Pediatrics, 168*(8), 729–736.

Meador, K. J., Baker, G. A., Browning, N., Cohen, M. J., Clayton-Smith, J., Kalayjian, L. A., et al. (2011). Foetal antiepileptic drug exposure and verbal versus non-verbal abilities at three years of age. *Brain, 134*(2), 396–404.

Meador, K. J., Baker, G. A., Finnell, R. H., Kalayjian, L. A., Liporace, J. D., Loring, D. W., et al. (2006). In utero antiepileptic drug exposure: Fetal death and malformations. *Neurology, 67*(3), 407–412.

Mejías-Aponte, C. A., Jiménez-Rivera, C. A., & Segarra, A. C. (2002). Sex differences in models of temporal lobe epilepsy: Role of testosterone. *Brain Research*, *944*(1–2), 210–218. https://doi.org/10.1016/s0006-8993(02)02691-4.

Morrell, M. J., Flynn, K. L., Doñe, S., Flaster, E., Kalayjian, L., & Pack, A. M. (2005). Sexual dysfunction, sex steroid hormone abnormalities, and depression in women with epilepsy treated with antiepileptic drugs. *Epilepsy & Behavior*, *6*(3), 360–365.

Morrell, M. J., Giudice, L., Flynn, K. L., Seale, C. G., Paulson, A. J., Doñe, S., et al. (2002). Predictors of ovulatory failure in women with epilepsy. *Annals of Neurology: Official Journal of the American Neurological Association and the Child Neurology Society*, *52*(6), 704–711.

Morrow, J. I., Hunt, S. J., Russell, A. J., Smithson, W. H., Parsons, L., Robertson, I., et al. (2009). Folic acid use and major congenital malformations in offspring of women with epilepsy: A prospective study from the UK epilepsy and pregnancy register. *Journal of Neurology, Neurosurgery & Psychiatry*, *80*(5), 506–511.

Ocek, L., Tarhan, H., Uludağ, F.İ., Sarıteke, A., Köse, C., Colak, A., et al. (2018). Evaluation of sex hormones and sperm parameters in male epileptic patients. *Acta Neurologica Scandinavica*, *137*(4), 409–416.

Oettel, M., & Mukhopadhyay, A. K. (2004). Progesterone: The forgotten hormone in men? *The Aging Male: The Official Journal of the International Society for the Study of the Aging Male*, 7(3), 236–257. https://doi.org/10.1080/13685530400004199.

Pack, A. M. (2011). Treatment of epilepsy to optimize bone health. *Current Treatment Options in Neurology*, *13*(4), 346–354.

Pack, A. M., Morrell, M. J., Randall, A., McMahon, D. J., & Shane, E. (2008). Bone health in young women with epilepsy after one year of antiepileptic drug monotherapy. *Neurology*, *70*(18), 1586–1593. https://doi.org/10.1212/01.wnl.0000310981.44676.de.

Panelli, D. M., Leonard, S. A., Kan, P., Meador, K. J., McElrath, T. F., Darmawan, K. F., et al. (2021). Association of Epilepsy and Severe Maternal Morbidity. *Obstetrics & Gynecology*, *138*(5), 747–754.

Peebles, C. T., McAuley, J. W., Moore, J. L., Malone, H. J., & Reeves, A. L. (2000). Hormone replacement therapy in a postmenopausal woman with epilepsy. *The Annals of Pharmacotherapy*, *34*(9), 1028–1031. https://doi.org/10.1345/aph.19417.

Pennell, P. B., French, J. A., Harden, C. L., Davis, A., Bagiella, E., Andreopoulos, E., et al. (2018). Fertility and birth outcomes in women with epilepsy seeking pregnancy. *JAMA Neurology*, *75*(8), 962–969.

Pennell, P. B., French, J. A., May, R. C., Gerard, E., Kalayjian, L., Penovich, P., et al. (2020). Changes in seizure frequency and antiepileptic therapy during pregnancy. *New England Journal of Medicine*, *383*(26), 2547–2556.

Pennell, P. B., Klein, A. M., Browning, N., Baker, G. A., Clayton-Smith, J., Kalayjian, L. A., et al. (2012). Differential effects of antiepileptic drugs on neonatal outcomes. *Epilepsy & Behavior*, *24*(4), 449–456.

Pesce, M. E., Acevedo, X., Bustamante, D., Miranda, H. E., & Pinardi, G. (2000). Progesterone and testosterone modulate the convulsant actions of pentylenetetrazol and strychnine in mice. *Pharmacology & Toxicology*, *87*(3), 116–119. https://doi.org/10.1111/j.0901-9928.2000.870303.x.

Rättyä, J., Pakarinen, A. J., Knip, M., RepoOutakoski, M., Myllylä, V. V., & Isojärvi, J. I. (2001). Early hormonal changes during valproate or carbamazepine treatment: A 3-month study. *Neurology*, *57*(3), 440–444.

Reddy, D. S. (2014). Neurosteroids and their role in sex-specific epilepsies. *Neurobiology of Disease*, *72*(Pt B), 198–209. https://doi.org/10.1016/j.nbd.2014.06.010.

Reimers, A., Brodtkorb, E., & Sabers, A. (2015). Interactions between hormonal contraception and antiepileptic drugs: Clinical and mechanistic considerations. *Seizure*, *28*, 66–70.

Reisinger, T. L., Newman, M., Loring, D. W., Pennell, P. B., & Meador, K. J. (2013). Antiepileptic drug clearance and seizure frequency during pregnancy in women with epilepsy. *Epilepsy & Behavior*, *29*(1), 13–18.

Sahoo, S., & Klein, P. (2005). Maternal complex partial seizure associated with fetal distress. *Archives of Neurology*, *62*(8), 1304–1305.

Scelzo, E., Mehrkens, J. H., Bötzel, K., Krack, P., Mendes, A., Chabardès, S., et al. (2015). Deep brain stimulation during pregnancy and delivery: Experience from a series of "DBS babies". *Frontiers in Neurology*, *6*, 191.

Scharfman, H. E., & MacLusky, N. J. (2006). The influence of gonadal hormones on neuronal excitability, seizures, and epilepsy in the female. *Epilepsia*, *47*(9), 1423–1440. https://doi.org/10.1111/j.1528-1167.2006.00672.x.

Shih, J. J., Whitlock, J. B., Chimato, N., Vargas, E., Karceski, S. C., & Frank, R. D. (2017). Epilepsy treatment in adults and adolescents: Expert opinion, 2016. *Epilepsy & Behavior*, *69*, 186–222.

Siniscalchi, A., Murphy, S., Cione, E., Piro, L., Sarro, G., & Gallelli, L. (2020). Antiepileptic drugs and bone health: Current concepts. *Psychopharmacology Bulletin*, *50*(2), 36–44.

Stefanidou, M., & Montouris, G. (2020). Reproductive and sexual health concerns in transition-age adolescents and young adults with epilepsy. *Seminars in Pediatric Neurology*, *36*, 100855. https://doi.org/10.1016/j.spen.2020.100855.

Stephen, L. J., Harden, C., Tomson, T., & Brodie, M. J. (2019). Management of epilepsy in women. *The Lancet Neurology*, *18*(5), 481–491.

Suller Marti, A., Mirsattari, S. M., Steven, D. A., Parrent, A. G., MacDougall, K. W., McLachlan, R. S., et al. (2019). Experience on the use of Vagus nerve stimulation during pregnancy. *Epilepsy Research*, *156*, 106186. https://doi.org/10.1016/j.eplepsyres.2019.106186.

Svalheim, S., Røste, L. S., Nakken, K. O., & Taubøll, E. (2011). Bone health in adults with epilepsy. *Acta Neurologica Scandinavica. Supplementum*, *191*, 89–95. https://doi.org/10.1111/j.1600-0404.2011.01551.x.

Svalheim, S., Sveberg, L., Mochol, M., & Taubøll, E. (2015). Interactions between antiepileptic drugs and hormones. *Seizure*, *28*, 12–17.

Sveberg, L., Svalheim, S., & Taubøll, E. (2015). The impact of seizures on pregnancy and delivery. *Seizure*, *28*, 35–38. https://doi.org/10.1016/j.seizure.2015.02.020.

Sveinsson, O., & Tomson, T. (2014). Epilepsy and menopause: Potential implications for pharmacotherapy. *Drugs & Aging*, *31*(9), 671–675. https://doi.org/10.1007/s40266-014-0201-5.

Taubøll, E., Isojärvi, J. I., & Herzog, A. G. (2021). The interactions between reproductive hormones and epilepsy. *Handbook of Clinical Neurology*, *182*, 155–174.

Tomson, T., Battino, D., Bonizzoni, E., Craig, J., Lindhout, D., Perucca, E., et al. (2018). Comparative risk of major congenital malformations with eight different antiepileptic drugs: A prospective cohort study of the EURAP registry. *The Lancet Neurology*, *17*(6), 530–538.

Tomson, T., Battino, D., Bromley, R., Kochen, S., Meador, K., Pennell, P., et al. (2019). Management of epilepsy in pregnancy: A report from the international league against epilepsy task force on women and pregnancy. *Epileptic Disorders*, *21*(6), 497–517.

Turner, K., Piazzini, A., Franza, A., Canger, R., Canevini, M. P., & Marconi, A. M. (2008). Do women with epilepsy have more fear of childbirth during pregnancy compared with women without epilepsy? A case-control study. *Birth*, *35*(2), 147–152.

Veiby, G., Daltveit, A. K., Engelsen, B. A., & Gilhus, N. E. (2009). Pregnancy, delivery, and outcome for the child in maternal epilepsy. *Epilepsia*, *50*(9), 2130–2139.

Veiby, G., Daltveit, A. K., Engelsen, B. A., & Gilhus, N. E. (2014). Fetal growth restriction and birth defects with newer and older antiepileptic drugs during pregnancy. *Journal of Neurology*, *261*(3), 579–588.

Veiby, G., Engelsen, B. A., & Gilhus, N. E. (2013). Early child development and exposure to antiepileptic drugs prenatally and through breastfeeding: A prospective cohort study on children of women with epilepsy. *JAMA Neurology*, *70*(11), 1367–1374.
Vélez-Ruiz, N. J., & Pennell, P. B. (2016). Issues for women with epilepsy. *Neurologic Clinics*, *34*(2), 411–425.
Velísková, J., Velísek, L., Galanopoulou, A. S., & Sperber, E. F. (2000). Neuroprotective effects of estrogens on hippocampal cells in adult female rats after status epilepticus. *Epilepsia*, *41*(Suppl 6), S30–S35. https://doi.org/10.1111/j.1528-1157.2000.tb01553.x.
Veroniki, A. A., Cogo, E., Rios, P., Straus, S. E., Finkelstein, Y., Kealey, R., et al. (2017). Comparative safety of anti-epileptic drugs during pregnancy: A systematic review and network meta-analysis of congenital malformations and prenatal outcomes. *BMC Medicine*, *15*(1), 1–20.
Verreault, N., Da Costa, D., Marchand, A., Ireland, K., Dritsa, M., & Khalifé, S. (2014). Rates and risk factors associated with depressive symptoms during pregnancy and with postpartum onset. *Journal of Psychosomatic Obstetrics and Gynaecology*, *35*(3), 84–91. https://doi.org/10.3109/0167482X.2014.947953.
Viale, L., Allotey, J., Cheong-See, F., Arroyo-Manzano, D., Mccorry, D., Bagary, M., et al. (2015). Epilepsy in pregnancy and reproductive outcomes: A systematic review and meta-analysis. *The Lancet*, *386*(10006), 1845–1852.
Voinescu, P. E., Park, S., Chen, L. Q., Stowe, Z. N., Newport, D. J., Ritchie, J. C., et al. (2018). Antiepileptic drug clearances during pregnancy and clinical implications for women with epilepsy. *Neurology*, *91*(13), e1228–e1236.
Voinescu, P. E. (2019). Catamenial Epilepsy. In *Neurology and psychiatry of women* (pp. 85–94). Springer International Publishing. https://doi.org/10.1007/978-3-030-04245-5_9.
Weston, J., Bromley, R., Jackson, C. F., Adab, N., Clayton-Smith, J., Greenhalgh, J., et al. (2016). Monotherapy treatment of epilepsy in pregnancy: Congenital malformation outcomes in the child. *Cochrane Database of Systematic Reviews*, *11*.
Woolley, C. S. (2000). Estradiol facilitates kainic acid-induced, but not flurothyl-induced, behavioral seizure activity in adult female rats. *Epilepsia*, *41*(5), 510–515. https://doi.org/10.1111/j.1528-1157.2000.tb00203.x.
Woolley, D. E., & Timiras, P. S. (1962). The gonad-brain relationship: Effects of female sex hormones on electroshock convulsions in the rat. *Endocrinology*, *70*(2), 196–209.
Wu, Y. V., & Burnham, W. M. (2018). Progesterone, 5a-dihydropogesterone and allopregnanolone's effects on seizures: A review of animal and clinical studies. *Seizure*, *63*, 26–36.
Xu, T., Yu, X., Ou, S., Liu, X., Yuan, J., Huang, H., et al. (2017). Risk factors for posttraumatic epilepsy: A systematic review and meta-analysis. *Epilepsy & Behavior*, *67*, 1–6.
Zupanc, M. L., & Haut, S. (2008). Epilepsy in women: Special considerations for adolescents. *International Review of Neurobiology*, *83*, 91–111.

Further reading

Bui E., Women's issues in epilepsy. Continuum (Minneap Minn). 2022 Apr 1;28(2):399-427. doi: 10.1212/CON.0000000000001126. PMID: 35393964.
Li Y., Meador K.J., Epilepsy and pregnancy. Continuum (Minneap Minn). 2022 Feb 1;28 (1):34-54. doi: 10.1212/CON.0000000000001056. PMID: 35133310.
Keni R., Mostacci B., Kiteva-Trenchevska G., Licchetta L., Ignjatova L., Thomas S., Meador K.J., Women's issues. Epileptic Disorders 2020 Aug 1;22(4):355-363. doi: 10.1684/epd.2020.1173. PMID: 32723702; PMCID: PMC7755427.

CHAPTER EIGHT

Sex and gender differences in pain

Natalie R. Osborne[a,b] and Karen D. Davis[a,b,c,*]
[a]Krembil Brain Institute, Krembil Research Institute, University Health Network, Toronto, Canada
[b]Institute of Medical Science, University of Toronto, Toronto, Canada
[c]Department of Surgery, University of Toronto, Toronto, Canada
*Corresponding author: e-mail address: karen.davis@uhnresearch.ca

Contents

Abstract

Chronic pain affects 20% of adults and is one of the leading causes of disability worldwide. Women and girls are disproportionally affected by chronic pain. About half of chronic pain conditions are more common in women, with only 20% having a higher prevalence in men. There are also sex and gender differences in acute pain sensitivity. Pain is a subjective experience made up of sensory, cognitive, and emotional components. Consequently, there are multiple dimensions through which sex and gender can influence the pain experience. Historically, most preclinical pain research was conducted exclusively in male animals. However, recent studies that included females have revealed significant sex differences in the physiological mechanisms underlying pain, including sex specific involvement of different genes and proteins as well as distinct interactions between hormones and the immune system that influence the transmission of pain signals. Human neuroimaging has revealed sex and gender differences in the neural circuitry associated with pain, including sex specific brain alterations in chronic pain conditions. Clinical pain research suggests that gender can affect how an individual contextualizes and copes with pain. Gender may also influence the susceptibility to develop chronic pain. Sex and gender biases can impact how pain

International Review of Neurobiology, Volume 164
ISSN 0074-7742
https://doi.org/10.1016/bs.irn.2022.06.013

is perceived and treated clinically. Furthermore, the efficacy and side effects associated with different pain treatments can vary according to sex and gender. Therefore, preclinical and clinical research must include sex and gender analyses to understand basic mechanisms of pain and its relief, and to develop personalized pain treatment.

1. Introduction

One in five adults experience pain and 1 in 10 are diagnosed with chronic pain each year, according to the International Association for the Study of Pain (IASP) (Goldberg & McGee, 2011). The 2016 Global Burden of Disease Study reported that chronic pain conditions, specifically low back pain and migraine, are among the five leading causes of disability and disease burden (Rice, Smith, & Blyth, 2016). Pain is also the most common reason for seeking medical care (Fishman, 2007), although the World Health Organization estimates that 80% of patients with severe pain do not experience adequate pain relief, particularly in developing nations (Taylor, 2007).

The IASP defines pain as "an unpleasant sensory and emotional experience associated with, or resembling that associated with, actual or potential tissue damage" (Raja et al., 2020). Pain is an inherently subjective and multidimensional experience comprised of sensory, emotional, and cognitive components. Consequently, there are many levels at which the biology of sex and the psychosocial attributes of gender can contribute to pain, from sex divergent molecular mechanisms and the actions of sex hormones, to differences in pain-related brain circuitry and cognitive coping strategies.

A better understanding of sex and gender effects on pain begins with a more comprehensive understanding of the concepts of sex and gender themselves. While science often strives to categorize, it is important to remember that many biological variables exist on a continuous spectrum. Although genetic sex is typically defined as XX (female) and XY (male), not all individuals will physically develop as the sex typically associated with their chromosomes, and other karyotypes exist, including X, XXY, XYY and XXXY (Fausto-Sterling, 1993). Furthermore, differences in the body's production of or response to sex hormones (e.g., androgen insensitivity, congenital adrenal hyperplasia) can influence secondary sex characteristics and genitalia (El-Maouche, Arlt, & Merke, 2017). Intersex individuals are estimated to make up 2% of the population, although that may be an underestimate as many people may not realize they are intersex if they have not undergone genotyping, internal imaging, or surgery (Blackless et al., 2000). Furthermore, a diverse spectrum of gender identities exists in humans.

It is important to note that many of the studies discussed in this chapter have evaluated only sex or gender or used the terms interchangeably. This is out of line with current concepts and distinctions between sex and gender. Here, we report the findings according to the term used in the studies, but otherwise refer to sex and gender using the definitions provided by the Canadian Institutes of Health Research (https://cihr-irsc.gc.ca/e/47830.html). That is, sex refers to "a set of biological attributes in humans and animals, primarily associated with physical and physiological features including chromosomes, gene expression, hormone levels and function, and reproductive/sexual anatomy." Gender is defined as "the socially constructed roles, behaviors, expressions and identities of girls, women, boys, men, and gender diverse people. Gender is usually conceptualized as a binary, yet there is considerable diversity in how individuals and groups understand, experience, and express it." Also of note, is that many of the findings discussed in this chapter that are assigned to a particular sex or gender are generally only valid at the group level, and individual differences exist within any group.

The incidence of many types of pain, how it is experienced, and its underlying mechanisms may differ for men and women. Sex and gender differences in pain can be studied in many ways, including documenting epidemiological data such as chronic pain prevalence, measuring physiological and psychophysical attributes such as hormone levels and pain thresholds, acquiring systems-level information using brain imaging, and measuring mechanistic cellular and molecule functions. It should be noted that any one approach needs to be informed by findings from other approaches. For example, a particular behavioral attribute of pain could arise from different nociceptive processing mechanisms in the body and brain in women and men, which is critically important for developing therapeutics. Because of the historical and present-day exclusion of female animals from preclinical pain research (Mogil & Chanda, 2005), and the lack of sex-specific and sex disaggregated analyses in human work, there is much still to be discovered about the influences of sex and gender on pain. We do know that there are demonstrable differences between women and men in pain sensitivity and coping, as well as an increased prevalence of chronic pain conditions in women.

Understanding how sex and gender interact to influence pain processing and coping could guide researchers in identifying optimal pain treatment strategies that are tailored for individuals. Part I of this chapter will discuss sex and gender differences in pain experiences. Part II will describe how these differences might arise, and Part III will explain why characterizing sex and gender differences in pain is important for the future of pain treatment.

2. Part I: How does pain differ by sex and gender?

Part I will provide an overview of sex and gender differences in the prevalence and severity of chronic pain, as well as experimental studies in healthy participants to contribute to our understanding of how sex and gender can influence pain-related behaviors and cognitions.

2.1 Sex and gender differences in chronic pain

Epidemiological studies have found that women are more likely to develop chronic pain than men (Unruh, 1996), throughout the world, in both developing and developed nations. For example, survey data from over 42,000 participants across 17 countries revealed a higher prevalence of chronic pain in the last 12 months in women compared to men (Tsang et al., 2008). This difference pertains to a variety of chronic pain conditions that affect different areas of the body. One study found that women are more likely than men to report pain in each of the 10 different body regions studied, and more likely to have widespread pain over multiple body regions (Gerdle et al., 2008). Among children and teens, girls also experience chronic pain more often than boys, with pain prevalence increasing in girls as they get older (King et al., 2011). These differences are particularly apparent in complex regional pain syndrome (CRPS), where girls represent 85–90% of patients (Tan, Zijlstra, Essink, Goris, & Severijnen, 2008), and juvenile arthritis, which has a prevalence of 20% among girls but only 11% in boys (Thierry, Fautrel, Lemelle, & Guillemin, 2014).

About 50% of chronic pain conditions are more prevalent in women, while just 20% are more common in men (Mogil, 2012). Migraine, musculoskeletal pain, temporomandibular disorder, fibromyalgia, rheumatoid and osteoarthritis, irritable bowel syndrome, and neuropathic pain are more common in women (Mogil, 2012; Unruh, 1996) and some studies find they are also associated with more severe clinical pain (Melchior, Poisbeau, Gaumond, & Marchand, 2016). Some chronic pain conditions that are more prevalent in men include episodic and chronic cluster headaches and spondyloarthritis, including ankylosing spondylitis (Dean et al., 2014). An examination of medical records from 11,000 patients compared disease-associated pain scores from men and women with the same diagnoses and revealed widespread sex differences. In fact, women had higher average pain scores than men for the majority (72–95%) of the diagnoses included in the study. For some conditions, women's pain scores were more than 20% higher than men's (Ruau, Liu, Clark, Angst, & Butte, 2012).

Symptoms of the same condition can also present differently in men and women. For example, women with spondyloarthritis often have different pain severity and symptoms than men, which the authors posit may affect clinician's judgments about disease severity and even influence management decisions (Barnabe et al., 2012). Significant disparities in care can occur when clinicians are only taught how a disease presents in one sex (e.g., the sex in which the condition is more prevalent), or when the "classical" definitions of the disease's manifestations are based on research from only one sex. Indeed, the sex difference in prevalence for spondyloarthritis, a male-dominant chronic pain condition, have been reported to be as high as 10:1 (West, 1949), although some have argued that women with AS have historically been underdiagnosed because they present with different symptoms, and recently the most commonly reported sex ratio has been estimated to be 3:1 (Rusman, van Vollenhoven, & van der Horst-Bruinsma, 2018).

2.2 Sex and gender differences in experimental pain in humans

Experimental pain sensitivity and reactions have been widely studied in healthy adults using a variety of pain modalities (e.g., thermal, electrical, ischemic) and behavioral paradigms (e.g., pain thresholds, short vs tonic stimuli, tests of pain tolerance), and many of these studies have investigated sex and gender differences. It is important to note that most studies and conclusions regarding pain sensitivity have been based on threshold measures of pain to short experimental stimuli, but that metrics of suprathreshold responses (e.g., tolerance, adaptation, habituation, coping, etc.) and the use of prolonged, tonic stimuli—which may be more relevant to chronic pain—can paint a different picture that is often not recognized (Hashmi & Davis, 2014).

Reviews of the studies focused primarily on threshold measures of pain mostly conclude that women tend to be more sensitive, that is, have lower pain thresholds, to experimental pain stimuli compared to men (Fillingim, King, Ribeiro-Dasilva, Rahim-Williams, & Riley, 2009; Greenspan et al., 2007), although one systematic review contends that a clear pattern of sex-differences in pain sensitivity has not been demonstrated (Racine et al., 2012a). While the statistical significance and effect sizes of some of the reported sex differences may vary between stimuli and across studies (Hashmi & Davis, 2014), qualitatively, women seem to generally exhibit greater pain sensitivity than men (Mogil, 2012). Interestingly, one study found that greater pain sensitivity (and stronger cortical responses) in women compared to men were not present when they controlled for participants' trait anxiety, suggesting this psychological variable is a major contributor to the observed sex-differences in pain thresholds (Goffaux et al., 2011).

Previous learning can also influence pain sensitivity in a sex-dependent manner. For example, a recent paradigm that paired tonic pain with a specific context found that re-exposure to that context produced conditioned pain hypersensitivity in males only (Martin et al., 2019). This was found for both rodents and human participants and was linked with testosterone and stress. Responses to suprathreshold and tonic stimuli, pain tolerance, and pain coping present a more complicated picture with respect to sex. One study found that women showed greater adaption and habituation to prolonged and repeated heat pain stimuli compared to men (Hashmi & Davis, 2009), while another found that women and men habituated to heat pain similarly, but showed differing activation of their sympathetic nervous system during pain (Breimhorst, Hondrich, Rebhorn, May, & Birklein, 2012). A common behavioral measure of the brain's endogenous pain modulation system is conditioned pain modulation (CPM, also referred to as diffuse noxious inhibitory control, or DNIC in experimental animal studies), whereby the perceived intensity of a painful "test" stimulus can be reduced when a second painful "conditioning" stimulus is applied to another part of the body (Nir & Yarnitsky, 2015). An inefficient or absent CPM effect has been implicated in the poor therapeutic response for some types of chronic pain and is proposed by some to represent a potential susceptibility to develop chronic pain. Some studies (Arendt-Nielsen, Sluka, & Nie, 2008), including a systematic literature review (Popescu, LeResche, Truelove, & Drangsholt, 2010) suggest that CPM is more efficient in men. However, this is not consistently found across studies, which may in part be due to the variation in how CPM is evaluated (Pud, Granovsky, & Yarnitsky, 2009). A recent (albeit small) study found that women had greater CPM responses than men, and stronger CPM responses were associated with higher serum levels of brain-derived neurotrophic factor (BDNF, a factor associated with neuroplasticity) in women but not men (Gasparin et al., 2020). Furthermore, CPM in women appears to be more efficient at specific phases of the menstrual cycle (i.e., ovulatory) (Tousignant-Laflamme & Marchand, 2009). A recent study found that both personal attributes, including resilience and sex, as well as attributes of the CPM test (i.e., the conditioning stimulus) combine to influence the CPM effect (Firouzian et al., 2020). Another study found that inter-individual differences (which could be due to genetic, epigenetic, behavioral or psychological traits) accounted for 24–34% of the variance in CPM effect in healthy adults, whereas sex, age, and conditioning stimulus combined accounted for only 3–12% (Graeff, Itter, Wach, & Ruscheweyh, 2021).

Pain sensitivity and tolerance seem to be influenced in part by gender roles and norms, defined as the role or behavior learned by a person as appropriate to their gender, as determined by social and cultural constructs (Keogh, 2021). One study reported that male participants who strongly identified with idealized "masculine" gender norms showed higher pain tolerance, while men who rated conforming to gender norms as not highly important to them had similar pain tolerance to women (Pool, Schwegler, Theodore, & Fuchs, 2007). A meta-analysis similarly concluded that participants who considered themselves more "masculine" and with higher pain thresholds than typical men do show decreased pain sensitivity and increased pain tolerance (Alabas, Tashani, Tabasam, & Johnson, 2012). Importantly, one study found that men who identified strongly with masculine gender traits, including higher levels of aggression and competitiveness, were more likely to volunteer for pain experiments, suggesting a potential bias in study samples which may not represent the diversity within the male population (Mattos Feijo et al., 2018). Interestingly, the gender of the experimenter can also influence participants' pain ratings. Some studies have found that men report less pain when tested by women, particularly when they found them attractive (Levine & Lee De Simone, 1991), while women reported more pain sensitivity when tested by attractive male experimenters (Gijsbers & Nicholson, 2005). But another study found both men and women tolerated pain longer when tested by the opposite sex (Kallai, Barke, & Voss, 2004).

There is very little experimental pain research that focusses on or even includes individuals who are non-binary, intersex, or transgender. One study of individuals living with HIV and chronic pain found that both cis and transgender women reported more severe chronic pain scores and showed greater thermal and mechanical temporal summation of pain (TSP), a pain paradigm thought to reflect central sensitization, compared to cisgender men (Strath et al., 2020). Furthermore, transgender women had greater mechanical TSP than cisgender women. The authors proposed that gender identity, rather than genetic sex, may be a more significant contributor to pain sensitivity. At present, almost nothing is known about how pain is processed in gender diverse individuals, and how that might contribute to their experience of clinical pain. One study found an increased prevalence of fibromyalgia in transgender individuals (Levit et al., 2021), and a recent review proposed that gender minority stress may affect pain experience in gender diverse people with migraine (Pace, Barber, Ziplow, Hranilovich, & Kaiser, 2021). Pain researchers should strive to incorporate measures of both sex and gender in their study designs and consider the diverse spectrum of gender identity and roles when evaluating differences in pain behavior.

3. Part II: What causes sex and gender differences in pain?

The previous section highlighted the many sex and gender differences that exist in experimental and chronic pain in humans. The following section will describe how the physiological processes underlying pain, as well as psycho-social factors such as gender, may contribute to these differences.

3.1 Sex differences in pain physiology

Preclinical work in animals has identified sex differences in the physiological mechanisms of pain, from genes and molecular messengers to the differential involvement of the immune system (for a comprehensive review see Mogil (2018, 2020)). In humans, chronic pain is associated with sex-specific difference in gene expression, and some different genes have been linked with pain in women and men (Mogil, 2020). For example, a certain rare allele for a gene associated with opioid receptors is associated with increased pain sensitivity in women and decreased pain sensitivity in men (Fillingim et al., 2005), and this same allele predicted worse recovery in women and better recovery in men following resolution of low back pain (Olsen et al., 2012). Women with two or more variants of a gene that encodes melanocortin-1 receptors (and is also associated with red hair and freckles) showed greater pain relief when given the drug pentazocine compared to women without the variant, but the same relationship was not seen in men (Mogil et al., 2003).

The immune system is an integral part of the body's response to tissue injury, and as such also plays a role in the experience of pain (Kavelaars & Heijnen, 2021). Sexual dimorphism in the immune system is well-established and believed to contribute to sex differences in pain (Rosen, Ham, & Mogil, 2017). One mechanism thought to lead to the development of neuropathic pain involves microglial activation and subsequent disinhibition of spinal cord neurons. However, studies in mice indicate that this is the primary mechanism in males, whereas a parallel T-cell mechanism is typically at play in female mice with neuropathic pain (Beggs & Salter, 2010). Recently, another study demonstrated a female-specific mechanism underlying the development of mechanical pain, where peripheral inflammatory mediators act on macrophages to release a specific cytokine that then activates nociceptors (Luo et al., 2021). This process was mediated by estrogen and does not appear to occur in male mice. Sex differences in the action of peripheral immune cells, which then infiltrate the dorsal root ganglion and modulate pain related neurons, have also been found (Lopes et al., 2017).

3.2 Sex hormones and pain

Communication between the neuroimmune system and the endocrine system can influence pain signaling mechanisms in ways that are sex-specific (Lenert, Avona, Garner, Barron, & Burton, 2021). From an epidemiological perspective, there are several clues that some chronic pain conditions are linked with sex hormones. First, sex differences in chronic pain prevalence largely begin after puberty, when circulating sex hormone levels first rise and fluctuate substantially (Vincent & Tracey, 2008). Second, the severity of chronic pain conditions can differ across the menstrual cycle (Vetvik, Macgregor, Lundqvist, & Russell, 2014) and throughout pregnancy (Craft, 2007). Third, decreases in sex hormones that accompany perimenopause have been linked with some chronic pain conditions, although it is difficult to rule out the contribution of advancing age (Dugan et al., 2006).

There are mixed findings from studies of naturally-cycling women on whether pain sensitivity fluctuates with the menstrual cycle (Iacovides, Avidon, & Baker, 2015; Riley, Robinson, Wise, & Price, 1999; Sherman & LeResche, 2006) and whether experimental or clinical pain is influenced by the suppression of endogenous hormone levels via hormonal oral contraceptives (Kowalczyk et al., 2010; LeResche, Saunders, Von Korff, Barlow, & Dworkin, 1997). A functional neuroimaging study of naturally cycling women and women using the oral contraceptive pill found that estradiol and progesterone were related to pain-evoked activity in emotional regulation networks, whereas testosterone was related to activity in the descending antinociceptive system, and that the sex hormones appeared to be generally antinociceptive (Vincent, Moore, Kennedy, & Tracey, 2014). Lower testosterone levels are associated with musculoskeletal pain in women (Kaergaard, Hansen, Rasmussen, & Andersen, 2000), and testosterone administration improves rheumatoid arthritis symptoms in older women and men (Morales, Nolan, Nelson, & Yen, 1994). Fluctuations in one type of estrogen, estradiol, may contribute to increased pain, while stable estradiol levels may reduce pain signaling (Lenert et al., 2021). Estrogen and progesterone fluctuations across the menstrual cycle can modulate the brain's functional architecture and structure (Pritschet, Taylor, Santander, & Jacobs, 2021). Women with primary dysmenorrhea, a cyclic pain condition linked with menstruation, show hypertrophic gray matter changes in some brain regions (e.g., medial orbitofrontal cortex, primary somatosensory cortex and hypothalamus) and atrophic changes in others (e.g., secondary somatosensory cortex, anterior/dorsoposterior cingulate cortex) during menstruation compared to healthy women (Tu et al., 2013).

Studies on the effects of hormone replacement therapy (HRT) on pain in postmenopausal women report mixed results, with some finding greater pain sensitivity (Fillingim & Edwards, 2001) and increased risk for chronic pain (Fillingim et al., 2009), and others finding HRT has no effect on pain (Stening et al., 2011) or may even protect against developing chronic pain (Al-Rousan et al., 2018; Jung et al., 2018). Research including individuals receiving hormone treatment as part of gender-affirming medical care provides another unique opportunity to better understand the action of hormones on the pain experience. Although there are only a few small studies to date, one study of 73 trans individuals reported that about a third of trans women receiving estrogen and androgens developed chronic pain, while those who did not develop clinical pain did report decreased tolerance and increased sensitivity to pain (Aloisi et al., 2007). Conversely, trans men in the study with a history of chronic pain reported an improvement in their pain symptoms while receiving testosterone. However, defining a specific sex hormone as "pro" or "anti" nociceptive is an oversimplification of the dynamic influences these hormones can have on pain. Furthermore, the actions of sex hormones can be immediate and short-term (e.g., binding to a receptor in the cell membrane), or long-lasting, such as modifying a cell's protein production through epigenetic modulation (McCarthy et al., 2009).

3.3 Sex and gender differences in pain and the brain

Sex and gender differences are seen in the structure and function of the brain of healthy individuals throughout the lifespan, from childhood through adolescence (Kurth, Gaser, & Luders, 2021) and into adulthood (McCarthy, Arnold, Ball, Blaustein, & De Vries, 2012; Ritchie et al., 2018). Sex differences in functional brain organization revealed in fetal (Wheelock et al., 2019) neuroimaging suggests some of these differences may be related to what are often referred to as "biological" factors associated with sex, including genetics and intra-uterine exposure to androgens during neurodevelopment. However, psychosocial factors known to be influenced by gender, such as socialization, cognition, and lived experience, also contribute to differences seen in the brain. For example, psychological trauma can affect central nervous and immune system function through epigenetic modifications in a sex-specific manner (Uddin, Sipahi, Li, & Koenen, 2013).

Recently, machine learning algorithms have been used to accurately classify individuals who identify as male or female based on their brain's functional organization (Weis et al., 2020). The existence of sexual dimorphism in the brain has been challenged (Eliot, Ahmed, Khan, & Patel, 2021), and some propose the brain is more likely made up of a mosaic of male and female-typical features (Joel, 2021). Sex differences in the brain have been proposed to underlie sex and gender differences in the prevalence and presentation of various psychiatric conditions, including depression and anxiety (Rubinow & Schmidt, 2019), and sex differences in how the brain ages are thought to underly women's increased risk for Alzheimer's disease and dementia (Mosconi et al., 2017). Whether sex and gender differences in pain-related brain regions could contribute to differences in chronic pain prevalence and severity has not been explored.

Studies in healthy participants indicate that several brain regions important for processing the sensory, emotional, and cognitive components of pain show sex and gender differences in their structural and functional connectivity. While structural connectivity measures (e.g., diffusion tensor imaging) assess the white matter tracts connecting two brain regions, functional connectivity represents the synchrony of oscillatory activity between brain areas and can be a measure of how these regions function together. The periaqueductal gray (PAG) is a key node in the brain's descending pain modulation system (Hemington & Coulombe, 2015) that shows sex differences in structural and functional connectivity in healthy adults (Coulombe, Erpelding, Kucyi, & Davis, 2016; Linnman, Beucke, Jensen, Gollub, & Kong, 2012). Preclinical studies suggest sexual dimorphism in the PAG may explain sex-differences in opioidergic modulation of pain (Averitt, Eidson, Doyle, & Murphy, 2019). Another brain region important in descending pain modulation, the subgenual anterior cingulate cortex (sgACC), also shows sex differences in structural and functional connectivity within the descending antinociceptive system and with other brain regions associated with pain (Kim et al., 2021; Osborne, Cheng, et al., 2021; Wang, Erpelding, & Davis, 2014). Sex and gender differences in the amygdala, mediated in part by corticosteroids and sex hormones, have also been well established (Kogler et al., 2016). The amygdala is thought to play an important role in processing the emotional component of pain (Neugebauer, 2015). A study using a large dataset from the human connectome project revealed sex differences in the brain's functional and structural organization starting after age 25 and including the amygdala, hippocampus, prefrontal cortex, temporal lobe, inferior parietal lobule, posterior cingulate and precuneus (Conrin et al., 2018). The latter two structures represent important nodes of the default mode network, which is implicated in pain-associated rumination and inward focused attention and often functions abnormally in chronic pain (Baliki, Mansour, Baria, & Apkarian, 2014). Below, we outline the brain areas that respond differently to experimentally induced pain in men and women. Then, we will describe sex and gender differences that have been observed in chronic pain-related plasticity throughout the brain.

3.3.1 Brain responses to experimental pain in humans

Neuroimaging studies that measure brain responses to painful stimuli frequently find common patterns of activity in women and men. There appear to be some sex and gender differences in how experimental pain is processed in the brain, but the findings have been mixed and can vary across pain stimuli and methodologies (see Table 1). Interestingly, sex differences

Table 1 Mixed findings of sex/gender differences in brain responses to experimental pain.

Study	Method & subjects	Pain stimuli & site	Brain activation findings (≥greater brain activity)
Paulson, Minoshima, Morrow, and Casey (1998)	PET $N=20$ (10F, 10M)	Thermal Skin (arm)	F > M: contralateral PFC, insula & thalamus (↑ activation in thalamus & anterior insula in females with ↑ pain intensity ratings) M > F: ipsilateral PFC
Derbyshire, Nichols, Firestone, Townsend, and Jones (2002)	PET $N=21$ (11F, 10M)	Laser Skin (hand)	F > M: ipsilateral perigenual cingulate cortex M > F: contralateral PFC, S1, S2, parietal cortex & insula
Henderson, Gandevia, and Macefield (2008)	fMRI $N=22$ (11F, 11M)	Hypertonic saline SubQ & intramuscular (leg)	F only: ↑ activity in MCC, ↓ activity in cerebellum; activity in hippocampus ↑ with subQ pain & ↓ with muscular pain M only: ↑ activity in lateral cerebellum, ↓ activity in DLPFC
Straube, Schmidt, Weiss, Mentzel, and Miltner (2009)	fMRI $N=36$ (18F, 18M)	Electrical SubQ (finger)	F > M: medial PFC M > F: ipsilateral insula F & M: Pain intensity correlated with activity in the ant. Insula, ACC, post. Insula/S2 & S1, and in the thalamus for M only
(Benson et al. (2012)	fMRI $N=30$ (15F, 15M)	Visceral (pressure) Rectum	F & M: pain-related activity in S1, insula & DLPFC F only: thalamus M only: VLPFC, MCC, brainstem F > M: right cerebellum, medial frontal gyrus

This table presents some examples of findings from neuroimaging studies of experimental pain. N = number of participants; M = men; F = female; ↑ = increased; fMRI = functional magnetic resonance imaging; PET = positron emission tomography; subQ = subcutaneous; S1 = primary somatosensory cortex; S2 = secondary somatosensory cortex; PFC = prefrontal cortex; DL = dorsolateral; VL = ventrolateral; ACC = anterior cingulate cortex; MCC = mid cingulate cortex.

in pain-related brain responses have even been observed in preterm babies. Female infants are more likely to show a more spatially distributed nociceptive event-related potential (nERP) in response to a heel-lance procedure than male infants (Verriotis et al., 2018). Several studies have proposed that sex and gender differences in pain are driven by differences in how emotional and arousal/attentional aspects of the pain experience are processed in the brains of men and women (Henderson et al., 2008; Labus et al., 2008; Straube et al., 2009). For example, subjective ratings of the unpleasantness of pain stimuli were associated with increased activity in the perigenual ACC in women and decreased activity in the ventromedial PFC in men (Girard-Tremblay et al., 2014). More research is needed to investigate sex/gender differences in pain across the lifespan, and whether pain is processed differently in the brains of gender diverse people.

3.3.2 Sex and gender differences in brain plasticity related to chronic pain

Chronic pain is associated with functional and structural brain reorganization (Kuner & Flor, 2017; Mansour, Farmer, Baliki, & Apkarian, 2014). Considering the preclinical evidence for sexually dimorphic pain mechanisms and demonstrable sex-differences in chronic pain prevalence in humans, there has been surprisingly little research examining sex and gender differences in chronic pain-related brain plasticity. This represents a serious knowledge gap in the field, as sex and gender differences in the neural circuitry of chronic pain are important to understand to develop a personalized approach to manage pain, including treatments that target the brain using neuromodulation, cognitive therapy, and virtual reality. To our knowledge there is only one review paper on the influence of sex in chronic pain-related brain plasticity. This review analyzed 143 single-sex studies, as well as 15 studies that directly compared sex differences in brain changes associated with various chronic pain conditions (Gupta et al., 2017). There is a dearth of chronic pain neuroimaging research that reports sex-disaggregated data or analyzes (rather than simply controls for) sex. This makes drawing conclusions about sex and gender's impacts on neuroplasticity in chronic pain difficult. However, the review by Gupta and colleagues did suggest some sex-specific commonalities in brain changes across conditions, which are summarized in Table 2.

Primary somatosensory and motor cortex abnormalities have been reported in chronic pain (Chang et al., 2018; Eto et al., 2011; Kim et al., 2020), and these regions are also the target for some non-invasive brain

Table 2 Summary of sex differences and sex-specific brain changes in chronic pain.

Brain network	**Interim conclusions from review**	
Sensorimotor (SMN)	Women	Women and girls with migraine, IBS, fibromyalgia, carpal tunnel syndrome & dysmenorrhea show abnormal S1 cortical thickness (vs healthy women)
	Men	Some abnormal sensorimotor function (e.g., ↑ FC with insula) has been reported in men, but more studies are needed
	General	Abnormal SMN structure & function appear more pronounced in women than men
Salience (SN)	Women	Abnormal insular FC with default mode network & ↑ dACC FC with limbic system in women with chronic pain (vs men)
	Men	↑ pain-induced insula activity in men with chronic pain (vs women)
	General	↑ pain-related activity in SN in both women & men with chronic pain (vs healthy controls) Sex differences in chronic pain-related changes to the insula & dACC may impact pain vigilance & attentional capacity
Limbic	Women	Women with chronic pain have ↑ amygdala activity & FC abnormalities during visceral pain (vs men)
	Men	Men may show ↑ activity in amygdala, hippocampus & ACC vs women in specific contexts (e.g., when viewing negative valence images)
	General	Both women & men with chronic pain show ↑ activity and abnormal FC in the limbic system in response to pain (vs healthy controls) Conflicting findings of hyper/hypo limbic activity in women vs men with chronic pain may reflect sex-specific triggers

In their review, Gupta et al. (2017) analyzed 143 single-sex studies and 15 sex difference studies on neuroplasticity in chronic pain. This table presents a summary of some of the authors' observations about the relationship between sex and brain changes in chronic pain. IBS = irritable bowel syndrome; S1 = primary somatosensory cortex; ↑ = greater; FC = functional connectivity; dACC = dorsal anterior cingulate cortex.

From Gupta, A., Mayer, E. A., Fling, C., Labus, J. S., Naliboff, B. D., Hong, J. Y., & Kilpatrick, L. A. (2017). Sex-based differences in brain alterations across chronic pain conditions. Journal of Neuroscience Research, 95(1–2), 604–616.

stimulation pain-treatments (Antal et al., 2008; Lima & Fregni, 2008). In general, chronic pain-related alterations in sensorimotor network structure and function appear to be more pronounced in women than men based on differences in single-sex studies, although more studies of male-specific conditions are needed (Gupta et al., 2017). A study of migraineurs that directly compared the sexes found that women have more sensorimotor dysfunction than men (Liu et al., 2011). This study also found that the sensorimotor regions in women had decreased centrality, a graph theory measure of how important that "node" is for the information flow in that network. However, sensorimotor abnormalities can also exist in men with chronic pain. A study of men and women with a male-dominant chronic pain condition (ankylosing spondylitis) found abnormal cross-network functional connectivity between key nodes of the sensorimotor, salience, and DMN that was associated with increased disease severity (Hemington, Wu, Kucyi, Inman, & Davis, 2016).

The salience network, including the anterior insula, temporoparietal junction, dorsolateral prefrontal cortex, and mid-cingulate cortex, plays an integral role in the pain experience as it is normally engaged when one directs their attention to a painful stimulus (Borsook, Edwards, Elman, Becerra, & Levine, 2013; Kucyi, Salomons, & Davis, 2013). While both men and women with chronic pain typically show increased salience network activity in response to pain compared with healthy adults, there are significant sex differences in the way chronic pain influences the structure and functional connectivity of two key nodes: the insula and the ACC (Gupta et al., 2017). The limbic system, particularly the amygdala and hippocampus, is a frequent focus of sex-difference studies on fear, emotional reactivity, and memory. Sex differences in the amygdala are proposed to underly sex-specific susceptibility to mood disorders (Seney & Sibille, 2014), and the hippocampus' dense concentration of estrogen receptors are studied in the context of sex-differences in stoke, memory, and aging (Choleris, Galea, Sohrabji, & Frick, 2018; Koss & Frick, 2017). Men and women with chronic pain generally have increased activity and altered connectivity of the amygdala and hippocampus in reaction to pain when compared to healthy controls, but there are conflicting reports of whether women or men with chronic pain show hyper-reactivity of the limbic system (Gupta et al., 2017). The role of the hippocampus in acute and chronic pain is currently a topic of interest (Ayoub et al., 2019), and a recent study reported that women with chronic low back pain, but not men, have deformations in hippocampal shape (Reckziegel et al., 2021).

Abnormalities in the sgACC, an important node in the descending pain modulation system, are also seen across chronic pain conditions (Fauchon et al., 2021). A study of ankylosing spondylitis, a chronic pain condition more common in men, found sex-differences in the sgACC functional connectivity of patients. However, sgACC connectivity was only abnormal in the women, who had greater functional connectivity with the default mode and salience networks and weaker with the left inferior frontal gyrus and dorsolateral PFC compared to healthy women (Osborne, Cheng, et al., 2021). In contrast, abnormally low sgACC functional connectivity with the medial PFC and temporal lobes in carpal tunnel syndrome, a condition more prevalent in women, seemed to be driven by men rather than women (Osborne, Anastakis, et al., 2021).

Neuroimaging analysis techniques such as graph theory can provide detailed characterizations of the functional organization of the entire brain. A recent graph theory analysis of individuals with low back pain found that there was atypical modular segregation in the mid-cingulate cortex and sgACC and lower connectivity with the default mode and frontoparietal networks in women, and stronger connectivity with the sensorimotor network in men (Fauchon et al., 2021). These graph metrics were also shown in a machine learning model to be able to classify a participant's sex, as well as whether or not they had chronic pain (or were a healthy control). Advanced computational techniques and shared neuroimaging databases for chronic pain could provide the large sample sizes and sensitive analyses required to investigate the effect sex and gender has on chronic pain-related neuroplasticity within and possibly across conditions.

3.4 Influence of psycho-social factors and gender on pain experience

Characterizing the psychosocial and gender-based contributions to differences in pain experience is as complex a task as disentangling the various genetic, hormonal, and physiological traits underlying sex differences in the pain system. Some studies suggest that men and women recruit different cognitive and emotional coping strategies to manage pain, with men preferring distraction and problem-focused techniques, and women using emotion-focused tactics, attentional focus, cognitive reframing and social support (Fillingim et al., 2009; Racine et al., 2012b). But this is likely an oversimplification and generalization that may not pertain to all individuals, and more research is needed to explain gender related variation in pain (see the gender context model of pain) (Keogh, 2021). Pain catastrophizing describes a way of thinking about one's pain, with specific focus on the pain's magnitude and threat, as well as rumination on pain (Melchior et al., 2016), and is associated with pain and its related disability (Keefe et al., 2000). Some studies report women are more likely to engage in a pain catastrophizing strategy to cope with pain than men (Edwards, Haythornthwaite, Sullivan, & Fillingim, 2004), while others find no gender difference (Ramirez-Maestre & Esteve, 2014).

Rollman and colleagues proposed that past pain experiences can influence current pain ratings, as those with more extensive pain histories also show lower pain tolerance experimentally (Rollman, Abdel-Shaheed, Gillespie, & Jones, 2004). They suggest that the generally increased prevalence of pain experience in women could contribute to the gender differences seen in experimental pain studies. Similarly, traumatic experiences can also influence pain perception in sex- and gender-dependent ways. Childhood abuse has been linked with chronic pain in adulthood in women (Walsh, Jamieson, Macmillan, & Boyle, 2007), and is also associated with decreased pain sensitivity (Fillingim & Edwards, 2005). Adverse childhood events are associated with sex-specific alterations in brain connectivity in IBS (Gupta et al., 2014) and chronic pelvic pain (Gupta et al., 2019). Some small studies of adults with post-traumatic stress disorder report increased sensitization of the central nervous system in men (Moeller-Bertram et al., 2014) and increased pain unpleasantness ratings in women (Gomez-Perez & Lopez-Martinez, 2013). Sociocultural factors may also influence chronic pain in gender specific ways. For example, an analysis of over 11,000 chronic pain patients found that two factors—age and being in a relationship—highly impacted men's pain intensity, anxiety, depression, pain catastrophizing, and the degree the pain interfered with their life, while these factors had no or lesser impact for women (Waardenburg, de Meij, Brouwer, van Zundert, & van Kuijk, 2021). One chronic pain study reported gender-specific associations between gray matter morphology throughout the brain and psychosocial characteristics, including pain catastrophizing, fear of movement (kinesiophobia), and perceived chronicity and emotional responses related to pain (Malfliet et al., 2019).

Gender stereotypes about pain likely influence its clinical treatment (Leresche, 2011). Work by Robinson and colleagues created the Gender Role Expectations of Pain (GREP) measure to assess people's thoughts about men's and women's pain sensitivity, tolerance, and willingness to report pain, and found that individuals of both sexes believe men are generally less willing to report pain than women (Robinson et al., 2001). Gender stereotypes about pain may partially explain why men report less pain in experiments and could also influence judgments about the severity of pain a person is experiencing. A recent study asked participants to watch videos of patients with a shoulder injury who were performing painful maneuvers and to rate the pain they believed the patients were experiencing (Zhang, Losin, Ashar, Koban, & Wager, 2021). Participants rated women as being in less pain than men, even when their pain intensity (from patient's own reports) and behavior (analyzed with a computerized pain scoring method) were equal. These judgments were partially related to endorsements of beliefs that women express their pain more than men. Furthermore, women were believed to be more likely to benefit from psychotherapy treatment, whereas men were thought to be more likely to benefit from medicine for their pain. These attitudes are reflected in real-world data that find clinicians recommend psychological treatment to women more often than men (Hamberg, Risberg, Johansson, & Westman, 2002), and analgesics to men more often than women (Schafer, Prkachin, Kaseweter, & Williams, 2016). However, other studies have found no difference in analgesic prescribing (Safdar et al., 2009), or variability in whether women (Alqudah, Hirsh, Stutts, Scipio, & Robinson, 2010) or men (Barr & Eilat-Tsanani, 2022) are more likely to receive strong opioids.

4. Part III: Why do sex and gender differences in pain matter?

Here, we review how sex and gender influence the efficacy of common pain treatments. We propose that characterizing sex and gender differences in pain and analgesia is vital to improving pain management. We conclude the section with recommendations on research study designs that can disentangle the complex contributions of sex and gender to the pain experience.

4.1 Sex and gender differences in pain treatment

Infrequent reporting of sex effects or sex disaggregated data in randomized controlled trials and systematic reviews makes it difficult to establish whether sex and gender differences in analgesic responses to a given pain treatment exist (Duan-Porter et al., 2016). Whether opioids are more or less effective in one sex has been the focus of many studies, with mixed results.

Some studies suggest mu opioids provide greater pain relief in women (Niesters et al., 2010), others suggest kappa opioids are more potent in women (Gear et al., 1996), and some find no sex differences in opioid relief (Bijur et al., 2008). A small study reported that the non-steroidal anti-inflammatory drug (NSAID) ibuprofen reduced experimental pain in men more than women (Butcher & Carmody, 2012), but no sex difference was seen in its relief of dental pain (Averbuch & Katzper, 2000). There are also sex differences in the efficacy of drugs that suppress the inflammatory response, such as tumor necrosis factor (TNF) inhibitors. A review of patients with ankylosing spondylitis reported worse treatment response and adherence in women compared to men (Rusman et al., 2018), and an evaluation of four clinical trials ($n = 1283$) revealed that women improved less than men after 12 weeks of treatment with TNF inhibitors (van der Horst-Bruinsma, Zack, Szumski, & Koenig, 2013).

Therapeutic spinal cord stimulation appears to produce similar levels of pain relief in men and women. However, one study found more women (13%) than men (6%) had the device removed due to inadequate pain relief (Bretherton et al., 2022), while another study reported that removal due to discomfort from the device itself (perhaps related to its size and placement) as well as higher rates of untreated depression was more common in women (Slyer et al., 2020). Physical therapy for acute back pain resulted in similar pain relief in women and men, although predictors of treatment outcome were sex-specific (George, Fritz, Childs, & Brennan, 2006). The effectiveness of chronic pain management intervention programs based on psychological treatments such as cognitive behavioral therapy (CBT) and acceptance and commitment therapy (ACT) also show gender differences. Some studies find women benefit more from such interventions (Racine et al., 2020), while others find more sustained long-term pain relief in men (Keogh, McCracken, & Eccleston, 2005). Here, we have illustrated just a few examples of how the same treatment can elicit different pain relief in men and women. Understanding how sex and gender influence the biological, psychological, and sociological components of pain can inform and guide treatments that optimize pain relief in every individual.

4.2 Future directions: How can we better understand the role of sex and gender in pain?

Sex and gender are some of the many factors that contribute to the inter-individual differences seen in pain. An important step towards better understanding sex and gender effects on pain is to design experiments that are more inclusive (Shansky & Murphy, 2021). At the preclinical level, this means including female animals (Mogil & Chanda, 2005), which have traditionally been excluded due to inaccurate beliefs that female data will be inherently more variable than male data due to estrous. An analysis of preclinical behavioral pain studies between 1996 and 2005 found that 79% used only male animals (Mogil, 2018). When the sex of tissues and cells used in preclinical biomedical research is reported (which only a quarter of studies do), they are also overwhelmingly male (71%) (Yoon et al., 2014). Encouragingly there is evidence that pain researchers are including females more often in their experiments and analyses (Mogil, 2020). Consideration of sex in preclinical research is important not only for the characterization of pain mechanisms, which have been shown to differ by sex, but also in the early development of novel pain treatments. Many examples exist where drugs have different pharmacokinetics, side-effects, and efficacy in women and men which were never explored in preclinical trials (LeGates, Kvarta, & Thompson, 2019). Moreover, drug discovery may be hampered when only one sex is studied or data from both sexes are pooled, as some treatments may work well in one sex but not the other. For example, since the 1980s calcitonin gene-related peptide (CGRP) was thought to play a role in migraine, but studies administering CGRP to the brain's meninges did not elicit migraine in rodents (Avona et al., 2019). However, when researchers used female animals, they found CGRP created a significant pain response at doses that had no effect on males (Avona et al., 2019). Multiple drugs targeting CGRP have recently become available for migraine treatment.

Inclusion of female-identifying participants in human biomedical research has increased since the 1990s, when funding agencies began mandating inclusion of females in studies and clinical trials (White, Tannenbaum, Klinge, Schiebinger, & Clayton, 2021). While most human studies may be recruiting relatively equal numbers of women and men, few are reporting data by sex or gender, with the majority either not commenting on sex/gender effects or, counter-productively, controlling for it in their analyses. Controlling for this important variable masks whether the relationship between the predictors and outcomes being studied differ by sex or gender (Shapiro, Klein, & Morgan, 2021). Instead, data should be disaggregated by sex/gender first to investigate sex/gender-specific findings for the mechanism or treatment of interest before (or instead) of being pooled, and claims of sex- or gender-differences should be made when direct comparisons between the two are made (Clayton & Tannenbaum, 2016).

There are several helpful guides with best practices for including sex and gender in research design (Clayton, 2018; Garcia-Sifuentes & Maney, 2021; Rich-Edwards, Kaiser, Chen, Manson, & Goldstein, 2018). Clinical studies should endeavor to collect information on both the sex and gender of their participants (Duchesne, Pletzer, Pavlova, Lai, & Einstein, 2020), ideally using measurement tools that include sex and gender options beyond the binary of male/female. Data on pain in intersex and gender diverse individuals is severely lacking, and increased collection and sharing of this data within the pain research community could help build databases that provide the sample sizes necessary to provide sufficient statistical power to answer questions about pain processing in these gender minority populations.

Finally, given the multiple levels at which sex and gender can exert influence over the experience of pain, from genes and signaling proteins to brain circuitry and pain-coping strategies, interdisciplinary research teams could be a useful tool for advancing and promoting sex and gender-based research questions in experimental and clinical pain. Greater knowledge exchange across scientific fields, including immunology, endocrinology, neurology, cognitive and behavioral neuroscience, and psychology will help piece together a comprehensive picture of the role that sex and gender play in pain. Equipped with this knowledge, researchers and clinicians will be one step closer to designing precision pain medicine strategies that consider the complexities of a person's sex and gender.

References

Alabas, O. A., Tashani, O. A., Tabasam, G., & Johnson, M. I. (2012). Gender role affects experimental pain responses: A systematic review with meta-analysis. *European Journal of Pain*, *16*(9), 1211–1223.

Aloisi, A. M., Bachiocco, V., Costantino, A., Stefani, R., Ceccarelli, I., Bertaccini, A., et al. (2007). Cross-sex hormone administration changes pain in transsexual women and men. *Pain*, *132*(Suppl 1), S60–S67.

Alqudah, A. F., Hirsh, A. T., Stutts, L. A., Scipio, C. D., & Robinson, M. E. (2010). Sex and race differences in rating Others' pain, pain-related negative mood, pain coping, and recommending medical help. *The Journal of Cybertherapy and Rehabilitation*, *3*(1), 63–70.

Al-Rousan, T., Sparks, J. A., Pettinger, M., Chlebowski, R., Manson, J. E., Kauntiz, A. M., et al. (2018). Menopausal hormone therapy and the incidence of carpal tunnel syndrome in postmenopausal women: Findings from the Women's Health Initiative. *PLoS One*, *13*(12), e0207509.

Antal, A., Brepohl, N., Poreisz, C., Boros, K., Csifcsak, G., & Paulus, W. (2008). Transcranial direct current stimulation over somatosensory cortex decreases experimentally induced acute pain perception. *The Clinical Journal of Pain*, *24*(1), 56–63.

Arendt-Nielsen, L., Sluka, K. A., & Nie, H. L. (2008). Experimental muscle pain impairs descending inhibition. *Pain*, *140*(3), 465–471.

Averbuch, M., & Katzper, M. (2000). Baseline pain and response to analgesic medications in the postsurgery dental pain model. *Journal of Clinical Pharmacology*, *40*(2), 133–137.

Averitt, D. L., Eidson, L. N., Doyle, H. H., & Murphy, A. Z. (2019). Neuronal and glial factors contributing to sex differences in opioid modulation of pain. *Neuropsychopharmacology*, *44*(1), 155–165.

Avona, A., Burgos-Vega, C., Burton, M. D., Akopian, A. N., Price, T. J., & Dussor, G. (2019). Dural calcitonin gene-related peptide produces female-specific responses in rodent migraine models. *The Journal of Neuroscience*, *39*(22), 4323–4331.

Ayoub, L. J., Barnett, A., Leboucher, A., Golosky, M., McAndrews, M. P., Seminowicz, D. A., et al. (2019). The medial temporal lobe in nociception: A meta-analytic and functional connectivity study. *Pain*, *160*(6), 1245–1260.

Baliki, M. N., Mansour, A. R., Baria, A. T., & Apkarian, A. V. (2014). Functional reorganization of the default mode network across chronic pain conditions. *PLoS One*, *9*(9), e106133.

Barnabe, C., Bessette, L., Flanagan, C., Leclercq, S., Steiman, A., Kalache, F., et al. (2012). Sex differences in pain scores and localization in inflammatory arthritis: A systematic review and metaanalysis. *The Journal of Rheumatology*, *39*(6), 1221–1230.

Barr, A., & Eilat-Tsanani, S. (2022). Prescribing analgesics for low Back pain: Is there a gender difference? *Journal of Women's Health*, *31*(1), 79–83.

Beggs, S., & Salter, M. W. (2010). Microglia-neuronal signalling in neuropathic pain hypersensitivity 2.0. *Current Opinion in Neurobiology*, *20*(4), 474–480.

Benson, S., Kotsis, V., Rosenberger, C., Bingel, U., Forsting, M., Schedlowski, M., et al. (2012). Behavioural and neural correlates of visceral pain sensitivity in healthy men and women: Does sex matter? *European Journal of Pain*, *16*(3), 349–358.

Bijur, P. E., Esses, D., Birnbaum, A., Chang, A. K., Schechter, C., & Gallagher, E. J. (2008). Response to morphine in male and female patients: Analgesia and adverse events. *The Clinical Journal of Pain*, *24*(3), 192–198.

Blackless, M., Charuvastra, A., Derryck, A., Fausto-Sterling, A., Lauzanne, K., & Lee, E. (2000). How sexually dimorphic are we? Review and synthesis. *American Journal of Human Biology*, *12*(2), 151–166.

Borsook, D., Edwards, R., Elman, I., Becerra, L., & Levine, J. (2013). Pain and analgesia: The value of salience circuits. *Progress in Neurobiology*, *104*, 93–105.

Breimhorst, M., Hondrich, M., Rebhorn, C., May, A., & Birklein, F. (2012). Sensory and sympathetic correlates of heat pain sensitization and habituation in men and women. *European Journal of Pain*, *16*(9), 1281–1292.

Bretherton, B., de Ridder, D., Crowther, T., Black, S., Whelan, A., & Baranidharan, G. (2022). Men and women respond equally well to spinal cord and dorsal root ganglion stimulation. *Neuromodulation: Technology at the Neural Interface*. In press.

Butcher, B. E., & Carmody, J. J. (2012). Sex differences in analgesic response to ibuprofen are influenced by expectancy: A randomized, crossover, balanced placebo-designed study. *European Journal of Pain*, *16*(7), 1005–1013.

Chang, W. J., O'Connell, N. E., Beckenkamp, P. R., Alhassani, G., Liston, M. B., & Schabrun, S. M. (2018). Altered primary motor cortex structure, organization, and function in chronic pain: A systematic review and meta-analysis. *The Journal of Pain*, *19*(4), 341–359.

Choleris, E., Galea, L. A. M., Sohrabji, F., & Frick, K. M. (2018). Sex differences in the brain: Implications for behavioral and biomedical research. *Neuroscience and Biobehavioral Reviews*, *85*, 126–145.

Clayton, J. A. (2018). Applying the new SABV (sex as a biological variable) policy to research and clinical care. *Physiology & Behavior*, *187*, 2–5.

Clayton, J. A., & Tannenbaum, C. (2016). Reporting sex, gender, or both in clinical research? *JAMA*, *316*(18), 1863–1864.

Conrin, S. D., Zhan, L., Morrissey, Z. D., Xing, M., Forbes, A., Maki, P., et al. (2018). From default mode Network to the basal configuration: Sex differences in the resting-state brain connectivity as a function of age and their clinical correlates. *Frontiers in Psychiatry*, *9*, 365.

Coulombe, M. A., Erpelding, N., Kucyi, A., & Davis, K. D. (2016). Intrinsic functional connectivity of periaqueductal gray subregions in humans. *Human Brain Mapping*, *37*(4), 1514–1530.

Craft, R. M. (2007). Modulation of pain by estrogens. *Pain*, *132*(Suppl 1), S3–S12.

Dean, L. E., Jones, G. T., MacDonald, A. G., Downham, C., Sturrock, R. D., & Macfarlane, G. J. (2014). Global prevalence of ankylosing spondylitis. *Rheumatology (Oxford)*, *53*(4), 650–657.

Derbyshire, S. W., Nichols, T. E., Firestone, L., Townsend, D. W., & Jones, A. K. (2002). Gender differences in patterns of cerebral activation during equal experience of painful laser stimulation. *The Journal of Pain*, *3*(5), 401–411.

Duan-Porter, W., Goldstein, K. M., McDuffie, J. R., Hughes, J. M., Clowse, M. E., Klap, R. S., et al. (2016). Reporting of sex effects by systematic reviews on interventions for depression, diabetes, and chronic pain. *Annals of Internal Medicine*, *165*(3), 184–193.

Duchesne, A., Pletzer, B., Pavlova, M. A., Lai, M. C., & Einstein, G. (2020). Editorial: Bridging gaps between sex and gender in neurosciences. *Frontiers in Neuroscience*, *14*, 561.

Dugan, S. A., Powell, L. H., Kravitz, H. M., Everson Rose, S. A., Karavolos, K., & Luborsky, J. (2006). Musculoskeletal pain and menopausal status. *The Clinical Journal of Pain*, *22*(4), 325–331.

Edwards, R. R., Haythornthwaite, J. A., Sullivan, M. J., & Fillingim, R. B. (2004). Catastrophizing as a mediator of sex differences in pain: Differential effects for daily pain versus laboratory-induced pain. *Pain*, *111*(3), 335–341.

Eliot, L., Ahmed, A., Khan, H., & Patel, J. (2021). Dump the "dimorphism": Comprehensive synthesis of human brain studies reveals few male-female differences beyond size. *Neuroscience and Biobehavioral Reviews*, *125*, 667–697.

El-Maouche, D., Arlt, W., & Merke, D. P. (2017). Congenital adrenal hyperplasia. *The Lancet*, *390*(10108), 2194–2210.

Eto, K., Wake, H., Watanabe, M., Ishibashi, H., Noda, M., Yanagawa, Y., et al. (2011). Inter-regional contribution of enhanced activity of the primary somatosensory cortex to the anterior cingulate cortex accelerates chronic pain behavior. *The Journal of Neuroscience*, *31*(21), 7631–7636.

Fauchon, C., Meunier, D., Rogachov, A., Hemington, K. S., Cheng, J. C., Bosma, R. L., et al. (2021). Sex differences in brain modular organization in chronic pain. *Pain*, *162*(4), 1188–1200.

Fausto-Sterling, A. (1993). The five sexes: Why male and female are not enough. *The Sciences*, *33*, 20.

Fillingim, R. B., & Edwards, R. R. (2001). The association of hormone replacement therapy with experimental pain responses in postmenopausal women. *Pain*, *92*(1–2), 229–234.

Fillingim, R. B., & Edwards, R. R. (2005). Is self-reported childhood abuse history associated with pain perception among healthy young women and men? *The Clinical Journal of Pain*, *21*(5), 387–397.

Fillingim, R. B., Kaplan, L., Staud, R., Ness, T. J., Glover, T. L., Campbell, C. M., et al. (2005). The A118G single nucleotide polymorphism of the mu-opioid receptor gene (OPRM1) is associated with pressure pain sensitivity in humans. *The Journal of Pain*, *6*(3), 159–167.

Fillingim, R. B., King, C. D., Ribeiro-Dasilva, M. C., Rahim-Williams, B., & Riley, J. L., 3rd. (2009). Sex, gender, and pain: A review of recent clinical and experimental findings. *The Journal of Pain*, *10*(5), 447–485.

Firouzian, S., Osborne, N. R., Cheng, J. C., Kim, J. A., Bosma, R. L., Hemington, K. S., et al. (2020). Individual variability and sex differences in conditioned pain modulation and the impact of resilience, and conditioning stimulus pain unpleasantness and salience. *Pain, 161*(8), 1847–1860.

Fishman, S. M. (2007). Recognizing pain management as a human right: A first step. *Anesthesia and Analgesia, 105*(1), 8–9.

Garcia-Sifuentes, Y., & Maney, D. L. (2021). Reporting and misreporting of sex differences in the biological sciences. *eLife, 10*, e70817.

Gasparin, A., Zortea, M., Dos Santos, V. S., Carvalho, F., Torres, I. L. S., de Souza, A., et al. (2020). Brain-derived neurotrophic factor modulates the effect of sex on the descending pain modulatory system in healthy volunteers. *Pain Medicine, 21*(10), 2271–2279.

Gear, R. W., Miaskowski, C., Gordon, N. C., Paul, S. M., Heller, P. H., & Levine, J. D. (1996). Kappa-opioids produce significantly greater analgesia in women than in men. *Nature Medicine, 2*(11), 1248–1250.

George, S. Z., Fritz, J. M., Childs, J. D., & Brennan, G. P. (2006). Sex differences in predictors of outcome in selected physical therapy interventions for acute low back pain. *The Journal of Orthopaedic and Sports Physical Therapy, 36*(6), 354–363.

Gerdle, B., Bjork, J., Coster, L., Henriksson, K., Henriksson, C., & Bengtsson, A. (2008). Prevalence of widespread pain and associations with work status: A population study. *BMC Musculoskeletal Disorders, 9*(1), 1–10.

Gijsbers, K., & Nicholson, F. (2005). Experimental pain thresholds influenced by sex of experimenter. *Perceptual and Motor Skills, 101*(3), 803–807.

Girard-Tremblay, L., Auclair, V., Daigle, K., Leonard, G., Whittingstall, K., & Goffaux, P. (2014). Sex differences in the neural representation of pain unpleasantness. *The Journal of Pain, 15*(8), 867–877.

Goffaux, P., Michaud, K., Gaudreau, J., Chalaye, P., Rainville, P., & Marchand, S. (2011). Sex differences in perceived pain are affected by an anxious brain. *Pain, 152*(9), 2065–2073.

Goldberg, D. S., & McGee, S. J. (2011). Pain as a global public health priority. *BMC Public Health, 11*(1), 1–5.

Gomez-Perez, L., & Lopez-Martinez, A. E. (2013). Association of trauma, posttraumatic stress disorder, and experimental pain response in healthy young women. *The Clinical Journal of Pain, 29*(5), 425–434.

Graeff, P., Itter, A., Wach, K., & Ruscheweyh, R. (2021). Inter-individual differences explain more variance in conditioned pain modulation than age, sex and conditioning stimulus intensity combined. *Brain Sciences, 11*(9), 1186.

Greenspan, J. D., Craft, R. M., LeResche, L., Arendt-Nielsen, L., Berkley, K. J., Fillingim, R. B., et al. (2007). Studying sex and gender differences in pain and analgesia: A consensus report. *Pain, 132*(Suppl 1), S26–S45.

Gupta, A., Bhatt, R. R., Naliboff, B. D., Kutch, J. J., Labus, J. S., Vora, P. P., et al. (2019). Impact of early adverse life events and sex on functional brain networks in patients with urological chronic pelvic pain syndrome (UCPPS): A MAPP research Network study. *PLoS One, 14*(6), e0217610.

Gupta, A., Kilpatrick, L., Labus, J., Tillisch, K., Braun, A., Hong, J. Y., et al. (2014). Early adverse life events and resting state neural networks in patients with chronic abdominal pain: Evidence for sex differences. *Psychosomatic Medicine, 76*(6), 404–412.

Gupta, A., Mayer, E. A., Fling, C., Labus, J. S., Naliboff, B. D., Hong, J. Y., et al. (2017). Sex-based differences in brain alterations across chronic pain conditions. *Journal of Neuroscience Research, 95*(1–2), 604–616.

Hamberg, K., Risberg, G., Johansson, E. E., & Westman, G. (2002). Gender bias in physicians' management of neck pain: A study of the answers in a Swedish national examination. *Journal of Women's Health & Gender-Based Medicine, 11*(7), 653–666.

Hashmi, J. A., & Davis, K. D. (2009). Women experience greater heat pain adaptation and habituation than men. *Pain, 145*(3), 350–357.

Hashmi, J. A., & Davis, K. D. (2014). Deconstructing sex differences in pain sensitivity. *Pain, 155*(1), 10–13.

Hemington, K. S., & Coulombe, M. A. (2015). The periaqueductal gray and descending pain modulation: Why should we study them and what role do they play in chronic pain? *Journal of Neurophysiology, 114*(4), 2080–2083.

Hemington, K. S., Wu, Q., Kucyi, A., Inman, R. D., & Davis, K. D. (2016). Abnormal cross-network functional connectivity in chronic pain and its association with clinical symptoms. *Brain Structure & Function, 221*(8), 4203–4219.

Henderson, L. A., Gandevia, S. C., & Macefield, V. G. (2008). Gender differences in brain activity evoked by muscle and cutaneous pain: A retrospective study of single-trial fMRI data. *NeuroImage, 39*(4), 1867–1876.

Iacovides, S., Avidon, I., & Baker, F. C. (2015). Does pain vary across the menstrual cycle? A review. *European Journal of Pain, 19*(10), 1389–1405.

Joel, D. (2021). Beyond the binary: Rethinking sex and the brain. *Neuroscience and Biobehavioral Reviews, 122*, 165–175.

Jung, J. H., Bang, C. H., Song, G. G., Kim, C., Kim, J. H., & Choi, S. J. (2018). Knee osteoarthritis and menopausal hormone therapy in postmenopausal women: A nationwide cross-sectional study. *Menopause, 26*(6), 598–602.

Kaergaard, A., Hansen, A. M., Rasmussen, K., & Andersen, J. H. (2000). Association between plasma testosterone and work-related neck and shoulder disorders among female workers. *Scandinavian Journal of Work, Environment & Health, 26*(4), 292–298.

Kallai, I., Barke, A., & Voss, U. (2004). The effects of experimenter characteristics on pain reports in women and men. *Pain, 112*(1–2), 142–147.

Kavelaars, A., & Heijnen, C. J. (2021). Immune regulation of pain: Friend and foe. *Science Translational Medicine, 13*(619). eabj7152.

Keefe, F. J., Lefebvre, J. C., Egert, J. R., Affleck, G., Sullivan, M. J., & Caldwell, D. S. (2000). The relationship of gender to pain, pain behavior, and disability in osteoarthritis patients: The role of catastrophizing. *Pain, 87*(3), 325–334.

Keogh, E. (2021). The gender context of pain. *Health Psychology Review, 15*(3), 454–481.

Keogh, E., McCracken, L. M., & Eccleston, C. (2005). Do men and women differ in their response to interdisciplinary chronic pain management? *Pain, 114*(1–2), 37–46.

Kim, J. A., Bosma, R. L., Hemington, K. S., Rogachov, A., Osborne, N. R., Cheng, J. C., et al. (2021). Sex-differences in network level brain dynamics associated with pain sensitivity and pain interference. *Human Brain Mapping, 42*(3), 598–614.

Kim, H., Mawla, I., Lee, J., Gerber, J., Walker, K., Kim, J., et al. (2020). Reduced tactile acuity in chronic low back pain is linked with structural neuroplasticity in primary somatosensory cortex and is modulated by acupuncture therapy. *NeuroImage, 217*, 116899.

King, S., Chambers, C. T., Huguet, A., MacNevin, R. C., McGrath, P. J., Parker, L., et al. (2011). The epidemiology of chronic pain in children and adolescents revisited: A systematic review. *Pain, 152*(12), 2729–2738.

Kogler, L., Muller, V. I., Seidel, E. M., Boubela, R., Kalcher, K., Moser, E., et al. (2016). Sex differences in the functional connectivity of the amygdalae in association with cortisol. *NeuroImage, 134*, 410–423.

Koss, W. A., & Frick, K. M. (2017). Sex differences in hippocampal function. *Journal of Neuroscience Research, 95*(1–2), 539–562.

Kowalczyk, W. J., Sullivan, M. A., Evans, S. M., Bisaga, A. M., Vosburg, S. K., & Comer, S. D. (2010). Sex differences and hormonal influences on response to mechanical pressure pain in humans. *The Journal of Pain, 11*(4), 330–342.

Kucyi, A., Salomons, T. V., & Davis, K. D. (2013). Mind wandering away from pain dynamically engages antinociceptive and default mode brain networks. *Proceedings of the National Academy of Sciences of the United States of America, 110*(46), 18692–18697.

Kuner, R., & Flor, H. (2017). Structural plasticity and reorganisation in chronic pain. *Nature Reviews. Neuroscience, 18*(2), 20–30.

Kurth, F., Gaser, C., & Luders, E. (2021). Development of sex differences in the human brain. *Cognitive Neuroscience, 12*(3–4), 155–162.

Labus, J. S., Naliboff, B. N., Fallon, J., Berman, S. M., Suyenobu, B., Bueller, J. A., et al. (2008). Sex differences in brain activity during aversive visceral stimulation and its expectation in patients with chronic abdominal pain: A network analysis. *NeuroImage, 41*(3), 1032–1043.

LeGates, T. A., Kvarta, M. D., & Thompson, S. M. (2019). Sex differences in antidepressant efficacy. *Neuropsychopharmacology, 44*(1), 140–154.

Lenert, M. E., Avona, A., Garner, K. M., Barron, L. R., & Burton, M. D. (2021). Sensory neurons, neuroimmunity, and pain modulation by sex hormones. *Endocrinology, 162*(8).

Leresche, L. (2011). Defining gender disparities in pain management. *Clinical Orthopaedics and Related Research, 469*(7), 1871–1877.

LeResche, L., Saunders, K., Von Korff, M. R., Barlow, W., & Dworkin, S. F. (1997). Use of exogenous hormones and risk of temporomandibular disorder pain. *Pain, 69*(1–2), 153–160.

Levine, F. M., & Lee De Simone, L. (1991). The effects of experimenter gender on pain report in male and female subjects. *Pain, 44*(1), 69–72.

Levit, D., Yaish, I., Shtrozberg, S., Aloush, V., Greenman, Y., & Ablin, J. N. (2021). Pain and transition: Evaluating fibromyalgia in transgender individuals. *Clinical and Experimental Rheumatology, 39*(*Suppl 130*(3)), 27–32.

Lima, M. C., & Fregni, F. (2008). Motor cortex stimulation for chronic pain: Systematic review and meta-analysis of the literature. *Neurology, 70*(24), 2329–2337.

Linnman, C., Beucke, J. C., Jensen, K. B., Gollub, R. L., & Kong, J. (2012). Sex similarities and differences in pain-related periaqueductal gray connectivity. *Pain, 153*(2), 444–454.

Liu, J., Qin, W., Nan, J., Li, J., Yuan, K., Zhao, L., et al. (2011). Gender-related differences in the dysfunctional resting networks of migraine suffers. *PLoS One, 6*(11), e27049.

Lopes, D. M., Malek, N., Edye, M., Jager, S. B., McMurray, S., McMahon, S. B., et al. (2017). Sex differences in peripheral not central immune responses to pain-inducing injury. *Scientific Reports*, 7(1), 1–8.

Luo, X., Chen, O., Wang, Z., Bang, S., Ji, J., Lee, S. H., et al. (2021). IL-23/IL-17A/TRPV1 axis produces mechanical pain via macrophage-sensory neuron crosstalk in female mice. *Neuron, 109*(17), 2691–2706.

Malfliet, A., De Pauw, R., Kregel, J., Coppieters, I., Meeus, M., Roussel, N., et al. (2019). Gender differences in the Association of Brain Gray Matter and Pain-Related Psychosocial Characteristics. *Pain Physician, 22*(3), E191–E203.

Mansour, A. R., Farmer, M. A., Baliki, M. N., & Apkarian, A. V. (2014). Chronic pain: The role of learning and brain plasticity. *Restorative Neurology and Neuroscience, 32*(1), 129–139.

Martin, L. J., Acland, E. L., Cho, C., Gandhi, W., Chen, D., Corley, E., et al. (2019). Male-specific conditioned pain hypersensitivity in mice and humans. *Current Biology, 29*(2), 192–201.

Mattos Feijo, L., Tarman, G. Z., Fontaine, C., Harrison, R., Johnstone, T., & Salomons, T. (2018). Sex-specific effects of gender identification on pain study recruitment. *The Journal of Pain, 19*(2), 178–185.

McCarthy, M. M., Arnold, A. P., Ball, G. F., Blaustein, J. D., & De Vries, G. J. (2012). Sex differences in the brain: The not so inconvenient truth. *The Journal of Neuroscience, 32*(7), 2241–2247.

McCarthy, M. M., Auger, A. P., Bale, T. L., De Vries, G. J., Dunn, G. A., Forger, N. G., et al. (2009). The epigenetics of sex differences in the brain. *The Journal of Neuroscience*, *29*(41), 12815–12823.
Melchior, M., Poisbeau, P., Gaumond, I., & Marchand, S. (2016). Insights into the mechanisms and the emergence of sex-differences in pain. *Neuroscience*, *338*, 63–80.
Moeller-Bertram, T., Strigo, I. A., Simmons, A. N., Schilling, J. M., Patel, P., & Baker, D. G. (2014). Evidence for acute central sensitization to prolonged experimental pain in posttraumatic stress disorder. *Pain Medicine*, *15*(5), 762–771.
Mogil, J. S. (2012). Sex differences in pain and pain inhibition: Multiple explanations of a controversial phenomenon. *Nature Reviews. Neuroscience*, *13*(12), 859–866.
Mogil, J. S. (2018). Sex-based divergence of mechanisms underlying pain and pain inhibition. *Current Opinion in Behavioral Sciences*, *23*, 113–117.
Mogil, J. S. (2020). Qualitative sex differences in pain processing: Emerging evidence of a biased literature. *Nature Reviews. Neuroscience*, *21*(7), 353–365.
Mogil, J. S., & Chanda, M. L. (2005). The case for the inclusion of female subjects in basic science studies of pain. *Pain*, *117*(1–2), 1–5.
Mogil, J. S., Wilson, S. G., Chesler, E. J., Rankin, A. L., Nemmani, K. V., Lariviere, W. R., et al. (2003). The melanocortin-1 receptor gene mediates female-specific mechanisms of analgesia in mice and humans. *Proceedings of the National Academy of Sciences of the United States of America*, *100*(8), 4867–4872.
Morales, A. J., Nolan, J. J., Nelson, J. C., & Yen, S. S. (1994). Effects of replacement dose of dehydroepiandrosterone in men and women of advancing age. *The Journal of Clinical Endocrinology and Metabolism*, *78*(6), 1360–1367.
Mosconi, L., Berti, V., Quinn, C., McHugh, P., Petrongolo, G., Varsavsky, I., et al. (2017). Sex differences in Alzheimer risk: Brain imaging of endocrine vs chronologic aging. *Neurology*, *89*(13), 1382–1390.
Neugebauer, V. (2015). Amygdala pain mechanisms. *Handbook of Experimental Pharmacology*, *227*, 261–284.
Niesters, M., Dahan, A., Kest, B., Zacny, J., Stijnen, T., Aarts, L., et al. (2010). Do sex differences exist in opioid analgesia? A systematic review and meta-analysis of human experimental and clinical studies. *Pain*, *151*(1), 61–68.
Nir, R. R., & Yarnitsky, D. (2015). Conditioned pain modulation. *Current Opinion in Supportive and Palliative Care*, *9*(2), 131–137.
Olsen, M. B., Jacobsen, L. M., Schistad, E. I., Pedersen, L. M., Rygh, L. J., Roe, C., et al. (2012). Pain intensity the first year after lumbar disc herniation is associated with the A118G polymorphism in the opioid receptor mu 1 gene: Evidence of a sex and genotype interaction. *The Journal of Neuroscience*, *32*(29), 9831–9834.
Osborne, N. R., Anastakis, D. J., Kim, J. A., El-Sayed, R., Cheng, J. C., Rogachov, A., et al. (2021). Sex-specific abnormalities and treatment-related plasticity of subgenual anterior cingulate cortex functional connectivity in chronic pain. *Frontiers in Pain Research*, *2*(9), 673538.
Osborne, N. R., Cheng, J. C., Rogachov, A., Kim, J. A., Hemington, K. S., Bosma, R. L., et al. (2021). Abnormal subgenual anterior cingulate circuitry is unique to women but not men with chronic pain. *Pain*, *162*(1), 97–108.
Pace, A., Barber, M., Ziplow, J., Hranilovich, J. A., & Kaiser, E. A. (2021). Gender minority stress, psychiatric comorbidities, and the experience of migraine in transgender and gender-diverse individuals: A narrative review. *Current Pain and Headache Reports*, *25*(12), 1–9.
Paulson, P. E., Minoshima, S., Morrow, T. J., & Casey, K. L. (1998). Gender differences in pain perception and patterns of cerebral activation during noxious heat stimulation in humans. *Pain*, *76*(1–2), 223–229.

Pool, G. J., Schwegler, A. F., Theodore, B. R., & Fuchs, P. N. (2007). Role of gender norms and group identification on hypothetical and experimental pain tolerance. *Pain, 129*(1–2), 122–129.

Popescu, A., LeResche, L., Truelove, E. L., & Drangsholt, M. T. (2010). Gender differences in pain modulation by diffuse noxious inhibitory controls: A systematic review. *Pain, 150*(2), 309–318.

Pritschet, L., Taylor, C. M., Santander, T., & Jacobs, E. G. (2021). Applying dense-sampling methods to reveal dynamic endocrine modulation of the nervous system. *Current Opinion in Behavioral Sciences, 40*, 72–78.

Pud, D., Granovsky, Y., & Yarnitsky, D. (2009). The methodology of experimentally induced diffuse noxious inhibitory control (DNIC)-like effect in humans. *Pain, 144*(1–2), 16–19.

Racine, M., Sole, E., Sanchez-Rodriguez, E., Tome-Pires, C., Roy, R., Jensen, M. P., et al. (2020). An evaluation of sex differences in patients with chronic pain undergoing an interdisciplinary pain treatment program. *Pain Practice, 20*(1), 62–74.

Racine, M., Tousignant-Laflamme, Y., Kloda, L. A., Dion, D., Dupuis, G., & Choiniere, M. (2012a). A systematic literature review of 10 years of research on sex/gender and experimental pain perception - part 1: Are there really differences between women and men? *Pain, 153*(3), 602–618.

Racine, M., Tousignant-Laflamme, Y., Kloda, L. A., Dion, D., Dupuis, G., & Choiniere, M. (2012b). A systematic literature review of 10 years of research on sex/gender and pain perception - part 2: Do biopsychosocial factors alter pain sensitivity differently in women and men? *Pain, 153*(3), 619–635.

Raja, S. N., Carr, D. B., Cohen, M., Finnerup, N. B., Flor, H., Gibson, S., et al. (2020). The revised International Association for the Study of Pain definition of pain: Concepts, challenges, and compromises. *Pain, 161*(9), 1976–1982.

Ramirez-Maestre, C., & Esteve, R. (2014). The role of sex/gender in the experience of pain: Resilience, fear, and acceptance as central variables in the adjustment of men and women with chronic pain. *The Journal of Pain, 15*(6), 608–618.

Reckziegel, D., Abdullah, T., Wu, B., Wu, B., Huang, L., Schnitzer, T. J., et al. (2021). Hippocampus shape deformation: A potential diagnostic biomarker for chronic back pain in women. *Pain, 162*(5), 1457–1467.

Rice, A. S. C., Smith, B. H., & Blyth, F. M. (2016). Pain and the global burden of disease. *Pain, 157*(4), 791–796.

Rich-Edwards, J. W., Kaiser, U. B., Chen, G. L., Manson, J. E., & Goldstein, J. M. (2018). Sex and gender differences research Design for Basic, clinical, and population studies: Essentials for investigators. *Endocrine Reviews, 39*(4), 424–439.

Riley, J. L., 3rd, Robinson, M. E., Wise, E. A., & Price, D. (1999). A meta-analytic review of pain perception across the menstrual cycle. *Pain, 81*(3), 225–235.

Ritchie, S. J., Cox, S. R., Shen, X., Lombardo, M. V., Reus, L. M., Alloza, C., et al. (2018). Sex differences in the adult human brain: Evidence from 5216 UK biobank participants. *Cerebral Cortex, 28*(8), 2959–2975.

Robinson, M. E., Riley, J. L., 3rd, Myers, C. D., Papas, R. K., Wise, E. A., Waxenberg, L. B., et al. (2001). Gender role expectations of pain: Relationship to sex differences in pain. *The Journal of Pain, 2*(5), 251–257.

Rollman, G. B., Abdel-Shaheed, J., Gillespie, J. M., & Jones, K. S. (2004). Does past pain influence current pain: Biological and psychosocial models of sex differences. *European Journal of Pain, 8*(5), 427–433.

Rosen, S., Ham, B., & Mogil, J. S. (2017). Sex differences in neuroimmunity and pain. *Journal of Neuroscience Research, 95*(1–2), 500–508.

Ruau, D., Liu, L. Y., Clark, J. D., Angst, M. S., & Butte, A. J. (2012). Sex differences in reported pain across 11,000 patients captured in electronic medical records. *The Journal of Pain, 13*(3), 228–234.

Rubinow, D. R., & Schmidt, P. J. (2019). Sex differences and the neurobiology of affective disorders. *Neuropsychopharmacology*, *44*(1), 111–128.

Rusman, T., van Vollenhoven, R. F., & van der Horst-Bruinsma, I. E. (2018). Gender differences in axial Spondyloarthritis: Women are not so lucky. *Current Rheumatology Reports*, *20*(6), 1–12.

Safdar, B., Heins, A., Homel, P., Miner, J., Neighbor, M., DeSandre, P., et al. (2009). Impact of physician and patient gender on pain management in the emergency department- -a multicenter study. *Pain Medicine*, *10*(2), 364–372.

Schafer, G., Prkachin, K. M., Kaseweter, K. A., & Williams, A. C. (2016). Health care providers' judgments in chronic pain: The influence of gender and trustworthiness. *Pain*, *157*(8), 1618–1625.

Seney, M. L., & Sibille, E. (2014). Sex differences in mood disorders: Perspectives from humans and rodent models. *Biology of Sex Differences*, *5*(1), 1–10.

Shansky, R. M., & Murphy, A. Z. (2021). Considering sex as a biological variable will require a global shift in science culture. *Nature Neuroscience*, *24*(4), 457–464.

Shapiro, J. R., Klein, S. L., & Morgan, R. (2021). Stop "controlling" for sex and gender in global health research. *BMJ Global Health*, *6*(4), e005714.

Sherman, J. J., & LeResche, L. (2006). Does experimental pain response vary across the menstrual cycle? A methodological review. *American Journal of Physiology. Regulatory, Integrative and Comparative Physiology*, *291*(2), R245–R256.

Slyer, J., Scott, S., Sheldon, B., Hancu, M., Bridger, C., & Pilitsis, J. G. (2020). Less pain relief, more depression, and female sex correlate with spinal cord stimulation explants. *Neuromodulation*, *23*(5), 673–679.

Stening, K. D., Eriksson, O., Henriksson, K. G., Brynhildsen, J., Lindh-Astrand, L., Berg, G., et al. (2011). Hormonal replacement therapy does not affect self-estimated pain or experimental pain responses in post-menopausal women suffering from fibromyalgia: A double-blind, randomized, placebo-controlled trial. *Rheumatology (Oxford)*, *50*(3), 544–551.

Strath, L. J., Sorge, R. E., Owens, M. A., Gonzalez, C. E., Okunbor, J. I., White, D. M., et al. (2020). Sex and gender are not the same: Why identity is important for people living with HIV and chronic pain. *Journal of Pain Research*, *13*, 829–835.

Straube, T., Schmidt, S., Weiss, T., Mentzel, H. J., & Miltner, W. H. (2009). Sex differences in brain activation to anticipated and experienced pain in the medial prefrontal cortex. *Human Brain Mapping*, *30*(2), 689–698.

Tan, E. C., Zijlstra, B., Essink, M. L., Goris, R. J., & Severijnen, R. S. (2008). Complex regional pain syndrome type I in children. *Acta Paediatrica*, *97*(7), 875–879.

Taylor, A. L. (2007). Addressing the global tragedy of needless pain: Rethinking the United Nations single convention on narcotic drugs. *The Journal of Law, Medicine & Ethics*, *35*(4), 556–570 (511).

Thierry, S., Fautrel, B., Lemelle, I., & Guillemin, F. (2014). Prevalence and incidence of juvenile idiopathic arthritis: A systematic review. *Joint, Bone, Spine*, *81*(2), 112–117.

Tousignant-Laflamme, Y., & Marchand, S. (2009). Excitatory and inhibitory pain mechanisms during the menstrual cycle in healthy women. *Pain*, *146*(1–2), 47–55.

Tsang, A., Von Korff, M., Lee, S., Alonso, J., Karam, E., Angermeyer, M. C., et al. (2008). Common chronic pain conditions in developed and developing countries: Gender and age differences and comorbidity with depression-anxiety disorders. *The Journal of Pain*, *9*(10), 883–891.

Tu, C. H., Niddam, D. M., Yeh, T. C., Lirng, J. F., Cheng, C. M., Chou, C. C., et al. (2013). Menstrual pain is associated with rapid structural alterations in the brain. *Pain*, *154*(9), 1718–1724.

Uddin, M., Sipahi, L., Li, J., & Koenen, K. C. (2013). Sex differences in DNA methylation may contribute to risk of PTSD and depression: A review of existing evidence. *Depression and Anxiety*, *30*(12), 1151–1160.

Unruh, A. M. (1996). Gender variations in clinical pain experience. *Pain*, *65*(2–3), 123–167.

van der Horst-Bruinsma, I. E., Zack, D. J., Szumski, A., & Koenig, A. S. (2013). Female patients with ankylosing spondylitis: Analysis of the impact of gender across treatment studies. *Annals of the Rheumatic Diseases*, *72*(7), 1221–1224.

Verriotis, M., Jones, L., Whitehead, K., Laudiano-Dray, M., Panayotidis, I., Patel, H., et al. (2018). The distribution of pain activity across the human neonatal brain is sex dependent. *NeuroImage*, *178*, 69–77.

Vetvik, K. G., Macgregor, E. A., Lundqvist, C., & Russell, M. B. (2014). Prevalence of menstrual migraine: A population-based study. *Cephalalgia*, *34*(4), 280–288.

Vincent, K., Moore, J., Kennedy, S., & Tracey, I. (2014). Steroid hormones and pain-related brain activity and functional connectivity in healthy women. *The Lancet*, *383*, S104.

Vincent, K., & Tracey, I. (2008). Hormones and their interaction with the pain experience. *Reviews in Pain*, *2*(2), 20–24.

Waardenburg, S., de Meij, N., Brouwer, B., van Zundert, J., & van Kuijk, S. M. (2021). Cross-sectional study on sex differences in chronic pain patients using the DATAPAIN cohort. *Regional Anesthesia and Pain Medicine*, *46*(11), 997–999.

Walsh, C. A., Jamieson, E., Macmillan, H., & Boyle, M. (2007). Child abuse and chronic pain in a community survey of women. *Journal of Interpersonal Violence*, *22*(12), 1536–1554.

Wang, G., Erpelding, N., & Davis, K. D. (2014). Sex differences in connectivity of the subgenual anterior cingulate cortex. *Pain*, *155*(4), 755–763.

Weis, S., Patil, K. R., Hoffstaedter, F., Nostro, A., Yeo, B. T. T., & Eickhoff, S. B. (2020). Sex classification by resting state brain connectivity. *Cerebral Cortex*, *30*(2), 824–835.

West, H. F. (1949). Aetiology of ankylosing spondylitis. *Annals of the Rheumatic Diseases*, *8*(2), 143–148.

Wheelock, M. D., Hect, J. L., Hernandez-Andrade, E., Hassan, S. S., Romero, R., Eggebrecht, A. T., et al. (2019). Sex differences in functional connectivity during fetal brain development. *Developmental Cognitive Neuroscience*, *36*, 100632.

White, J., Tannenbaum, C., Klinge, I., Schiebinger, L., & Clayton, J. (2021). The integration of sex and gender considerations into biomedical research: Lessons from international funding agencies. *The Journal of Clinical Endocrinology and Metabolism*, *106*(10), 3034–3048.

Yoon, D. Y., Mansukhani, N. A., Stubbs, V. C., Helenowski, I. B., Woodruff, T. K., & Kibbe, M. R. (2014). Sex bias exists in basic science and translational surgical research. *Surgery*, *156*(3), 508–516.

Zhang, L., Losin, E. A. R., Ashar, Y. K., Koban, L., & Wager, T. D. (2021). Gender biases in estimation of Others' pain. *The Journal of Pain*, *22*(9), 1048–1059.

CHAPTER NINE

Sex and age differences in migraine treatment and management strategies

Elena R. Lebedeva[a,b,*]
[a]Department of Neurology, The Ural State Medical University, Yekaterinburg, Russia
[b]International Headache Centre "Europe-Asia", Yekaterinburg, Russia
*Corresponding author: e-mail address: cosmos@k66.ru

Contents

International Review of Neurobiology, Volume 164
ISSN 0074-7742
https://doi.org/10.1016/bs.irn.2022.07.003

Abstract

Migraine is one of the leading causes of disability worldwide, especially in women younger than 50 years old. Migraine has three times higher prevalence in women than in men and tends to decrease after the menopausal transition. Migraine has different clinical features in people of different ages. Clinical symptoms and factors associated with migraine can be various in women and men. Women have special types of migraine, such as pure menstrual migraine and menstrually related migraine. Besides, clinical symptoms of migraine can change during pregnancy, postpartum and lactation. Women are significantly more often than men consulting a doctor because of migraine. These features of migraine lead to different treatment and management strategies in females and males of different ages. Migraine therefore is a disorder that demonstrates the necessity of a personalization of healthcare—ensuring the proper treatment for the right patient, at the right time. Considering all the available literature and guidelines, in this chapter several strategies for management of acute and prophylactic treatments of migraine, according to sex and age differences, are discussed. The purpose of this chapter is to provide a useful piece of information improving the treatment and management of migraine.

1. Introduction

Migraine is one of the leading causes of disability worldwide, especially in women younger than 50 (Steiner et al., 2020). Among females aged 15–49 years, headache disorders are second only to gynecological diseases, according to the Global Burden of Disease study 2019 (Steiner et al., 2020).

One-year prevalence of migraine in women is 18%, and one-year prevalence in men is 6% (Ashina, Buse, et al., 2021; Ashina, Katsarava, et al., 2021). Overall, and in females, migraine is most prevalent between 20 and 64 years of age, whereas in males it was somewhat higher from 10 to 19 (Stovner, Hagen, Linde, & Steiner, 2022). Migraine can affect anyone, at any age, including young children. About 4–5 out of 100 children have migraines before puberty. Boys and girls are almost equally affected. Migraines become more common during puberty, especially in girls. In general, migraine has three times higher prevalence in women than in men and 3- to 4-fold higher prevalence in women after puberty compared to men. The largest 1-year prevalence ratio between sexes was observed at 30.2 years of age (Al-Hassany et al., 2020). Active migraine tends to decrease after the menopausal transition.

Migraine manifests clinically as recurrent attacks of headache with a range of accompanying symptoms. In approximately one-third of individuals with migraine, headache is sometimes preceded or accompanied by transient neurological disturbances, referred to as migraine aura (ICHD 3, 2018). Furthermore, a minority of those affected develop chronic migraine, in which attacks become highly frequent (Eigenbrodt et al., 2021).

Migraine has clinical features in people of different ages, especially in children. Typical migraine symptoms such as nausea, vomiting, photo- and phonophobia are not constantly prevalent in child sufferers. Besides, children sometimes cannot describe what they are feeling, which makes the diagnosis of headaches even more difficult. These children may go undiagnosed for years or have the wrong diagnosis, leading into adulthood, affecting their psychological and mental well-being. The main differences between childhood migraine and those of adults include:

- migraine in children has usually shorter duration and low frequency
- in children, the pain experience tends to be more bilateral, localized across the forehead, rather than unilateral
- while typical symptoms of migraine maybe not be reported, a lot can be obtained from a behavior of a child, such as wanting to sleep or lay down in a dark, quiet room
- specific early indicators of migraine in children include car or motion sickness, cyclic vomiting periods, and infant colic.

Clinical symptoms of migraine can be a little different in women and men. Women have more severe, prolonged, and frequent attacks, leading to more significant disability and a longer recovery (Al-Hassany et al., 2020). A 1.4-fold increased risk for high-frequency headache was found during perimenopause compared to premenopause (Martin et al., 2016). This effect might be confounded or mediated by medication overuse or depression. Accompanying symptoms including photo and/or phonophobia, nausea, vomiting occur more often in women (Lebedeva, Kobzeva, Gilev, & Olesen, 2016a, 2016b, 2016c; Lipton et al., 2018). Men with migraine generally have less severe attacks and disability, and tend to have more extended remission periods than women. Men have fewer monthly headache days, and they have chronic migraine rarely and migraine with aura. Women, as distinct from men, have special types of migraine, such as pure menstrual migraine and menstrually related migraine. Women with pure menstrual migraine are more impaired due to the attacks and have the overall highest burden. Besides, clinical symptoms of migraine can change during pregnancy, postpartum and lactation.

Factors associated with migraine can differ in women and men. For example, consumption of light alcoholic beverages (OR 3.49; 2.03–6.02) and low physical activity was associated with migraine in females (OR 1.9; 95% CI 1.1–3.2) (Lebedeva, Kobzeva, Gilev, & Olesen, 2016a, 2016b, 2016c). Common factors associated with migraine in both sexes included a history of head trauma, a family history of headache, and arterial hypertension (Lebedeva, Kobzeva, Gilev, & Olesen, 2016a, 2016b, 2016c). Being a student was associated with the highest risk of migraine (OR 6.6; 95% CI 4.2–10.4) (Lebedeva, Kobzeva, Gilev, & Olesen, 2016a, 2016b, 2016c). Stress and lifestyle synergistic relationship between female sex and high levels of daily stress on risk of migraine has been described. Females exposed to stress seem to have a higher 1-year prevalence of migraine, indicating an interplay between biological, sex-related factors and environmental stress in migraine progression (Lipton et al., 2018).

Women are significantly more often than men consulting a doctor because of migraine (Lebedeva, Kobzeva, Gilev, & Olesen, 2016a, 2016b, 2016c; Lipton et al., 2018). Women are more likely to report a healthcare professional diagnosis. Males more often underreport their symptoms and receive a migraine diagnosis if they are consulted. Women frequently use all prescriptions, which can lead to medication overuse headache.

These features of migraine lead to different treatment and management strategies in females and males of different ages. Migraine is a unique disorder that demonstrates the necessity of the personalization of healthcare—ensuring the proper treatment for the right patient at the right time. Based on available literature and guidelines on treatment of migraine, strategies for management and treatment of migraine are reported, according to sex and age differences.

(1) Management of migraine in girls and boys 3–17 years old
(2) Management of migraine in women 18–40 years old
(3) Management of menstrual migraine
(4) Management of migraine in pregnancy
(5) Management of migraine during lactation
(6) Management of migraine in perimenopause
(7) Management of migraine in menopause and post-menopause
(8) Management of migraine in men 18–50 years old
(9) Management of migraine in men >50 years old

For each group, a table, including acute and prophylactic treatments of migraine, consisting of first-line, second-line, and third-line medication, and also strategies of management is reported.

2. Part I: Management of migraine in girls and boys 3–17 years old

Migraine is often underdiagnosed in children. The diagnosis of migraine is not supported by laboratory or imaging findings and it is based only on the history information and a thorough physical examination. The lack of studies regarding its treatment in this particular population makes it harder to enlarge the choices of treatment (Teleanu, Vladacenco, Teleanu, & Epure, 2016). However, recent trials made it easier to utilize newer compounds that improve the outcome of the disease.

This subchapter describes the treatment and management of pediatric migraine and divided therapeutic methods into two broad areas: treatment of the acute attack—used both in the emergency room and as home options and prophylactic treatment (Table 1).

2.1 Strategy of management for acute treatment of migraine in girls and boys 3–17 years old

- Non-pharmacological treatment should always be attempted before initiating drug treatment (Schytz et al., 2021).
- Minimize or eliminate triggers, e.g., stress or poor posture during schoolwork.
- Exclude other underlying disorders, e.g., pain of any localization, stress, psychogenic factors (problems at home, school or/and among peers), depression, anxiety, refraction anomalies, strabismus, over-strained eyes (computer work/games), oromandibular dysfunction, sinusitis, posture anomaly, passive/active smoking and inappropriate lifestyle (Practice Guideline Update Summary, 2020).

Table 1 Acute and preventive treatment of migraine in girls and boys 3–17 years old.

Acute treatment	Preventive treatment
First-line medication • *Nonpharmacologic treatment* ✓ Bed rest alone can be sufficient ✓ Identification and elimination/reduction of provocative headache trigger factors ✓ Improvement of sleep duration and quality ✓ Maintaining regular meals and good hydration ✓ The moderate use of caffeine and caffeinated soft drink ✓ Good schedule of physical activity • *Pharmacological treatment:* ✓ NSAIDs (ibuprofen 7.5–10.0 mg/kg/dose, maximum dose of 1000 mg/dose) Second-line medication • *Triptans* ✓ Sumatriptan 10 mg nasal spray (approved for adolescents aged 12–17 years) ✓ Zolmitriptan tablets/nasal spray ✓ Almotriptan tablets ✓ Rizatriptan tablets • Sumatriptan/naproxen in combination • *Acetaminophen* (10–20 mg/kg, maximum dose of 1000 mg) *Adjunct medications for nausea and/or vomiting* • Prokinetic antiemetics ✓ Domperidone (only for children older than 12 y.o. with weight >35 kg) ✓ Metoclopramide *Contraindications* ✓ Aspirin-containing products should be avoided in children under the age of 16 years to avoid the risk of Reye syndrome. ✓ Before prescription, it is necessary to check contraindications for drugs in each patient using their instructions; besides, a patient should read a medicine label.	Beta-blockers: ✓ Propranolol ✓ Metoprolol Antidepressants: ✓ Amitriptyline (the target dose of is 1 mg/kg/day) Antiepileptics: ✓ Topiramate (the target dose is 2–4 mg/kg/day) ✓ Valproate (15–20 mg/kg/day) ✓ Levetiracetam Flunarizine *Nonpharmacologic treatment* • Vitamin D • Magnesium • Riboflavin (vitamin B2) • Coenzyme Q10 *Contraindications* ✓ Before prescription, it is necessary to check contraindications for drugs in each patient using their instructions; besides, a patient should read a medicine label.

- Exclude medication-overuse headache.
- Inform about disease mechanisms so that both child and parents understand it.
- Thorough information of children/youth and their parents regarding conclusion and plan for treatment. In addition, it is essential to minimize their possible concern about serious illness—when there is no evidence of this.
- Triptans should be prescribed with caution for patients <18 years. However, Sumatriptan 10 mg nasal spray is approved for adolescents aged 12–17 years (Schytz et al., 2021).
- Monotherapy with other NSAIDs (naproxen, ketoprofen, diclofenac and indomethacin) has not been tested in children/adolescents in RCTs (Kacperski, Kabbouche, O'Brien, & Weberding, 2016).
- To avoid the development of medication overuse headache (MOH), abortive medications should be used no more than three days/week. Migraine-specific drugs, particularly triptans, should be used fewer than nine times per month.
- In practice, NSAIDs are used to treat mild to moderate cases of migraine, with triptans reserved for moderate to severe headaches unresponsive to over-the-counter therapy.
- Clinicians should offer an alternate triptan, if one triptan fails to provide pain relief, to find the most effective agent to reduce migraine symptoms.
- In adolescents whose migraine is incompletely responsive to a triptan, clinicians should offer ibuprofen or naproxen in addition to a triptan to improve migraine relief.
- Clinicians must not prescribe triptans to those with a history of ischemic vascular disease or accessory conduction pathway disorders to avoid the morbidity and mortality associated with aggravating these conditions.
- Due to lower efficacy, acetaminophen should be reserved as a treatment choice for patients with documented hypersensitivity to nonsteroidal anti-inflammatory drugs (NSAIDs), upper gastrointestinal disease, renal impairment, bleeding disorders, or current oral anticoagulant use (Kacperski et al., 2016).
- For patients who experience nausea and vomiting with their migraines, the sublingual, intranasal, or subcutaneous injection preparations may be prescribed (Kacperski et al., 2016).
- For younger children who are unable to swallow pills, rizatriptan and zolmitriptan are available in melt preparations or zolmitriptan and sumatriptan can be given as a nasal spray.

2.2 Strategy of management for preventive treatment of migraine in girls and boys 3–17 years old

- Non-pharmacological treatment should always be attempted before initiating drug treatment, but the evidence base is extremely limited.
- There is only an indication for pharmacological preventive treatment of migraine in children if the non-pharmacological treatment and acute treatment are ineffective and/or the headache attacks are frequent (more than 3–4 days per month), long and/or so severe that the attacks significantly influence the quality of life and/or functional level (Schytz et al., 2021).
- Currently, no standardized guidelines for choosing a preventive for pediatric patients exist. Clinicians are frequently guided by extrapolation from adult studies, as well as a limited number of pediatric studies. Medications frequently used in children include tricyclic antidepressants, antiepileptic medications, and antiserotonergic agents (Schytz et al., 2021).
- Topiramate has gained approval from the FDA for use in childhood migraine.
- When selecting an agent for preventive treatment, any comorbid conditions, including anxiety and depression, should be considered (Kacperski et al., 2016).
- Slow titration of doses over a period of 4–12 weeks is common to ensure that the child tolerates the medication with minimal adverse effects. When an effective dose is reached, relief must be sustained for 2–3 months before considering alternative medication. Once this is achieved, the child may slowly be weaned off the therapy.
- Treatment should not be abandoned until it has been given an adequate trial of at least 6–8 weeks unless the child experiences intolerable side effects (Kacperski et al., 2016).
- Clear instructions should be given to families regarding the medication's mechanism of action, possible adverse effects, and a clear titration schedule should be provided. It is essential to discuss the time it will take, often several weeks, for the preventive to become effective (Kacperski et al., 2016).
- The risk of fertility-related adverse effects needs to be discussed with females of child-bearing age prior to initiation of therapy with Valproic acid and should not be prescribed to females who are at risk for pregnancy.

3. Part II: Management of migraine in women 18–40 years old

In females of childbearing age, migraine accounts for 11.2% of total YLDs (Al-Hassany et al., 2020). Sex hormones, especially fluctuations of estrogen and progesterone, are believed to impact the pathogenesis of migraine. Although the exact etiology of migraine remains unknown, research efforts over the past decade have greatly expanded our scientific knowledge concerning migraine pathophysiology, especially in relation to the roles of serotonin receptors and calcitonin gene-related peptide (CGRP). This knowledge has greatly contributed to the development of new drugs that are more selective and specific to migraine pain mechanisms.

In this subchapter, we aim to discuss the drug classes available for the symptomatic and preventive treatment of migraine and the considerations that physicians must bear in mind prescribing the best therapeutic option in females of childbearing age (Table 2).

3.1 Strategies of management for acute treatment of migraine in women 18–40 years old

- To rule out medication overuse headache (Schytz et al., 2021).
- Treatment should be initiated as early as possible (Schytz et al., 2021).
- Triptans should be taken at the onset of head pain for maximum efficacy.
- In migraine with aura, however, triptans should be taken only when the headache starts (Schytz et al., 2021).
- Analgesics often has a better effect when combined with rest and, possibly, sleep. If the patient has difficulty calming down, benzodiazepine may be given, e.g. 5 mg diazepam (Schytz et al., 2021).
- Simple analgesics should be used for a maximum of 14 days per month to avoid medication overuse headaches (Eigenbrodt et al., 2021; Schytz et al., 2021).
- Triptans or combined analgesics should be used for a maximum of 9 days per month to avoid medication overuse headaches (Eigenbrodt et al., 2021; Schytz et al., 2021).
- Patients who have no effect on one particular triptan may benefit from another triptan. Before the effect of any triptan is ruled out, the patient should, as a rule of thumb, have tried three different triptans, each during three attacks (Schytz et al., 2021).

Table 2 Acute and preventive treatment of migraine in women 18–40 years old.

Acute treatment	Preventive treatment
First-line medication • *Simple analgesics:* ✓ Ibuprofen 400–600 mg ✓ Naproxen 500 mg ✓ Paracetamol 1.000 mg ✓ Aspirin/caffeine 500/50 mg • *Antiemetic:* ✓ Metoclopramide 10 mg ✓ Domperidone 10 mg Second-line medication • *Triptans:* ✓ Sumatriptan 50–100 mg ✓ Zolmitriptan 2.5 mg ✓ Naratriptan 2.5 mg ✓ Rizatriptan 5–10 mg ✓ Almotriptan 12.5 mg ✓ Eletriptan 40 mg ✓ Frovatriptan 2.5 mg • *Gepants* ✓ Rimegepant 75 mg ✓ Ubrogepant 50–100 mg *Contraindication* ✓ Triptans are contraindicated in stroke, uncontrolled hypertension, ischemic heart disease, previous cerebral infarction, peripheral vascular diseases, severe hepatic impairment, and those taking monoamine oxidase inhibitors. ✓ Before prescription, it is necessary to check contraindications for drugs in each patient using their instructions; besides, a patient should read a medicine label.	First-line medication • Beta-blockers: ✓ Metoprolol is 50 mg × 1 for 1 week and then 100 mg × 1, increased to 150–200 mg per day in case of lack of effect ✓ Propranolol is 40 mg × 2 increased at a weekly interval to a maximum of 120 mg × 2 • Angiotensin II receptor antagonist: ✓ Candesartan 8 mg × 1 for 1 week and then 16 mg × 1, possibly increasing to 24–32 mg × 1 Second-line medication • Anti-epileptics: ✓ Topiramate 25–100 (200) mg ✓ Valproate sodium 500–1800 mg ✓ Divalproex sodium • Antidepressants: ✓ Amitriptyline 10 mg × 1 increasing by 10 mg at one-week intervals to 10–100 mg daily (1–2 h before bedtime) • Botulinum type A toxin 155–195 units i.m. (every 12 weeks) • CGRP-antibodies: ✓ Erenumab 70–140 mg s.c. every 4 weeks ✓ Fremanezumab 225 mg s.c. every month or 675 mg s.c. every 3 months ✓ Galcanezumab Start dose 240 mg s.c. followed by 120 mg s.c. every month ✓ Eptinezumab 100–300 mg i.v. every 3 months *Nonpharmacologic treatment* • Riboflavin • Vitamin D supplementation • Acupuncture • Physical exercise • Cognitive behavioral therapy *Contraindications* ✓ Before prescription, it is necessary to check contraindications for drugs in each patient using their instructions; besides, a patient should read a medicine label.

- Ergot alkaloids (ergotamine derivate) are rarely used due to the risk of serious side effects. These should only be used by specialists (Schytz et al., 2021).
- Nasal spray and subcutaneous injection of triptans act more rapidly than tablets.
- A combination of triptan and NSAIDs may be more effective in some patients than each drug alone (Schytz et al., 2021).
- Patients should also be advised that triptans should not be used within a 24h period of another triptan or with ergot-containing medications (Schytz et al., 2021).
- Women with migraines with aura should be informed that they have a slightly increased risk of an ischemic stroke in the brain, but that the risk is very small if there are no other risk factors and if they refrain from smoking and taking estrogen-containing birth control pills.
- Women with migraine should be counselled about a healthy lifestyle such as regular physical activity, smoking cessation, and modification of other traditional CVD risks (e.g., blood pressure control and weight loss).
- It is advisable to make persistent foramen ovale (PFO) closure for women with migraine for prevention of stroke and improvement in migraine symptoms.

3.2 Strategies of management for preventive treatment of migraine in women 18–40 years old

- At ≥15 headache days per month, medication overuse should be ruled out (Schytz et al., 2021).
- Preventive treatment should be considered if (Schytz et al., 2021):
  - **(1)** the number of migraine days per month is four or higher or two days per month of attacks with severe disability.
  - **(2)** attacks significantly interfere with patients' daily routines despite acute treatment and the patient's quality of life is reduced considerably due to the migraine
  - **(3)** attack medications have a poor effect
  - **(4)** frequent or very long-lasting cases of aura occur
- For better choice and increasing of effectiveness of preventive treatment, it is necessary to diagnose and manage comorbid diseases: hypertension, asthma, severe obesity and depression.
- Females with migraine should be screened for endometriosis criteria and vice versa, allowing for more individualized treatment (Al-Hassany et al., 2020).

- Valproate and topiramate should be prescribed only to women with adequate contraception.
- The prevention treatment should attempt for a minimum of 2–3 months at full dose before it may be finally assessed whether there is an effect (unless it is not tolerated due to side effects) (Schytz et al., 2021).
- In case of effect, the preventive drug should be discontinued every 6–12 months to ensure that there is still a need for and effect of the medication (Schytz et al., 2021).
- Lack of the effect of one type of preventive does not preclude the effect of other types of prevention (Schytz et al., 2021).
- Amitriptyline is particularly suitable if the patient also suffers from frequent tension-type headache or chronic migraine, or insomnia.
- Clinicians should consider the use of a neuromodulatory device in patients who may benefit from preventive treatment but must limit or avoid medications due to comorbid and coexistent illness or concomitant medication (Ailani, Burch, & Robbins, 2021).
- Practitioners should seek alternatives to erenumab in patients with a latex allergy as well as constipation, as its use has been associated with cases of severe constipation. Because erenumab has been shown to precipitate or exacerbate hypertension, its use should be evaluated on an individual basis in patients with preexisting hypertension (Ailani et al., 2021).

4. Part III: Management of menstrual migraine

Hormonal factors play a relevant impact in migraine. Menstruation is a common trigger of migraine attacks among females. Migraine usually starts after menarche and occurs more frequently during menstruation. According to the "estrogen withdrawal" hypothesis, migraine is triggered by the sudden decline in estrogen levels occurring immediately before menses.

The International Classification of Headache Disorders (ICHD-3) (Headache Classification Committee of the International Headache Society (IHS), 2018) includes two types of menstrual migraine: migraine related to menstruation and pure menstrual migraine. Women with a diagnosis of menstrual migraine report that menstrual attacks are more painful, longer-lasting, more disabling and less responsive to treatment.

The aim of this subchapter—is to provide an updated overview of acute and preventive treatment of menstrual migraine and how to manage this type of debilitating migraine (Table 3).

Table 3 Acute and preventive treatment of menstrual migraine

Acute treatment	**Preventive treatment**
First-line medication • *Simple analgesics:* ✓ Ibuprofen 400–600 mg ✓ Naproxen 500 mg ✓ Paracetamol 1.000 mg • *Antiemetic:* ✓ Metoclopramide 10 mg ✓ Domperidone 10 mg	Nonhormonal treatment (1) *Perimenstrual therapy* • *Triptans* ✓ Frovatriptan 5 mg twice daily started two days before predicted menstrual migraine, with 2.5 mg twice daily for a further 5 days ✓ Naratriptan 1 mg or 2.5 mg twice daily for 5 days ✓ Zolmitriptan 2–3 days/week during the menstrual week
Second-line medication • *Triptans:* ✓ Frovatriptan 2.5 mg ✓ Naratriptan 2.5 mg ✓ Sumatriptan 50–100 mg ✓ Zolmitriptan 2.5 mg ✓ Rizatriptan 5–10 mg ✓ Almotriptan 12.5 mg ✓ Eletriptan 40 mg • *Gepants (rimegepant)* • *Ditans (lasmiditan)*	• *NSAIDs* ✓ Naproxen 500 mg twice daily 7 days ✓ Mefenamic acid ✓ Nimesulide 100 mg thrice daily 7 days • Combination of triptans and NSAIDs (sumatriptan plus naproxen or frovatriptan plus dexketoprofen) (2) *Regular therapy* • *Anti-epileptics:* ✓ Topiramate 25–100 (200) mg • *CGRP-antibodies:* ✓ Erenumab 70–140 mg s.c. every 4 weeks
Contraindication ✓ Triptans are contraindicated in stroke, uncontrolled hypertension, ischemic heart disease, previous cerebral infarction, peripheral vascular diseases, severe hepatic impairment, and those taking monoamine oxidase inhibitors. ✓ Before prescription, it is necessary to check contraindications for drugs in each patient using their instructions; besides, a patient should read a medicine label.	Hormonal treatment (for women whose menstrual cycle is irregular) • *Short-term hormonal supplementation* of low-dose estrogen (1.5-mg estradiol gel used during the menstrual week) or patches • *Continuous use of exogenous hormones in fixed doses, without hormone-free intervals,* especially for women needing contraception (contraceptive containing 20 μg ethinyl estradiol on days 1–21, supplemented with 0.9 mg conjugated equine estrogens on days 22–28) • *Progestogen-containing contraceptives* (particularly for women who have migraine with aura: the depot injection (medroxyprogesterone) or desogestrel

Continued

Table 3 Acute and preventive treatment of menstrual migraine—cont'd

Acute treatment	Preventive treatment
	Nonpharmacologic treatment • Phytoestrogens (soy isoflavones) • magnesium tablets 360–400 mg × 1 which is taken daily from day 15 of the cycle (calculated from the first day of menstruation) • vitamin E (400 daily units) • non-invasive Vagus nerve Stimulation *Contraindications* ✓ Before prescription, it is necessary to check contraindications for drugs in each patient using their instructions; besides, a patient should read a medicine label.
	Choice of contraception • *Migraine with aura* ✓ Hormonal contraception containing progestogens exclusively: mini-pills, the subdermal implant (contraceptive stick), depot injection and IUD • *Migraine without aura* ✓ Estrogen-containing contraceptive pills ✓ Mini-pills (containing desogestrel 75 μg/day only)

4.1 Strategies of management for acute treatment of menstrual migraine

- Be aware of the risk of medication overuse headaches. The number of days per month of triptan consumption should not exceed nine.
- The commonly used nonsteroidal anti-inflammatory drugs inhibit prostaglandin synthesis by inhibiting cyclooxygenase enzymes and can therefore be used in the acute treatment of menstrual migraine, especially in women with dysmenorrhea (Ornello et al., 2021).
- Menstrual migraine is associated with an increased occurrence of gastrointestinal symptoms, such as nausea and vomiting, using non-oral formulations can contribute to increasing the effectiveness of acute treatments.

- Triptans and common analgesics act on different pathways; therefore, combining triptans with analgesics is an alternative approach to treating menstrual migraine effectively. The available trials confirmed the efficacy of sumatriptan plus naproxen and of frovatriptan plus dexketoprofen (Ornello et al., 2021).
- Common analgesics can also be used as rescue treatment if triptans fail. It is advisable that patients adopting rescue treatments take a triptan first to maximize its efficacy; common analgesics can be taken soon after the triptan, maintaining gastrointestinal safety.
- Gepants, the oral CGRP receptor antagonists, can act both as acute and as preventative migraine treatments.

4.2 Strategies of management for preventive treatment of menstrual migraine

- The treating physician should consider the woman's comorbidities that might influence the hormonal status. Collaboration between headache physicians and gynecologists is encouraged to manage menstrual irregularities and reproductive conditions that can affect migraine. For example, obesity has a complex relationship with female sex hormones regarding appetite regulation and adipose tissue distribution (Ornello et al., 2021).
- Females with migraine should be screened for endometriosis criteria and vice versa, allowing for more individualized treatment (Al-Hassany et al., 2020).
- The choice of preventive treatment depends on the regularity of menstruation, presence or absence of menstrual disorders, presence or absence of vasomotor symptoms, sleep disturbance or depression, need for contraception, as well as a woman's personal preferences (MacGregor, 2020).
- It might be advisable to prescribe triptans as short-term prevention only to women with very regular menstrual cycles or under combined hormonal treatments in pure menstrual migraine to maximize efficacy while minimizing the risk of medication overuse (Ornello et al., 2021).
- Short-term cyclic prevention can start at the onset of menstrual migraine or menstruation, whichever comes first, continuing for up to 7 days or starting 2 days before the expected menstrual period and continuing for 3–5 days. The highest quality of evidence is available for frovatriptan and naratriptan (MacGregor, 2020).
- Women with frequent menstrual and non-menstrual attacks should consider standard migraine prophylaxis in the first instance.

- Short-term dose escalation of common preventive medication may be attempted.
- Women with migraines with aura should be informed that they have a slightly increased risk of an ischemic stroke in the brain, but that the risk is very small if there are no other risk factors and if they refrain from smoking and taking estrogen-containing birth control pills.

4.3 Strategies for choice of contraception in menstrual migraine

If contraception is needed in women with migraine with aura, birth control pills with the lowest possible estrogen content are preferred, and the patient must be informed of the increased risk of ischemic stroke. They should avoid using contraceptive pills with estrogen content ≥35 μg, estrogen patches and a vaginal ring.

In case of need for contraception in women with migraine without aura, where a worsening of migraine is experienced at the same time, the following can be tried:

(1) Use of estrogen-containing contraceptive pills, where there is no contraceptive pill breakthrough several cycles, e.g., by taking birth control pills continuously for 9 weeks (instead of the usual 3 weeks) followed by a 7-day pill-free period. If breakthrough bleeding occurs earlier, there is a pause at the time of breakthrough bleeding.

(2) Use of mini-pills.

5. Part IV: Management of migraine in pregnancy

Migraine is a common disorder among pregnant women with a prevalence as high as 35% (Hamilton & Robbins, 2019). While migraine usually improves during pregnancy, many women can have worsening course and may even experience first-time migraine attacks (Hamilton & Robbins, 2019). Twenty-six percent of women with migraine during the first trimester of pregnancy report moderate or severe headache-related disability (Hamilton & Robbins, 2019). Despite the high prevalence of migraine, treatment of migraine can be difficult during pregnancy, because of lack of medication trials and fears of teratogenicity. There are currently no guidelines for treating migraine during pregnancy.

This subchapter aims to characterize the treatment of pregnant women with migraine (Table 4).

Table 4 Acute and preventive treatment of migraine in pregnancy.

Acute treatment	**Preventive treatment**
First-line interventions • Optimization of lifestyle factors (improvement of sleep duration and quality, maintaining regular meals and good hydration, and a good schedule of physical activity) • Introduction of nonpharmacologic techniques and therapies (relaxation training and biofeedback, cognitive behavioral therapy, bed rest, ice packs) • Combination of nonpharmacologic techniques/ therapies and pharmacologic treatments First-line medication • Paracetamol • Acetaminophen (except the third trimester) • Diphenhydramine • Peripheral nerve blocks with lidocaine/ropivacaine Second-line medication • Triptans (Sumatriptan) Non-invasive devises Supraorbital nerve stimulation and transcranial magnetic stimulation *Adjunct medications for nausea and/or vomiting* First-line medication • Metoclopramide • IV fluids Second-line medication • Promethazine • Prochlorperazine *Contraindications* ✓ Contraindicated drugs: • Opioids • Ergotamine (contraindicated due to the uterine contracting effect) • Domperidone (not recommended) • NSAIDs (should be avoided in the third trimester) ✓ Before prescription, it is necessary to check contraindications for drugs in each patient using their instructions; besides, a patient should read a medicine label.	First-line treatment • Propranolol Second-line treatment • Cyclobenzaprine • Coenzyme Q10 • Memantine • Amitriptyline (only if there is a significant need for treatment that outweighs the potential risks) • Venlafaxine (only if there is a significant need for treatment that outweighs the potential risks) *Nonpharmacologic treatment* • Transcranial magnetic stimulation • Vagus nerve stimulation • Transcutaneous supraorbital nerve stimulation • Acupuncture *Contraindications* ✓ Contraindicated drugs • Valproate • Topiramate • Candesartan • Lisinopril ✓ Before prescription, it is necessary to check contraindications for drugs in each patient using their instructions; besides, a patient should read a medicine label.

5.1 The strategy of management for acute treatment of migraine in pregnancy

- Treatment of migraine should be considered a necessary part of prenatal care.
- Nonpharmacologic treatments, including relaxation training and biofeedback, have shown efficacy for the prevention of migraine and are most effective when practiced regularly. Starting these techniques prior to conception ensures that the patient is able to use them early during pregnancy (Burch, 2020).
- These behavioral techniques can be combined with pharmacologic treatments, and these approaches are more effective in combination (Burch, 2020).
- Many NSAIDs, including ibuprofen and naproxen, are considered safe for use during the second trimester only. Use during the first trimester may increase the risk of miscarriage. Use in the third trimester can lead to premature closure of the fetal ductus arteriosus (Burch, 2020).
- Several population-based and registry studies support the safety of triptans during pregnancy (Burch, 2020).
- While the butalbital/acetaminophen/caffeine combination medication has historically been favored as the second-line treatment for women during pregnancy, safety has not been rigorously studied. Acetaminophen has been associated with an increased risk of attention deficit hyperactivity disorder and wheezing, butalbital—with congenital heart defects.
- Opiates are generally avoided due to the risk of overuse and neonatal abstinence syndrome, but rare use in the second and early third trimester is unlikely to cause significant problems (Burch, 2020).
- Dihydroergotamine and other ergot derivatives are contraindicated during pregnancy as they cause decreased blood flow to the uterus and thus increase the risk of spontaneous abortion (Burch, 2020).
- The antiemetics promethazine and prochlorperazine have not been associated with fetal malformations in good quality studies.
- Pregnancy is a risk factor for many types of secondary headaches.
- Migraine is a risk factor for the development of cerebral venous thrombosis (CVT), pregnancy-associated stroke, and preeclampsia (Burch, 2020).
- Any new headache, change in headache type, or headache associated with hypertension, fever, or neurologic signs or symptoms should raise suspicion for secondary headache.
- Noncontrast magnetic resonance imaging (MRI) can be performed with little risk during pregnancy, while computed tomography (CT) is generally avoided when possible. Gadolinium should be avoided during pregnancy due to embryocidal effects (Burch, 2020).

- Over one-half of women with migraine will experience recurrence of headaches within the first month after delivery. Most of these headaches are due to migraine, but increased risk for reversible cerebral vasoconstriction syndrome and preeclampsia/eclampsia persists for around 6 weeks into the postpartum period. Postdural puncture headache is the most common cause of secondary headache in the puerperium (Burch, 2020).
- Women with migraine should be monitored for the development of complications during pregnancy, particularly vascular conditions.

5.2 The strategy of management for preventive treatment of migraine in pregnancy

- Preventive treatment should be avoided if possible.
- The medication with the best evidence for safety during pregnancy is propranolol.
- Beta-blockers can be used in the lowest possible dosage, but there is a risk of side effects in the newborn such as bradycardia, hypotension and hypoglycemia.
- Topiramate, lisinopril, and candesartan have all been conclusively linked to fetal malformations. Given the number of safer options, these medications are almost never continued through pregnancy.
- Sodium valproate was listed as an FDA pregnancy category X, meaning it should never be used for migraine prevention during pregnancy due to its teratogenic effects (Burch, 2020).
- Due to increasing evidence that intravenous magnesium sulfate for longer than five days increased the risk of fetal bone malformations, the FDA reclassified IV magnesium sulfate from a category A to a category D in 2013 (Burch, 2020). No statement was made about the safety of oral magnesium.
- There is no evidence to suggest that the teratogenic risk of any oral preventive medication extends beyond its typical half-life. There is, therefore, no need for a prolonged oral preventive washout period before attempting pregnancy.

6. Part V: Management of migraine during lactation

After childbirth, the recurrence rate of migraine attacks increases, especially during the first month. Breastfeeding usually exerts a protective effect against the reappearance of attacks. While the safety of migraine medications during breastfeeding is poorly studied, the available evidence suggests that there are more options for treatment during lactation than during pregnancy.

Since there is no evidence to suggest that breastfeeding worsens migraine, and because there are many safe treatment options available, migraine is almost never a reason to withhold breastfeeding (Burch, 2020).

This subchapter aims to characterize the treatment and management of migraine during lactation (Table 5).

6.1 The strategy of management for acute treatment of migraine during lactation

- Many frequently used acute medications for migraine are considered safe during lactation. Acetaminophen, ibuprofen, and diclofenac are considered safe; there is less evidence for naproxen and indomethacin.
- Aspirin should be avoided as a first-line treatment due to the risk of hemolysis and bleeding conditions in the newborn. However, a recent study found infant exposure to be very minimal with low-dose aspirin (Burch, 2020).
- Triptans have not been well studied during lactation but are categorized as "usually compatible" with breastfeeding by the American Academy of Pediatrics "Transfer of drugs and other chemicals into human milk" evaluation (Burch, 2020).
- Breast-feeding is not recommended for 12h after ingestion of triptans except eletriptan. Eletriptan is more highly protein-bound than other triptans and is theoretically less likely to be transferred into breastmilk (Burch, 2020).
- Metoclopramide may increase milk supply and is sometimes used as a galactagogue to stimulate milk production. It should be avoided in women who do not wish to breastfeed for the first eight weeks after delivery (Burch, 2020).
- Prochlorperazine may also support milk production by increasing prolactin levels (Burch, 2020).
- Dihydroergotamine and other ergots should never be used during breastfeeding, both due to effects on milk supply and due to reports of weakness in the infant (Burch, 2020).
- Opiates may cause sedation in infants and are therefore not the first choice for migraine treatment.
- Codeine and hydrocodone may cause a rare but severe opiate reaction in infants due to hypermetabolism in mothers and should therefore be avoided (Burch, 2020).

Table 5 Acute and preventive treatment of migraine during lactation.

Acute treatment	Preventive treatment
First-line medication • Paracetamol • Ibuprofen • Acetaminophen • Peripheral nerve blocks with lidocaine/ropivacaine Second-line medication • Triptans (eletriptan is preferable or sumatriptan) *Non-invasive devises* Supraorbital nerve stimulation and transcranial magnetic stimulation *Adjunct medications for nausea and/or vomiting* First-line medication • Metoclopramide (may increase plasma prolactin and lactation) • IV fluids • Second-line medication • Prochlorperazine *Contraindications* ✓ Contraindicated drugs: 1. Acetylsalicylic acid 2. Benzodiazepines 3. Ergotamine 4. Metoclopramide (should be avoided: it is absorbed into breast milk) 5. Codeine ✓ Before prescription, it is necessary to check contraindications for drugs in each patient using their instructions; besides, a patient should read a medicine label.	First-line treatment • Propranolol • Magnesium Second-line treatment • Amitriptyline • Topiramat (the infant needs to be observed for signs of sedation) • Candesartan (should be avoided in mothers of premature infants) *Nonpharmacologic treatment* • Riboflavin • Magnesium • Vitamin D supplementation • Acupuncture • Physical exercise • Cognitive behavioral therapy *Contraindications* ✓ Before prescription, it is necessary to check contraindications for drugs in each patient using their instructions; besides, a patient should read a medicine label.

- Infant exposure to acute migraine medications can be minimized by either pumping and discarding the milk once after medication use or taking the medication just before an anticipated longer stretch between feedings.
- If there are lingering concerns, the infant's pediatrician may also be a helpful resource.

6.2 Strategies of management for preventive treatment of migraine during lactation

- Avoid drug prevention as much as possible.
- While the safety of migraine medications during breastfeeding is poorly studied, the available evidence suggests that, on balance, there are more options for treatment during lactation than during pregnancy (Burch, 2020).
- Since there is no evidence to suggest that breastfeeding worsens migraine, and because there are many safe treatment options available, migraine is rarely a reason to withhold breastfeeding (Burch, 2020).
- Since the safety of on botulinum toxin or a CGRP monoclonal antibody has not been established during lactation, the need to use such treatment after pregnancy is one of the few clinical scenarios in which breastfeeding may need to be stopped due to migraines (Burch, 2020).
- Amitriptyline is considered a second-line treatment due to concerns about sedation in infants. This concern generally decreases as the infant grows older.

7. Part VI: Management of migraine in perimenopause

Migraine is triggered by the sudden decline in estrogen levels occurring during the transition toward menopause or in the early postmenopausal period (Ripa et al., 2015). The menopausal transition, known as perimenopause, is associated with fluctuating ovarian function. According to the theory of "estrogen withdrawal headache," migraine may be caused by the cessation of a course of exogenous estrogens. A link between hormonal imbalance and migraine may stem from the opposing roles of estrogen and progesterone within the brain. Indeed, while estrogen stimulates neural excitability, progesterone exhibits inhibitory actions in central neurons. Therefore, the imbalance between these neuromodulatory hormones may give rise to physiological conditions that alter susceptibility to migraine (Ripa et al., 2015).

This subchapter reviews existent possibilities for the treatment of migraine and its management in perimenopause (Table 6).

Table 6 Acute and preventive treatment of migraine in perimenopause.

Acute treatment	**Preventive treatment**
First-line medication • *Simple analgesics:* ✓ Ibuprofen 400–600 mg ✓ Naproxen 500 mg ✓ Paracetamol 1.000 mg ✓ Aspirin/caffeine 500/50 mg • *Antiemetic:* ✓ Metoclopramide 10 mg Second-line medication • *Triptans:* ✓ Sumatriptan 50–100 mg ✓ Zolmitriptan 2.5 mg ✓ Naratriptan 2.5 mg ✓ Rizatriptan 5–10 mg ✓ Almotriptan 12.5 mg ✓ Eletriptan 40 mg ✓ Frovatriptan 2.5 mg *Contraindications* ✓ Contraindicated drugs: (1) Triptans are contraindicated in uncontrolled hypertension, ischemic heart disease, previous cerebral infarction and peripheral vascular diseases. (2) Domperidone 10 mg should be avoided due to the risk of extrapyramidal side effects ✓ Before prescription, it is necessary to check contraindications for drugs in each patient using their instructions; besides, a patient should read a medicine label.	Non-hormonal treatment • *Beta-blockers:* ✓ Metoprolol is 50 mg × 1 for 1 week and then 100 mg × 1, increased to 150–200 mg per day in case of lack of effect ✓ Propranolol is 40 mg × 2 increased at a weekly interval to a maximum of 120 mg × 2 • *Angiotensin II receptor antagonist:* ✓ Candesartan 8 mg × 1 for 1 week and then 16 mg × 1, possibly increasing to 24–32 mg × 1 • *Anti-epileptics:* ✓ Topiramate 25–100 (200) mg ✓ Valproate 500–1800 mg • *Antidepressants:* ✓ Amitriptyline 10 mg × 1 increasing by 10 mg at 1-week intervals to 10–100 mg daily (1–2 h before bedtime) • *Botulinum type A toxin* 155–195 units i.m. (every 12 weeks) • *CGRP-antibodies:* ✓ Erenumab 70–140 mg s.c. every 4 weeks ✓ Fremanezumab 225 mg s.c. every month or 675 mg s.c. every 3 months ✓ Galcanezumab Start dose 240 mg s.c. followed by 120 mg s.c. every month ✓ Eptinezumab 100–300 mg i.v. every 3 months Hormonal treatment • Administration of estrogen (intrauterine or transdermal rather than oral: 52 mg levonorgestrel intrauterine system) • Transdermal estrogen together with progesterone • The administration of non-estrogen compounds (tibolone) • Preferably to avoid hormonal replacement therapy

Continued

Table 6 Acute and preventive treatment of migraine in perimenopause.—cont'd

Acute treatment	Preventive treatment
	Treatment of perimenopausal vasomotor symptoms • Escitalopram • Venlafaxine • Paroxetine *Nonpharmacologic treatment* • Phytoestrogens (soy isoflavones) • Vitamin E • Vitamin D supplementation • Acupuncture • Physical exercise • Cognitive behavioral therapy *Contraindications* ✓ Before prescription, it is necessary to check contraindications for drugs in each patient using their instructions; besides, a patient should read a medicine label.

7.1 The strategy of management for acute treatment of migraine in perimenopause

- New onset migraine in women over 40 years of age requires to exclude secondary headache (Filatova et al., 2020).
- To rule out medication overuse headache (Schytz et al., 2021).
- Treatment should be initiated as early as possible.
- Triptans should be taken at the onset of head pain for maximum efficacy.
- In migraine with aura, however, triptans should be taken only when the headache starts.
- Analgesics often has a better effect when combined with rest and, possibly, sleep. If the patient has difficulty calming down, benzodiazepine may be given, e.g., 5 mg diazepam (Schytz et al., 2021).
- Simple analgesics should be used for a maximum of 14 days per month to avoid medication overuse headache (Eigenbrodt et al., 2021; Schytz et al., 2021).
- Triptans or combined analgesics should be used for a maximum of 9 days per month to avoid medication overuse headache (Eigenbrodt et al., 2021; Schytz et al., 2021).

- Patients who have no effect of one particular triptan may benefit from another triptan. Before the effect of any triptan is ruled out, the patient should, as a rule of thumb, have tried three different triptans, each during three attacks.
- Nasal spray and subcutaneous injection of triptans act more rapidly than tablets.
- A combination of triptan and NSAIDs may be more effective in some patients than each drug alone.
- Patients should also be advised that triptans should not be used within a 24 h period of another triptan or with ergot-containing medications.
- Women with migraine should be counselled about a healthy lifestyle such as regular physical activity, smoking cessation, and modification of other traditional CVD risks (e.g., blood pressure control and weight loss).

7.2 The strategy of management for preventive treatment of migraine in perimenopause

- At $\geq$15 headache days per month, medication overuse should be ruled out (Schytz et al., 2021).
- Preventive treatment should be considered if:
  (1) the number of migraine days per month is four or higher or two days per month of attacks with severe disability (Ailani et al., 2021).
  (2) attack medications have a poor effect (Schytz et al., 2021)
  (3) attacks significantly interfere with patients' daily routines despite acute treatment and the patient's quality of life is reduced considerably due to the migraine (Schytz et al., 2021)
  (4) frequent or very long-lasting cases of aura occur (Schytz et al., 2021)
- There are no specific guidelines for the preventive treatment of migraine during the menopausal transition; nevertheless, some special considerations may apply to that population of women regarding both hormonal and non-hormonal therapies.
- As high-estrogen states can increase the susceptibility to migraine with aura using the lowest possible doses of estrogens can contribute to avoiding the worsening of migraine during the menopausal transition (Ornello et al., 2021).
- The use of products without androgenic properties or natural progesterone (109) or non-oral routes of administration might be beneficial to migraine (Ornello et al., 2021).

- When considering hormonal treatments for postmenopausal women, the possible benefits should be compared with the risk of cerebrovascular events, which is increased both by migraine itself and by estrogens even at the lowest doses (Ornello et al., 2021).
- Women with migraine, mainly those with migraine with aura, who enter the perimenopausal period should be advised to avoid HRT if possible due to the potential harm to their health.
- Women with migraines with aura should be informed that they have a slightly increased risk of an ischemic stroke but that the risk is minimal if there are no other risk factors and if they refrain from smoking
- In case of migraine with aura to make the correct differential diagnosis and exclude TIA.
- Women with migraine should be counselled about a healthy lifestyle such as regular physical activity, smoking cessation, and modification of other traditional CVD risks (e.g., blood pressure control and weight loss).
- It is advisable to make persistent foramen ovale (PFO) closure for women with migraine for prevention of stroke and improvement in migraine symptoms.

8. Part VII: Management of migraine in menopause and post-menopause

Menopause is a normal physiological event, defined as the final menstrual period and reflects a loss of ovarian follicular function and decline of estrogen levels. This may contribute to migraine course and menopause. A low level of estrogen is associated with decreased production of serotonin, increased serotonin reuptake, and raised serotonin degradation (Ripa et al., 2015). All these factors can influent the frequency of migraine and its comorbidity including depression and anxiety. Migraine shows variability in the clinical presentations in the menopausal period, as migraine attacks might improve, worsen or remain unchanged during this transition (Al-Hassany et al., 2020).

Menopause leads to a loss of vasoprotective functions of natural estrogens and increased thrombogenesis (Rathfoot, Edrissi, Sanders, et al., 2021). A recent study showed that females with migraine, atrial fibrillation (AF), and high levels of LDL -cholesterol had an increased risk of ischemic stroke (Rathfoot et al., 2021). Besides, if older women continue to have migraines and smoke at the same time, it seems to increase the risk of stroke.

This subchapter reviews existent possibilities for the treatment of migraine and its management in menopause and post-menopause (Table 7).

Table 7 Acute and preventive treatment of migraine in menopause and post-menopause.

Acute treatment	**Preventive treatment**
First-line medication • *Simple analgesics:* ✓ Ibuprofen 400–600 mg ✓ Naproxen 500 mg ✓ Paracetamol 1.000 mg ✓ Aspirin/caffeine 500/50 mg • *Antiemetic:* ✓ Metoclopramide 10 mg Second-line medication • *Triptans:* ✓ Sumatriptan 50–100 mg ✓ Zolmitriptan 2.5 mg ✓ Naratriptan 2.5 mg ✓ Rizatriptan 5–10 mg ✓ Almotriptan 12.5 mg ✓ Eletriptan 40 mg ✓ Frovatriptan 2.5 mg *Contraindication* *Contraindicated drugs* (1) Triptans are contraindicated in uncontrolled hypertension, ischemic heart disease, previous cerebral infarction and peripheral vascular diseases. (2) Domperidone 10 mg should be avoided due to the risk of extrapyramidal side effects ✓ Before prescription, it is necessary to check contraindications for drugs in each patient using their instructions; besides, a patient should read a medicine label.	Non-hormonal treatment • *Beta-blockers:* ✓ Metoprolol is 50 mg × 1 for 1 week and then 100 mg × 1, increased to 150–200 mg per day in case of lack of effect ✓ Propranolol is 40 mg × 2 increased at a weekly interval to a maximum of 120 mg × 2 • *Angiotensin II receptor antagonist:* ✓ Candesartan 8 mg × 1 for 1 week and then 16 mg × 1, possibly increasing to 24–32 mg × 1 • *Anti-epileptics:* ✓ Topiramate 25–100 (200) mg ✓ Valproate 500–1800 mg • *Antidepressants:* ✓ Amitriptyline 10 mg × 1 increasing by 10 mg at one-week intervals to 10–100 mg daily (1–2 h before bedtime) • *Botulinum type A toxin* 155–195 units i.m. (every 12 weeks) • *CGRP-antibodies:* ✓ Erenumab 70–140 mg s.c. every 4 weeks ✓ Fremanezumab 225 mg s.c. every month or 675 mg s.c. every 3 months ✓ Galcanezumab Start dose 240 mg s.c. followed by 120 mg s.c. every month ✓ Eptinezumab 100–300 mg i.v. every 3 months • Hormonal treatment *Non-hysterectomized women:* ✓ Continuous transdermal estrogen plus 52 mg levonorgestrel intrauterine system ✓ Tibolone 2.5 mg *Hysterectomized women:* ✓ Continuous transdermal estrogen

Continued

Table 7 Acute and preventive treatment of migraine in menopause and post-menopause.—cont'd

Acute treatment	Preventive treatment
	Treatment of menopausal vasomotor symptoms • Escitalopram • Venlafaxine • Paroxetine *Nonpharmacologic treatment* • Phytoestrogens (soy isoflavones) • Vitamin E • Vitamin D supplementation • Acupuncture • Physical exercise • Cognitive behavioral therapy *Contraindications* ✓ Combined hormonal contraception is contraindicated for contraceptive use in women with migraine with aura. ✓ Before prescription, it is necessary to check contraindications for drugs in each patient using their instructions; besides, a patient should read a medicine label.

8.1 The strategy of management for acute treatment of migraine in menopause and post-menopause

- New onset of migraine at this period requires excluding secondary headache and to do neuroimaging and other tests (Filatova et al., 2020).
- In case of migraine with aura it is necessary to make the correct differential diagnosis and exclude transient ischemic attack.
- To rule out medication overuse headache.
- Treatment should be initiated as early as possible.
- Triptans should be taken at the onset of head pain for maximum efficacy (Schytz et al., 2021).
- In migraine with aura, however, triptans should be taken only when the headache starts (Schytz et al., 2021).
- Analgesics often has a better effect when combined with rest and, possibly, sleep. If the patient has difficulty calming down, benzodiazepine may be given, e.g., 5 mg diazepam (Schytz et al., 2021).

- Simple analgesics should be used for a maximum of 14 days per month to avoid medication overuse headache (Eigenbrodt et al., 2021; Schytz et al., 2021).
- Triptans or combined analgesics should be used for a maximum of 9 days per month to avoid medication overuse headache (Eigenbrodt et al., 2021; Schytz et al., 2021).
- Patients who have no effect of one particular triptan may benefit from another triptan. Before the effect of any triptan is ruled out, the patient should, as a rule of thumb, have tried three different triptans, each during three attacks.
- Nasal spray and subcutaneous injection of triptans act more rapidly than tablets.
- A combination of triptan and NSAIDs may be more effective in some patients than each drug alone.
- Patients should also be advised that triptans should not be used within a 24h period of another triptan or with ergot-containing medications.
- Women with migraine should be counselled about a healthy lifestyle such as regular physical activity, smoking cessation, and modification of other traditional CVD risks (e.g., blood pressure control and weight loss).
- Using of oral anticoagulants should be considered for stroke prevention in women with migraine and atrial fibrillation.

8.2 The strategy of management for preventive treatment of migraine in menopause and post-menopause

- At ≥15 headache days per month, medication overuse should be ruled out.
- Preventive treatment should be considered if:
  - **(1)** the number of migraine days per month is four or higher or two days per month of attacks with severe disability (Ailani et al., 2021).
  - **(2)** attack medications have a poor effect (Schytz et al., 2021).
  - **(3)** attacks significantly interfere with patients' daily routines despite acute treatment, and the patient's quality of life is reduced considerably due to the migraine (Schytz et al., 2021).
  - **(4)** frequent or very long-lasting cases of aura occur (Schytz et al., 2021).
- For better choice and increasing of effectiveness of preventive treatment, it is necessary to diagnose and manage comorbid diseases: hypertension, sleep disorders, severe obesity, depression and anxiety (Schytz et al., 2021).

- Options for managing menopause symptoms are the same as for women without migraine but considering how the strategy might benefit or adversely affect migraine (Ornello et al., 2021).
- The most effective treatment for menopause symptoms is menopausal hormone therapy (MHT). However, MHT use adversely affects migraine compared to never use. Combined hormonal contraception (CHC) may be the better option. CHC is also an effective treatment for menopause symptoms (Ornello et al., 2021).
- CHC can be used by women with migraine without aura but is contraindicated for contraceptive use in women with migraine with aura since both CHC and aura are independent risk factors for ischemic stroke (Ornello et al., 2021).
- CHC should be prescribed in accordance with national guidelines. In particular, women should be screened for vascular risk factors such as hypertension, atherosclerosis, obesity, smoking, etc.
- While progestogen-only methods have the advantage that use is not associated with increased risk of ischemic stroke, data regarding the effect on migraine are limited to small observational studies with desogestrel 75 mcg daily. EHF and ESCRH recommend the use of desogestrel for CHC but with the additional use in women who develop new-onset or worsening migraine with CHCs (Ornello et al., 2021).
- Transdermal estrogen is preferred to oral estrogen and continuous combined regimens are better tolerated than cyclical regimens. In perimenopausal women, the 52 mg levonorgestrel intrauterine system is an appropriate option since it provides contraception, endometrial protection, and in women who become amenorrhoeic can also benefit migraine (MacGregor, 2020).
- It is necessary to titrate estrogen dose to control vasomotor symptoms.
- In postmenopausal women who favor oral MHT, tibolone may be a better option than other oral preparations, although data regarding migraine are limited to comparison with estradiol plus norethisterone and it is less effective for menopause symptoms than estrogen/progestogen MHT (MacGregor, 2020).
- For medically induced menopause, the addition of estradiol significantly benefits migraine. No studies have assessed the effects of progestogens on migraine, but the potential antinociceptive effects of allopregnanolone, a metabolite of progesterone, could benefit migraine (MacGregor, 2020).
- Women developing aura for the first time after starting MHT should be assessed for stroke risk factors. Once transient ischemic attacks are

excluded, the dose of transdermal estrogen is reduced to the lowest dose necessary to control menopause symptoms. If aura does not resolve or increases in frequency, consider stopping MHT in favor of non-hormonal management (MacGregor, 2020).

- Adverse events are all more likely in older women.
- Poor evidence base for all drugs in older women.

9. Part VIII: Management of migraine in men 18–50 years old

While migraine affects more women, migraines in men do happen. Gender differences in headache treatment and management are existing. As distinct from women, men don't have many specific forms of migraine during this period of life. Their treatment is simpler and hormone therapies don't prescribe at all. Men with migraine tend to utilize health resources less frequently than women. There needs to be greater awareness among men of the fact that migraine is a brain disorder that can cause severe disability.

There are currently no guidelines for treating migraine in men. This subchapter summarizes existing knowledge on the treatment and management of migraine in men 18–50 years old (Table 8).

9.1 Strategies of management for acute treatment of migraine in men 18–50 years old

- To rule out medication overuse headache.
- Treatment should be initiated as early as possible.
- Triptans should be taken at the onset of head pain for maximum efficacy (Schytz et al., 2021).
- In migraine with aura, however, triptans should be taken only when the headache starts (Schytz et al., 2021).
- Analgesics often has a better effect when combined with rest and, possibly, sleep. If the patient has difficulty calming down, benzodiazepine may be given, e.g., 5 mg diazepam (Schytz et al., 2021).
- Simple analgesics should be used for a maximum of 14 days per month to avoid medication overuse headache (Schytz et al., 2021).
- Triptans or combined analgesics should be used for a maximum of 9 days per month to avoid medication overuse headache (Schytz et al., 2021).
- Patients who have no effect of one particular triptan may benefit from another triptan. Before the effect of any triptan is ruled out, the patient should, as a rule of thumb, have tried three different triptans, each during three attacks.

Table 8 Acute and preventive treatment of migraine in men 18–50 years old.

Acute treatment	Preventive treatment
First-line medication • *Simple analgesics:* ✓ Ibuprofen 400–600 mg ✓ Naproxen 500 mg ✓ Paracetamol 1.000 mg ✓ Aspirin/caffeine 500/50 mg • *Antiemetic:* ✓ Metoclopramide 10 mg ✓ Domperidone 10 mg Second-line medication • *Triptans:* ✓ Sumatriptan 50–100 mg ✓ Zolmitriptan 2.5 mg ✓ Naratriptan 2.5 mg ✓ Rizatriptan 5–10 mg ✓ Almotriptan 12.5 mg ✓ Eletriptan 40 mg ✓ Frovatriptan 2.5 mg *Contraindications* ✓ Triptans are contraindicated in stroke, uncontrolled hypertension, ischemic heart disease, previous cerebral infarction, peripheral vascular diseases, severe hepatic impairment, and those taking monoamine-oxidase inhibitors. ✓ Before prescription, it is necessary to 0check contraindications for drugs in each patient using their instructions; besides, a patient should read a medicine label.	• Beta-blockers: ✓ Metoprolol is 50 mg × 1 for 1 week and then 100 mg × 1, increased to 150–200 mg per day in case of lack of effect ✓ Propranolol is 40 mg × 2 increased at a weekly interval to a maximum of 120 mg × 2 • Angiotensin II receptor antagonist: ✓ Candesartan 8 mg × 1 for 1 week and then 16 mg × 1, possibly increasing to 24–32 mg × 1 • Anti-epileptics: ✓ Topiramate 25–100 (200) mg ✓ Valproate 500–1800 mg • Antidepressants: ✓ Amitriptyline 10 mg × 1 increasing by 10 mg at one-week intervals to 10–100 mg daily (1–2 h before bedtime) • Botulinum type a toxin Botulinum type A toxin 155–195 units i.m. (every 12 weeks) • CGRP-antibodies: ✓ Erenumab 70–140 mg s.c. every 4 weeks ✓ Fremanezumab 225 mg s.c. every month or 675 mg s.c. every 3 months ✓ Galcanezumab Start dose 240 mg s.c. followed by 120 mg s.c. every month ✓ Eptinezumab 100–300 mg i.v. every 3 months *Nonpharmacologic treatment* • Riboflavin • Vitamin D supplementation • Acupuncture • Physical exercise • Cognitive behavioral therapy *Contraindications* ✓ Before prescription, it is necessary to check contraindications for drugs in each patient using their instructions; besides, a patient should read a medicine label.

- Nasal spray and subcutaneous injection of triptans act more rapidly than tablets.
- A combination of triptan and NSAIDs may be more effective in some patients than each drug alone.
- Patients should also be advised that triptans should not be used within a 24h period of another triptan or with ergot-containing medications.
- Men with migraine should be counselled about a healthy lifestyle such as regular physical activity, smoking cessation, and modification of other traditional CVD risks (e.g. blood pressure control and weight loss).
- It is advisable to make persistent foramen ovale (PFO) closure for men with migraine for prevention of stroke and improvement in migraine symptoms.

9.2 Strategies of management for preventive treatment of migraine in men 18–50 years old

- At ≥15 headache days per month, medication overuse should be ruled out.
- Preventive treatment should be considered if:

(1) the number of migraine days per month is four or higher or two days per month of attacks with severe disability (Ailani et al., 2021).
(2) attack medications have a poor effect (Schytz et al., 2021)
(3) attacks significantly interfere with patients' daily routines despite acute treatment and the patient's quality of life is reduced considerably due to the migraine (Schytz et al., 2021)
(4) frequent or very long-lasting cases of aura occur (Schytz et al., 2021)

- For better choice and increasing of effectiveness of preventive treatment it is necessary to diagnose and manage comorbid diseases: hypertension, sleep disorders, severe obesity, depression and anxiety.
- The prevention treatment should attempt for a minimum of 2–3 months at full dose before it may be finally assessed whether there is an effect (unless it is not tolerated due to side effects) (Schytz et al., 2021).
- In case of effect, the preventive drug should be discontinued every 6–12 months to ensure that there is still a need for and effect of the medication (Schytz et al., 2021).
- Lack of effect of one type of preventive treatment does not preclude the effect of other types of prevention.
- Amitriptyline is particularly suitable if the patient also suffers from frequent tension-type headache or chronic migraine or insomnia, but it can sometimes provoke erectile dysfunction, especially in high doses.

- When outcomes of preventive treatment are suboptimal, review the diagnosis, treatment strategy, dosing and adherence.
- When preventive treatment fails, re-evaluate before changing.

10. Part IX: Management of migraine in men >50 years old

Ongoing research has suggested that migraine is associated with an increased heart attack risk and stroke risk among active smokers, especially in older men with migraine with aura (Monteith, Gardener, Rundek, Elkind, & Sacco, 2015). Therefore, treatment and management of migraine in men >50 years old should include not only effective personalized treatment plan of migraine but also prevention of stroke.

This subchapter describes existing knowledge on the treatment and management of migraine in men >50 years old (Table 9).

10.1 Strategies of management for acute treatment of migraine in men >50 years old

- New onset of migraine at this period requires excluding secondary headache and to do neuroimaging and other tests (Filatova et al., 2020).
- In case of migraine with aura it is necessary to make correct differential diagnosis and exclude TIA.
- To rule out medication overuse headache.
- Treatment should be initiated as early as possible.
- Triptans should be taken at the onset of head pain for maximum efficacy (Schytz et al., 2021).
- In migraine with aura, however, triptans should be taken only when the headache starts (Schytz et al., 2021).
- Analgesics often has a better effect when combined with rest and, possibly, sleep. If the patient has difficulty calming down, benzodiazepine may be given, e.g., 5 mg diazepam (Schytz et al., 2021).
- Simple analgesics should be used for a maximum of 14 days per month to avoid medication overuse headache (Eigenbrodt et al., 2021; Schytz et al., 2021).
- Triptans or combined analgesics should be used for a maximum of 9 days per month to avoid medication overuse headache (Eigenbrodt et al., 2021; Schytz et al., 2021).

Table 9 Acute and preventive treatment of migraine in men >50 years old.

Acute treatment	Preventive treatment
First-line medication • *Simple analgesics:* ✓ Ibuprofen 400–600 mg ✓ Naproxen 500 mg ✓ Paracetamol 1.000 mg ✓ Aspirin/caffeine 500/50 mg • Antiemetic: ✓ Metoclopramide 10 mg Second-line medication • *Triptans:* ✓ Sumatriptan 50–100 mg ✓ Zolmitriptan 2.5 mg ✓ Naratriptan 2.5 mg ✓ Rizatriptan 5–10 mg ✓ Almotriptan 12.5 mg ✓ Eletriptan 40 mg ✓ Frovatriptan 2.5 mg *Contraindications* *Contraindicated drugs:* (1) Triptans are contraindicated in uncontrolled hypertension, ischemic heart disease, previous cerebral infarction and peripheral vascular diseases. (2) Domperidone 10 mg should be avoided due to the risk of extrapyramidal side effects ✓ Before prescription, it is necessary to check contraindications for drugs in each patient using their instructions; besides, a patient should read a medicine label.	Preventive treatment • Beta-blockers: ✓ Metoprolol is 50 mg × 1 for 1 week and then 100 mg × 1, increased to 150–200 mg per day in case of lack of effect ✓ Propranolol is 40 mg × 2 increased at a weekly interval to a maximum of 120 mg × 2 • Angiotensin II receptor antagonist: ✓ Candesartan 8 mg × 1 for 1 week and then 16 mg × 1, possibly increasing to 24–32 mg × 1 • Anti-epileptics: ✓ Topiramate 25–100 (200) mg ✓ Valproate 500–1800 mg • Antidepressants: ✓ Amitriptyline 10 mg × 1 increasing by 10 mg at one-week intervals to 10–100 mg daily (1–2 h before bedtime) • Calcium channel blockers: ✓ Flunarizin 5–10 mg • Other drugs: ✓ Lisinopril 20 mg ✓ Pitzotifen 1.5–3 mg • Botulinum type A toxin 155–195 units i.m. (every 12 weeks) • CGRP-antibodies: ✓ Erenumab 70–140 mg s.c. every 4 weeks ✓ Fremanezumab 225 mg s.c. every month or 675 mg s.c. every 3 months ✓ Galcanezumab Start dose 240 mg s.c. followed by 120 mg s.c. every month ✓ Eptinezumab 100–300 mg i.v. every 3 months *Nonpharmacologic treatment* • Riboflavin • Vitamin D supplementation • Acupuncture • Physical exercise • Cognitive behavioral therapy *Contraindications* ✓ Before prescription, it is necessary to check contraindications for drugs in each patient using their instructions; besides, a patient should read a medicine label.

- Patients who have no effect of one particular triptan may benefit from another triptan. Before the effect of any triptan is ruled out, the patient should, as a rule of thumb, have tried three different triptans, each during three attacks.
- Nasal spray and subcutaneous injection of triptans act more rapidly than tablets.
- A combination of triptan and NSAIDs may be more effective in some patients than each drug alone.
- Patients should also be advised that triptans should not be used within a 24h period of another triptan or with ergot-containing medications.
- Men with migraine should be counselled about screening for risk factors for stroke and about a healthy lifestyle such as regular physical activity, smoking cessation, and modification of other traditional CVD risks (e.g., blood pressure control and weight loss).

10.2 Strategies of management for preventive treatment of migraine in men >50 years old

- At $\geq$15 headache days per month, medication overuse should be ruled out.
- Preventive treatment should be considered if:
  1. the number of migraine days per month is four or higher or two days per month of attacks with severe disability (Ailani et al., 2021)
  2. attack medications have a poor effect (Schytz et al., 2021)
  3. attacks significantly interfere with patients' daily routines despite acute treatment, and the patient's quality of life is reduced considerably due to the migraine (Schytz et al., 2021)
  4. frequent or very long-lasting cases of aura occur (Schytz et al., 2021)
- For better choice and increasing of effectiveness of preventive treatment it is necessary to diagnose and manage comorbid diseases: hypertension, sleep disorders, severe obesity, depression and anxiety.
- Men developing aura for the first time should be assessed for stroke risk factors.
- The prevention treatment should attempt for a minimum of 2–3 months at full dose before it may be finally assessed whether there is an effect (unless it is not tolerated due to side effects) (Schytz et al., 2021).
- In case of effect, the preventive drug should be discontinued every 6–12 months to ensure there is still a need for and effect of the medication (Schytz et al., 2021).

- Lack of effect of one type of preventive does not preclude the effect of other kinds of prevention.
- Amitriptyline is particularly suitable if the patient also suffers from frequent tension-type headache or chronic migraine, or insomnia.
- When outcomes of preventive treatment are suboptimal, review the diagnosis, treatment strategy, dosing and adherence.
- When preventive treatment fails, re-evaluate before changing.
- Adverse events are all more likely in older men.
- Poor evidences for all drugs in older men.

11. Conclusions

✓ Females and males should have a different approach to the management of migraine depending on pathophysiology, clinical characteristics, risk factors and comorbidity.

✓ There is a specific treatment of migraine in women in different periods of life, including pregnancy, lactation, menstrual migraine, perimenopause, menopause and post-menopause. The options for treatment should study more in future randomized control trials. However, there is no specific treatment for men since it was not studied yet.

✓ At $\geq$15 headache days per month, medication overuse should be ruled out at any age.

✓ Non-pharmacological treatment should always be attempted before initiating drug treatment in children, in pregnant women and during lactation.

✓ Women and men with migraine with aura should be counselled about screening for risk factors for stroke.

✓ In case of migraine with aura it is necessary to make correct differential diagnosis and exclude TIA, especially in people $>$ 50 years old.

✓ Adverse events of treatment are all more likely in older women and men.

12. Key points

- The development of personalized health care strategies in migraine should be a priority in future because of significant burden of this disorder.
- Future guidelines on migraine management and treatment should also include sex and age differences.

- For better choice and increasing of effectiveness of preventive treatment, it is necessary to diagnose and manage comorbid diseases: depression, anxiety, hypertension, sleep disorders and severe obesity.
- It is necessary to provide standardized guidelines for choosing a preventive treatment of pediatric patients.
- Children and adults need to have different treatments and management strategies for migraine.

References

Ailani, J., Burch, R. C., & Robbins, M. S. (2021). Board of Directors of the American Headache Society. The American Headache Society Consensus Statement: Update on integrating new migraine treatments into clinical practice. *Headache*, *61*(7), 1021–1039. https://doi.org/10.1111/head.14153. Epub 2021 Jun 23. PMID: 34160823.

Al-Hassany, L., Haas, J., Piccininni, M., Kurth, T., Maassen Van Den Brink, A., & Rohmann, J. L. (2020). Giving researchers a headache—Sex and gender differences in migraine. *Frontiers in Neurology*, *22*(11), 549038. https://doi.org/10.3389/fneur.2020.549038. PMID: 33192977; PMCID: PMC7642465.

Ashina, M., Buse, D. C., Ashina, H., Pozo-Rosich, P., Peres, M. F. P., Lee, M. J., et al. (2021). Migraine: Integrated approaches to clinical management and emerging treatments. *Lancet*, *397*(10283), 1505–1518. https://doi.org/10.1016/S0140-6736(20)32342-4.

Ashina, M., Katsarava, Z., Do, T. P., Buse, D. C., Pozo-Rosich, P., Özge, A., et al. (2021). Migraine: Epidemiology and systems of care. *Lancet*, *397*(10283), 1485–1495. https://doi.org/10.1016/S0140-6736(20)32160-7. Epub 2021 Mar 25. PMID: 33773613.

Burch, R. (2020). Epidemiology and treatment of menstrual migraine and migraine during pregnancy and lactation: A narrative review. *Headache*, *60*(1), 200–216. https://doi.org/10.1111/head.13665. Epub 2019 Oct 3. PMID: 31579938.

Eigenbrodt, A. K., Ashina, H., Khan, S., Diener, H. C., Mitsikostas, D. D., Sinclair, A. J., et al. (2021). Diagnosis and management of migraine in ten steps. *Nature Reviews. Neurology*, *17*(8), 501–514. https://doi.org/10.1038/s41582-021-00509-5.

Filatova, E. G., Osipova, V. V., Tabeeva, G. R., Parfenov, V. A., Ekusheva, E. V., Azimova, Y. E., et al. (2020). Diagnosis and treatment of migraine: Russian experts' recommendations. *Neurology, Neuropsychiatry, Psychosomatics*, *12*(4), 4–14 (In Russ.) https://doi.org/10.14412/2074-2711-2020-4-4-14.

Hamilton, K. T., & Robbins, M. S. (2019). Migraine treatment in pregnant women presenting to acute care: A retrospective observational study. *Headache*, *59*(2), 173–179. https://doi.org/10.1111/head.13434.

Headache Classification Committee of the International Headache Society (IHS). (2018). The international classification of headache disorders, 3rd ed. *Cephalalgia*, *38*(1), 1–211. https://doi.org/10.1177/0333102417738202. PMID: 29368949.

Kacperski, J., Kabbouche, M. A., O'Brien, H. L., & Weberding, J. L. (2016). The optimal management of headaches in children and adolescents. *Therapeutic Advances in Neurological Disorders*, *9*(1), 53–68. https://doi.org/10.1177/1756285615616586. PMID: 26788131; PMCID: PMC4710107.

Lebedeva, E. R., Kobzeva, N. R., Gilev, D., & Olesen, J. (2016a). Prevalence of primary headache disorders diagnosed according to ICHD-3 beta in three different social groups. *Cephalalgia*, *36*(6), 579–588. https://doi.org/10.1177/0333102415596442. Epub 2015 Oct 6. PMID: 26442929.

Lebedeva, E. R., Kobzeva, N. R., Gilev, D. V., & Olesen, J. (2016b). Factors associated with primary headache according to diagnosis, sex, and social group. *Headache*, *56*(2), 341–356. https://doi.org/10.1111/head.12757. Epub 2016 Feb 1. PMID: 26833220.

Lebedeva, E. R., Kobzeva, N. R., Gilev, D. V., & Olesen, J. (2016c). The quality of diagnosis and management of migraine and tension-type headache in three social groups in Russia. *Cephalalgia*, *37*(3), 225–235. https://doi.org/10.1177/0333102416642603.

Lipton, R. B., Munjal, S., Alam, A., Buse, D. C., Fanning, K. M., Reed, M. L., et al. (2018). Migraine in America symptoms and treatment (mast) study: Baseline study methods, treatment patterns, and gender differences. *Headache*, *58*(9), 1408–1426. https://doi.org/10.1111/head.13407.

MacGregor, E. A. (2020). Menstrual and perimenopausal migraine: A narrative review. *Maturitas*, *142*, 24–30. https://doi.org/10.1016/j.maturitas.2020.07.005.

Martin, V. T., Pavlovic, J., Fanning, K. M., Buse, D. C., Reed, M. L., & Lipton, R. B. (2016). Perimenopause and menopause are associated with high frequency headache in women with migraine: Results of the American Migraine Prevalence and Prevention Study. *Headache*, *56*(2), 292–305. https://doi.org/10.1111/head.12763.

Monteith, T. S., Gardener, H., Rundek, T., Elkind, M. S., & Sacco, R. L. (2015). Migraine and risk of stroke in older adults: Northern Manhattan Study. *Neurology*, *85*(8), 715–721. https://doi.org/10.1212/WNL.0000000000001854.

Ornello, R., De Matteis, E., Di Felice, C., Caponnetto, V., Pistoia, F., & Sacco, S. (2021). Acute and preventive management of migraine during menstruation and menopause. *Journal of Clinical Medicine*, *10*(11), 2263. https://doi.org/10.3390/jcm10112263.

Practice Guideline Update Summary. (2020). Acute treatment of migraine in children and adolescents: Report of the Guideline Development, Dissemination, and Implementation Subcommittee of the American Academy of Neurology and the American Headache Society. *Neurology*, *94*(1), 50. https://doi.org/10.1212/WNL.0000000000008728.

Rathfoot, C., Edrissi, C., Sanders, C. B., et al. (2021). Gender differences in comorbidities and risk factors in ischemic stroke patients with a history of atrial fibrillation. *BMC Neurology*, *21*(1), 209. https://doi.org/10.1186/s12883-021-02214-8.

Ripa, P., Ornello, R., Degan, D., Tiseo, C., Stewart, J., Pistoia, F., et al. (2015). Migraine in menopausal women: A systematic review. *International Journal of Women's Health*, *20*(7), 773–782. https://doi.org/10.2147/IJWH.S70073. PMID: 26316824; PMCID: PMC4548761.

Schytz, H. W., Amin, F. M., Jensen, R. H., Carlsen, L., Maarbjerg, S., Lund, N., et al. (2021). Reference programme: Diagnosis and treatment of headache disorders and facial pain. Danish Headache Society, 3rd edition, 2020. *The Journal of Headache and Pain*, *22*(1), 22. https://doi.org/10.1186/s10194-021-01228-4.

Steiner, T. J., Stovner, L. J., Jensen, R., Uluduz, D., Katsarava, Z., & Lifting The Burden: the Global Campaign against Headache. (2020). Migraine remains second among the world's causes of disability, and first among young women: findings from GBD2019. *The Journal of Headache and Pain*, *21*(1), 137. https://doi.org/10.1186/s10194-020-01208-0.

Stovner, L. J., Hagen, K., Linde, M., & Steiner, T. J. (2022). The global prevalence of headache: an update, with analysis of the influences of methodological factors on prevalence estimates. *The Journal of Headache and Pain*, *23*(1), 34. https://doi.org/10.1186/s10194-022-01402-2.

Teleanu, R. I., Vladacenco, O., Teleanu, D. M., & Epure, D. A. (2016). Treatment of pediatric migraine: A review. *Maedica (Bucur)*, *11*(2), 136–143. PMID: 28461833; PMCID: PMC5394581.

CHAPTER TEN

Sex and gender differences in mild traumatic brain injury/concussion

Samaneh Chaychi[a,b], Eve Valera[c,d], and Maria Carmela Tartaglia[a,b,e,*]
[a]Memory Clinic, Krembil Brain Institute, University Health Network, Toronto, ON, Canada
[b]Canadian Concussion Centre, Krembil Brain Institute, University Health Network, Toronto, ON, Canada
[c]Department of Psychiatry, Harvard Medical School, Boston, MA, United States
[d]Department of Psychiatry, Massachusetts General Hospital, Charlestown, MA, United States
[e]Tanz Centre for Research in Neurodegenerative Diseases, University of Toronto, Toronto, ON, Canada
*Corresponding author: e-mail address: carmela.tartaglia@uhn.ca

Contents

Abstract

The high incidence of concussions/mild traumatic brain injury and the significant number of people with persisting concussion symptoms as well as the concern for delayed, neurodegenerative effects of concussions makes them a major public health concern. There is much to learn on concussions with respect to pathophysiology as well as vulnerability and resiliency factors. The heterogeneity in outcome after a concussion warrants a more personalized approach to better understand the biological and psychosocial factors that may affect outcome. In this chapter we address biological sex and gender as they impact different aspects of concussion including incidence, risk factors and outcome. As well, this chapter will provide a more fulsome overview of intimate partner violence, an often-overlooked cause of concussion in women. Applying the sex and gender lens to concussion/mild traumatic brain injury is imperative for discovery of its pathophysiology and moving closer to treatments.

International Review of Neurobiology, Volume 164
ISSN 0074-7742
https://doi.org/10.1016/bs.irn.2022.07.004

1. Mild traumatic brain injury/concussion

Mild traumatic brain injury (mTBI) is described as "an alteration in brain function caused by an external force" with a Glasgow Coma Scale (GCS) of 13–15 (Wagner et al., 2021). Approximately 100–300/100,000 individuals are seen for mTBI at a hospital annually (Cassidy et al., 2004). This figure, however, is likely an underestimate of number sustained, as many injured individuals do not seek medical attention, for this reason, the Centers for Disease Control and Prevention (CDC) defined mTBI as a "silent epidemic" (Langlois, Marr, Mitchko, & Johnson, 2005).

Concussions are a type of mTBI but the terms are often used interchangeably. Both are due to a biomechanical force that transiently disrupts normal brain function resulting in neurological signs and symptoms (Anderson, Heitger, & Macleod, 2006; Kamins & Giza, 2016). Mild TBI is typically considered more severe as GCS can be between 13 and 15 (Anderson et al., 2006) while GCS in concussion is usually 15.

According to studies, the majority of concussed patients (85%) recover from concussion within 7 to 10 days, but some cases may suffer from post-concussion symptoms for months or even years (Hiploylee et al., 2017; McCrea et al., 2013; Tator et al., 2016; Tator & Davis, 2014). This condition was previously known as post-concussion syndrome but given the negative connotations of the term, there has been a move away from using the word "syndrome" and towards simply describing the state such as "persisting or prolonged concussion symptoms (PCS)" (Doroszkiewicz et al., 2021). However, other terms such as "late," "chronic", "long-term," and "permanent" are inconsistently used to describe symptoms that persist from one month to several years after a concussion (King & Kirwilliam, 2013; Tsushima, Lum, & Geling, 2009). Previously, the diagnostic criteria for Post-concussion syndrome according to the Diagnostic and Statistical Manual of Mental Disorders (DSM-IV) and the International Statistical Classification of Diseases and Related Health Problems (ICD-10) included a history of head trauma leading to concussion, proof of unsatisfactory performance on the neuropsychological evaluation of memory and attention along with three or more of the following symptoms: fatigue, mood alterations, personality changes, headache, sleep disturbances, and cognitive issues, which last at least three months following injury and result in a significant attenuation in social and professional activities of the injured person (American Psychiatric Association, 1996; Association & Association, 1994;

McCrory, Meeuwisse, Aubry, et al., 2013). Of note, in DSM-V, PCS has been substituted by "Major or Mild Neurocognitive Disorder Due to Traumatic Brain Injury" the diagnostic criteria of which, however, does not include the majority of patients diagnosed with PCS according to DSM-IV or ICD-10 (Tator et al., 2016).

2. Sex-related differences in mild traumatic brain injury/concussion

A wide variety of factors have been proposed to influence outcome after a concussion or post-concussion symptoms. These factors include the presence of loss of consciousness, certain pre-morbid conditions, substance abuse, pre-injury mental health disorders, post-traumatic stress, age, and sex (Carroll et al., 2004; Gallagher et al., 2018). Over the past decade, the influence of sex and gender on the risk of concussion and the course of recovery after concussion has attracted increased attention (McCrory et al., 2013; Mollayeva, El-Khechen-Richandi, & Colantonio, 2018).

It is important to consider that sex and gender are two distinct terms. Sex refers to a collection of biological characteristics that are determined by chromosomes, hormone levels, and reproductive anatomy in humans and animals, while gender refers to socially created roles, expressions, and identities, as well as behavioral manifestations (Clayton & Tannenbaum, 2017; World Health Organization, 2015). However, these terms are frequently used interchangeably in the scientific literature.

Regarding the socio-cultural (gender) aspect, it has been reported that women have greater vulnerabilities to concussion and its consequences compared to men (Brook, Luo, Curry, & Matzkin, 2016; Covassin, Schatz, & Swanik, 2007; King, 2014a,b; Mikolic et al., 2021; Varriano et al., 2018). A recent scoping review reported that the majority of human mTBI studies show women have more severe post-concussive symptoms and of longer duration compared to men (Gupte, Brooks, Vukas, Pierce, & Harris, 2019). This gender-related disparity observed may be partially explained by a greater frequency of psychiatric disorders including mood and anxiety disorders in women, particularly young women (Gulland, 2016; McManus, Bebbington, Jenkins, & Brugha, 2016). Interestingly, a recent study revealed that history of psychiatric disorders was an important risk factor that could in part explain poor outcomes in women following concussion (Gulland, 2016). Another proposed difference to explain the gender disparity is that women are more likely to report concussion symptoms and seek medical attention than men (Daneshvar, Nowinski, McKee, & Cantu, 2011; Makdissi et al., 2013; Moser, Olek, & Schatz, 2019).

It has become increasingly clear that sex affects the physiology and function of neural cells and greatly impacts symptoms and presentation of neurological disease as well as recovery following the injury and also the response to treatment (Anto-Ocrah, Tiffany, Hasman, & van Wijngaarden, 2020). Epidemiological studies of concussion typically rely on records from emergency departments and hospitals and so constantly report higher incidence in males, which is often explained by the fact that males are more frequently involved in physical activities, military service, and contact sports where many concussions occur (Centers for Disease Control and Prevention, 2014, 2019; Frost, Farrer, Primosch, & Hedges, 2013; National Center for Injury Prevention and Control (US), 2003). This has largely ignored the millions of females exposed to recurring concussions caused by intimate partner violence (IPV), which typically remains undetected and underreported (Centers for Disease Control and Prevention, 2008; Monahan, Goldfine, & Biegon, 2017). Given the fact that in most observational and clinical studies of concussion, female participants constitute only around 30% or less of study subjects, the natural history, incidence, and outcome of concussion in females in general, and IPV victims in particular, has not been well-reported to date (Biegon, 2021). Even though in recent decades the number of females who serve in the military or participate in contact sports activities has increased, they still remain an understudied population in the context of brain injury.

Although the incidence of concussion in males and females, as mentioned above, is difficult to accurately determine, it has been reported that females have a 1.4 times higher overall concussion injury rate than males in sex-comparable sports (Covassin, Moran, & Elbin, 2016). As well, there exists some evidence that there is some heterogeneity in concussion symptoms amongst the sexes. It has been reported that females are more likely to report concussion symptoms including headache/migraine, difficulty concentrating, vision/hearing problems, low energy, sleep disturbances, and emotional instability compared to males (Broshek et al., 2005; Covassin et al., 2006; Preiss-Farzanegan, Chapman, Wong, Wu, & Bazarian, 2009; Schneider, Emery, Kang, Schneider, & Meeuwisse, 2010; Shehata et al., 2009). There are, however, some studies that did not find any sex-related differences between males and females in terms of concussion outcome and symptom severity (Black, Sergio, & Macpherson, 2017; Brooks et al., 2014, 2018; Kontos, Covassin, Elbin, & Parker, 2012; Zuckerman et al., 2012). In addition, a few studies demonstrated better neuropsychological results, superior execution function, and less disability among concussed females compared to males after concussion (Barker et al., 2017; Covassin, Elbin, Kontos, & Larson, 2010; Fakhran, Yaeger, Collins, & Alhilali, 2014; Léveillé, Guay, Blais, Scherzer, & De Beaumont, 2017). Interestingly, a meta-analysis of symptom prevalence after concussion revealed that females were 45% less likely to report confusion than males (Brown, Elsass, Miller, Reed, & Reneker, 2015). Another study evaluating concussion symptoms among high school athletes showed that male athletes more frequently reported amnesia and disorientation (Frommer et al., 2011). The data to date provide some evidence that there are sex-related discrepancies in reporting concussion symptoms but there are consistent post-concussion symptom profiles that distinguish females from males (Brook et al., 2016).

As for concussion recovery, which is defined as the number of days to resume daily activities or become asymptomatic after a concussion, some report that females need more days to completely recover and return to their baseline than males (Covassin et al., 2016; Henry, Elbin, Collins, Marchetti, & Kontos, 2016; Merritt, Padgett, & Jak, 2019; Zuckerman et al., 2014), and on average they are more likely to have symptoms that last longer than a month (Bauman, Ray, & Joseph, 2017; Covassin, Savage, Bretzin, & Fox, 2018; Eisenberg, Andrea, Meehan, & Mannix, 2013; Zemek et al., 2016). In a case-controlled study aimed at evaluating risk factors linked to the delayed recovery after a sports-related concussion in a pediatric population, female sex was one of the strongest risk factors associated with prolonged recovery lasting more than 28 days (Miller et al., 2016). Moreover, one study reported that female collegiate concussed athletes took longer to recover from concussions than male collegiate concussed athletes in all sports except lacrosse (McCrory et al., 2013). Among the elderly population, it has been shown that older females are more vulnerable to the development of chronic PCS than males (Farace & Alves, 2000a; King, 2014b). However, the data are conflicting since some studies reported no sex-related differences in post-concussion outcomes (Black et al., 2017; Henry et al., 2016).

It is noteworthy that the adverse effects of concussion are not limited to the acute or sub-acute post-injury phase but can be seen years following an injury as well. It has become widely recognized that head injuries including concussions have long-term consequences, and multiple studies have found greater cognitive impairment and higher incidences of dementia in those who have had a brain injury (Barnes et al., 2014, 2018; Fann et al., 2018; Gardner et al., 2014; Lee et al., 2013; Nordström, Michaëlsson, Gustafson, & Nordström, 2014; Nordström & Nordström, 2018; Yaffe et al., 2019). In a more recent study, it has been demonstrated that head injury including concussion was linked to an increased risk for dementia 25 years after follow-up, and this risk was dose-dependent and stronger for those who had more prior head injuries (Schneider et al., 2021). This study reported a greater risk of dementia as a sequelae of head injury among female participants compared to males.

3. Possible mechanisms underlying sex differences in concussion

The mechanisms underlying the proposed disparities between male and female outcomes after concussion are not well understood. Several theories have been proposed to explain the sex differences including sex hormones, brain anatomy and function, and neck muscle strength and head/neck dimensions (Farace & Alves, 2000b; Mollayeva et al., 2018; Solomito, Reuman, & Wang, 2019; Tierney et al., 2005).

3.1 Effects of gonadal steroid hormones

Gonadal steroid hormones exert multiple diverse effects on the structure and function of the brain in males and females. These effects can be categorized into irreversible (organizational) effects referring to permanent actions of sex steroids on the structural development of the brain which occurs during early brain development and reversible (activational) effects that occur after puberty and impact sexual differentiation (Biegon, 2021; Marrocco & McEwen, 2022). The role of gonadal steroids including estrogen and progesterone on the observed sex differences in outcome after a concussion is still not clearly understood. Some studies have shown better post-concussion outcomes for premenarche and postmenopausal females (Berry et al., 2009; Davis et al., 2006; Ley et al., 2013), compared with females of childbearing age with the highest levels of circulating sex hormones (Wunderle, Hoeger, Wasserman, & Bazarian, 2014). Moreover, the average PCS score amongst females is higher than those of males starting around menarche, reaching a peak during the reproduction ages, and then returning to the level of males after menopause (Bazarian, Blyth, Mookerjee, He, & McDermott, 2010). These findings support the hypothesis that higher concentrations of estrogen and progesterone are associated with worse results after concussion.

As for the effects of sex hormones on the severity of concussion symptoms, it has been reported that higher levels of estrogen and progesterone are associated with elevated pain perception (Choi et al., 2006; Fillingim et al., 1997; Ring, van Zanten, & Kavussanu, 2009). According to the evidence, gonadal hormones have different receptors located throughout the nervous system that could moderate pain (McEwen, Alves, Bulloch, & Weiland, 1998). Interestingly, females' experience of acute and chronic pain can be influenced by the level of circulating gonadal hormones and, subsequently, can differ across the menstrual phases, i.e. follicular and luteal (Riley, Robinson, Wise, Myers, & Fillingim, 1998; Riley, Robinson, Wise, & Price, 1999; Sherman & LeResche, 2006). Females report more pain compared to males in response to similar painful stimuli but they also report more pain during the luteal phase than the follicular phase indicating higher concentrations of estradiol and progesterone are associated with greater pain perception (Ring et al., 2009). There is controversy regarding whether consumption of hormonal contraceptives may affect the severity of concussive symptoms and duration of post-concussion recovery. It has been demonstrated that concussed females taking oral contraceptive pills (OCPs) experienced significantly lower symptom intensity and fewer symptoms compared to those not on OCPs (Mollayeva et al., 2018; Mihalik, Ondrak, Guskiewicz, & McMurray, 2009). Since OCPs adjust and generally lower the level of circulating female reproductive hormones, a decrement in the level of sex hormones in OCP users seems to be protective following a concussive incident (Gaspard et al., 1983; Mishell, Thorneycroft, Nakamura, Nagata, & Stone, 1972; Mollayeva et al., 2018). It has been reported that symptom severity is subjectively lower in females taking OCPs, compared to non-users (Mollayeva et al., 2018). Moreover, this study demonstrated that lower levels of estrogen and progesterone observed in females taking OCPs are associated with better recovery from concussion. Better outcomes from concussion in those taking OCPs may also be partially explained by protection from an abrupt drop in circulating gonadal hormones as seen in females getting a concussion while not on OCPs (Bazarian et al., 2010; Mollayeva et al., 2018).

There is, however, controversy regarding the role of sex hormones on the severity of post-concussion symptoms and outcome following concussion. Two clinical trials reported improvement in patients taking progesterone following head injury (Wright et al., 2007; Xiao, Wei, Yan, Wang, & Lu, 2008), which contradicts the worse outcomes seen in females with a higher concentration of progesterone. To explain this discrepancy in the effect of progesterone, a group of researchers proposed the "withdrawal hypothesis," which states that concussions occurring in a high progesterone environment, that is in the luteal phase of the menstrual cycle, results in an abrupt decrease of progesterone and so can consequently worsen outcomes compared with those occurring in the follicular phase (Wunderle et al., 2014). The reason for the change in hormonal levels could be due to injury of the anterior pituitary gland (Center, 2014; Schneider, Kreitschmann-Andermahr, Ghigo, Stalla, & Agha, 2007; Tanriverdi, Unluhizarci, & Kelestimur, 2010) whose proper functioning is pivotal to regulating and maintaining normal reproductive function in humans. Following a head injury, it has been shown that the hypothalamic-pituitary-gonadal axis is suppressed, which subsequently results in lower luteinizing hormone, follicle-stimulating hormone, testosterone, progesterone, and estrogen concentrations (Agha & Thompson, 2006). Post-injury deficits in sex hormones could explain why concussed females frequently miss regular menstrual cycles and even may experience amenorrhea following TBI (Ripley et al., 2008).

It is important to note that although over the past decade there has been some research on the influence of gonadal hormones to explain sex differences in recovery after concussion, further studies are required to further characterize the impact of these hormones on the pathophysiology of concussion and its functional outcomes.

3.2 Effects of brain anatomy and neural function

Steroid hormones play a critical role in the development of the vertebrate brain. By interacting with certain receptors, they influence myelination, neurogenesis, and brain maturation, as well as sex-specific features in the developing brain that will be present in adults (Bakker, 2021; Diviccaro, Melcangi, & Giatti, 2021; Stortz, Presman, Pecci, & Levi, 2021). Of note, the effects of gonadal sex hormones during fetal brain development are presumably linked to the sex differences in intracranial volume, and focal anatomy (Biegon, 2021). Accordingly, males possess a larger brain and whole-brain volume, approximately 8-15% greater than females (Ruigrok et al., 2014; Shin et al., 2005). Moreover, the number of neurons, total thickness of the cortex, and global cortical surface area are relatively larger in males (Escorial et al., 2015; Luders, Gaser, Narr, & Toga, 2009; Pakkenberg & Gundersen, 1997). It has been postulated that females may demonstrate a more severe reaction to concussion because of smaller brain size and less brain reserve compared to their male counterparts (Biegon, 2021).

One of the brain structures that is significantly different between the sexes is the corpus callosum (CC) (Allen, Richey, Chai, & Gorski, 1991; Menzler et al., 2011; Tierney et al., 2005), which is more cylindrical in males and a more bulbous shape in females (Aboitiz, Rodriguez, Olivares, & Zaidel, 1996; Luders, Toga, & Thompson, 2014). Moreover, the cross-sectional area of the CC is larger in males compared to females (Allen et al., 1991; Luders et al., 2014) reflecting a greater amount of large diameter (3 μm) myelinated axons in males (Aboitiz et al., 1996; Luders et al., 2006). Although males have thicker fibers, females have a greater number of fibers crossing the CC indicating more interhemispheric communication in females (Aboitiz et al., 1996; Allen et al., 1991; Zaidi, 2010). Furthermore, females employ a greater number of areas in their brain when doing certain activities implying that they have more integration and intricate connections than males do (Zaidi, 2010). According to computational models of concussive impacts, the CC is a region of the brain exposed to the highest amount of shear and strain and is a primary site of axonal damage after a concussion (Giordano & Kleiven, 2014; Giordano, Zappalà, & Kleiven, 2017; McAllister et al., 2012; Ng et al., 2017). Interestingly, it has been postulated that the neurobehavioral consequences of a concussion, notably the loss in attention, response speed, and working memory are likely due to the increased strain on the CC following a concussive injury (Ng et al., 2017). It is noteworthy that the part of the brain that displays the highest risk of axonal damage is also the one structure in the brain that differs the most between females and males (Tierney et al., 2005). Thus, following a head impact, males may be able to withstand greater axonal stresses than females because of the larger cross-sectional area, and fiber diameter, as well as a greater number of microtubules in the axon, contributing to less axonal damage (Dollé et al., 2018; Tierney et al., 2005).

Moreover, the CC also serves as the major interhemispheric connector and it has been demonstrated that females have more interhemispheric connections and consequently may be at greater risk of concussion symptoms and prolonged recovery given that damage to the CC will affect their ability to use interhemispheric communication (Ingalhalikar et al., 2014; Tierney et al., 2005).

In-vivo neural network connectivity has been widely investigated using resting-state functional magnetic resonance imaging (rs-fMRI). A recent rs-fMRI study revealed sex-specific differences in network-based connectivity amongst distinct brain regions. They showed that certain structural vulnerabilities especially in temporal regions may predispose females to greater injury than males following concussion (Shafi et al., 2020).

3.3 Biomechanical differences in head and neck anatomy

Several studies have suggested that head/neck anatomy differences between males and females may explain sex differences following a concussion (Mansell, Tierney, Sitler, Swanik, & Stearne, 2005; Tierney et al., 2005, 2008). A computer modeling simulation has shown that head size has a considerable effect on both linear and angular acceleration of the head; the larger the head mass of the athlete, the lower the linear and angular head acceleration (Queen, Weinhold, Kirkendall, & Yu, 2003). Interestingly, direct comparisons demonstrated that females' average head-neck segment size was 43% less than males, and they also had 23% less neck girth (Tierney et al., 2008). Moreover, females also displayed considerably higher peak angular acceleration and displacement in the head-neck region than males which may cause females to experience more head acceleration resulting in more severe head injuries following a concussion (Tierney et al., 2008).

Female neck musculature strength has been demonstrated to be 49% less compared to males, predisposing females to greater angular acceleration of the head/neck axis during a head blow (Tierney et al., 2005, 2008). Having less head mass, weaker neck muscles, and smaller neck girth may be considered as factors that make females more susceptible to concussion and its associated consequences compared to males (Merritt et al., 2019).

Table 1 summarizes the aforementioned factors that are possibly attributed to the higher vulnerability of females to concussion and its following sequels.

Table 1 Proposed factors explaining the higher susceptibility of females to concussion.

Sex hormones	Brain anatomy	Head-neck anatomy
– Higher levels of estrogen and progesterone increase pain perception leading to a greater reaction to concussion and more severe symptoms (particularly in the luteal phase) – More severe symptoms and worse outcome following concussion in non-OCP users propose unfavorable effects of circulating sex hormones which stabilize with OCPs	– Smaller brain size and brain reserve – Smaller axon fiber diameter – Less axonal microtubules. – Structural vulnerability (particularly in temporal cortices) – More fibers crossing corpus callosum indicating greater interhemispheric communication that makes females more vulnerable to effects of CC injury which sustains the greatest amount of axonal damage following concussion	– Smaller head size – Weaker Neck muscles – Smaller neck girth

OCP, oral contraceptive pills.

4. Different mechanisms of injury leading to mild traumatic brain injury/concussion

Concussion affects individuals of all ages and may occur due to a broad spectrum of mechanisms including sports, motor vehicle accidents, falls, military activities, occupational injuries, and intimate partner violence (Mollayeva et al., 2018; Shah, Ba, & Daniels, 2020; Varriano et al., 2018).

Sports and recreation are considered one of the main causes of concussion, particularly in adolescents and young adults (17–20 years) (Mollayeva et al., 2018). Not surprisingly, sport-related concussion (SRC) has become one of the major public health concerns worldwide. Athletes, particularly those who participate in contact sports, such as boxing, American football, ice hockey, etc. are more susceptible to concussions compared to those involved in non-contact sports including biking, tennis, and alpine skiing (Musumeci, Ravalli, Amorini, & Lazzarino, 2019). It is notable that horse racing and equestrian sports have a concussion rate of 13%, which is greater than that of American football and boxing (Connor et al., 2019; Rueda, Halley, & Gilchrist, 2010). Among school-aged children and youth, hockey/skating is the most common cause of head injury (Macpherson, Fridman, Scolnik, Corallo, & Guttmann, 2014). Football, boys' lacrosse, and girls' soccer have the greatest SRC in high school athletics, with 9.21, 6.65, and 6.11 concussions per 10,000 athlete-exposures, respectively (O'Connor et al., 2017). The incidence of concussions in sports also differs by sex. Boys are more likely to experience a concussion when playing football or hockey while females are more likely to suffer a concussion when playing soccer (Halstead & Walter, 2010; Schatz & Moser, 2011). Of interest, it has been shown that the most prevalent form of injury among male and female soccer players was player-to-player contact, which often occurred when players used their shoulder and neck in case of male players, and head-to-head contact, in case of female players (Barnes et al., 1998; Fuller, Junge, & Dvorak, 2005). As for ice hockey, it has been reported that the majority of male players sustained their concussion with player-player contact while females more often experienced concussion from surface contact (Lincoln, Hinton, Almquist, Lager, & Dick, 2007). According to a study on athletes of the National Collegiate Athletic Association, female athletes experienced concussions 1.4 times more than age-matched male athletes playing the same sport suggesting a greater vulnerability of female participants to SRC (Covassin et al., 2016).

Motor vehicle accidents (MVAs) are another significant cause of head trauma. It has been suggested that MVAs can result in more serious concussions than sports since MVA victims appear to show worse symptoms and require more days to recover (Seiger, Goldwater, & Deibert, 2015). Interestingly, one study aiming at better understanding the interaction between variables such as age, sex, mechanism of injury, and history of concussion in patients with PCS reported that following an MVA a higher proportion of females experienced PCS than males (Varriano et al., 2018). However, the cause of increased PCS in females from MVA is unclear because male drivers are more frequently involved in vehicle injuries (Ramage-Morin, 2008; Vehicle RSaM, Regulation TC, 2011). One possible reason is that males in MVAs end up with more serious injuries than concussions, or that MVA, like sport and recreational activities, puts females at a higher risk of developing PCS (Ramage-Morin, 2008; Varriano et al., 2018; Vehicle RSaM, Regulation TC, 2011).

Fall is another potential source of head injury particularly among the elderly and can result in concussion and its related complications

(Thompson, McCormick, & Kagan, 2006). As the population ages, the risk of falls and fall-related head injuries including concussion increases. Several prospective and retrospective studies in New Zealand, Finland, Australia, and the US have observed sex differences in fall risk and fall-related injuries among different populations (Campbell, Spears, & Borrie, 1990; Ellis & Trent, 2001; Kannus et al., 1999; Peel, Kassulke, & McClure, 2002). These studies reported a higher prevalence of falls among elderly females. Moreover, one study reported a greater number of females with PCS following mTBI resulting from a fall possibly due to a higher prevalence of falls among females compared to their male counterparts (Varriano et al., 2018).

More males participate in military services, but the involvement of females in military activities is growing globally, and combat-related concussions became a critical issue in recent wars in Iraq and Afghanistan (Zuckerman et al., 2014). Given that females comprise only 15% of active military personnel (Department of Defense, 2015), they have often been overlooked in military research. While there is a notable paucity of research on evaluating the effect of sex on concussion among military members and veterans, few studies have focused on the prevalence of PCS and psychological disorders amongst female veterans (Duhart, 2011; Hourani, Williams, Bray, & Kandel, 2015; Mattocks et al., 2010; McGraw, 2016; Merritt et al., 2019). It has been shown that female veterans experienced more neurobehavioral symptoms and were more likely to report greater levels of somatosensory, cognitive, and vestibular symptoms compared to male veterans after concussion (Brickell et al., 2017; Iverson et al., 2011).

Occupation-related concussions may be caused by different mechanisms at the workplace including moving or falling objects, falls, bodily movement collisions, and violence (Kim, Colantonio, & Chipman, 2006). Of note, it has been reported that the majority of work-related injuries among female workers occurred in the education and healthcare sectors (Colantonio et al., 2010). In the workplace, female employees account for over 40 percent of head injuries including all levels of severity, but catastrophic and deadly injuries occur significantly more often among males (Colantonio et al., 2010), which is easily explained by the greater numbers of males in high-risk workplaces such as mining, agriculture, and construction industries (Chang, Guerriero, & Colantonio, 2015). The Ontario Workplace Safety and Insurance Board (2016) reported that at the workplace, younger males aged 20–24 years and older females in the 50–54 age group were at the greatest risk of head injuries, mostly in the form of concussion (WSIB, n.d.).

Another very important cause of concussion with a significant sex and gender component is intimate partner violence (IPV) which is experienced by approximately 1 in 3 women globally (Devries et al., 2013). IPV is any violence perpetrated by a current or former partner, spouse, boyfriend, girlfriend or significant other. It is used to exert power and control over a partner and, as such, can include a wide range of abuses including physical, sexual, emotional/psychological, verbal, financial, reproductive, cultural, or spiritual. Although IPV occurs across all sexes and genders, women are disproportionately affected as rates of IPV are higher in women, and women are more likely to be seriously injured as a result of the violence (Smith et al., 2017). Women who experience IPV report being struck in the head with fists and hard objects, having their heads slammed into hard objects (e.g., walls, floors, car windows), being violently shaken and thrown to the ground or down stairs (Valera & Kucyi, 2017). All of these abuses have the potential to cause a concussion or a more severe traumatic brain injury. Women who are abused also report being strangled into unconsciousness which has the potential to lead to anoxic or hypoxic ischemic events that may have additive or interactive effects with other brain injuries (Monahan, Bannon, & Dams-O'Connor, 2020). It has been reported that up to 94% of abuse injures the head and neck (Arosarena, Fritsch, Hsueh, Aynehchi, & Haug, 2009; Wu, Huff, & Bhandari, 2010).

There are no good epidemiologically-based estimates of IPV-related BI in men or women but there are a number of findings that indicate the rate is higher than in other circumstances often considered to be high risk for BI. A recent review based on samples of convenience and varying "definitions" of both BI and IPV indicated estimates ranging from 28-100% of women reporting IPV-related BIs (Campbell, Joseph, Rothman, & Valera, 2022). Some definitions of IPV restricted samples to physical abuse (Valera & Berenbaum, 2003; Valera & Kucyi, 2017), which may artificially inflate BI estimates. Yet other definitions only included loss of consciousness, rather than all possible indicators of BI, likely leading to deflated estimates of IPV-related BI. Nonetheless, one nationally representative survey in the United States, found that approximately 6.2 million women reported being "knocked out after getting hit, slammed against something, or choked" from partner violence (Breiding, Chen, & Black, 2010). As being "knocked out," or losing consciousness, is only one of many indicators of a BI, the number of women sustaining BIs is likely significantly higher. These data underscore the need for epidemiological data aimed at understanding the true prevalence of IPV-related BI, while still indicating that this is a largely overlooked public health crisis.

Overall, our knowledge on IPV-related BI is relatively recent, scant, and based completely on female data. This is in stark contrast to what we know about brain injuries in other circumstances, for which nearly all data are based on males (in both human and animal research). The first studies considering the effects of brain injuries in IPV were published in the late 90s/early 2000s (Corrigan, Wolfe, Mysiw, Jackson, & Bogner, 2003; Jackson, Philp, Nuttall, & Diller, 2002; Monahan & O'Leary, 1999; Valera & Berenbaum, 2003). These reports provided preliminary prevalence estimates as well as correlates of IPV-related BI. Monahan and O'Leary (Monahan & O'Leary, 1999) reported higher rates of somatic, cognitive and emotional symptoms in women who had reported head traumas from IPV. Others, (Jackson et al., 2002; Smirl et al., 2019; Valera & Berenbaum, 2003) showed dose-response effects of IPV-related BI with a range of neurobehavioral symptoms. For example, Valera and Berenbaum (Valera & Berenbaum, 2003) generated a brain injury score based on number, recency and severity of partner related BIs (including strangulation related alterations in consciousness) and used neuropsychological measures to assess facets of cognitive functioning. They found that, compared to women with lower scores, women with higher brain injury scores tended to perform more poorly on measures of memory, learning and cognitive flexibility, and tended to have higher levels of depression, anxiety, worry, general distress and post-traumatic stress symptoms. Studies such as these, suggested early on, not only that IPV-related BI was a problem, but also that many women sustained repetitive brain injuries from partner violence, and the more BIs they experienced, the greater were their levels of emotional distress and cognitive difficulties.

Notably, some studies (Maldonado-Rodriguez et al., 2021; Valera & Berenbaum, 2003) controlled for variables such as psychopathology, substance use/dependence and/or abuse severity and still found relationships between BIs and cognitive and psychological outcomes. In doing so, these studies demonstrated that the difficulties women were reporting or experiencing were not merely the effects of being in a traumatic and psychologically stressful relationship. Rather, the BIs themselves were contributing factors.

Since that time a growing number of studies have provided additional evidence supporting the link between IPV, BIs and negative outcomes in a range of samples. For example, Iverson and colleagues examined women veterans who had experienced IPV. They found that women with IPV-related BIs demonstrated higher depression scores, PTSD scores, and poorer perceptions of physical health relative to women veterans with no

BIs (Iverson, Dardis, & Pogoda, 2017; Iverson & Pogoda, 2015). Other survey data conducted in primary care or family planning clinics showed a higher rate of central nervous system (CNS) symptoms (e.g., memory loss, ears ringing, dizzy spells, vision or hearing problems, difficulty concentrating) in women who reported IPV with a "probable" BI relative to women reporting IPV with no probable BI (Campbell et al., 2018). In one relatively novel study (Maldonado-Rodriguez et al., 2021), investigators used a device with movable robotic arms to assess cognitive-motor function in women who had experienced IPV. They found a relationship between a brain injury score and women's performance such that the higher the score (indicating a higher brain injury 'load') the more difficulty women had performing the task. Overall, a growing number of studies have found similar links between IPV-related BI and negative outcomes clearly indicating the importance of understanding the impact of BIs on a range of functions for women who experience IPV.

To date, there are only three studies that have used neuroimaging to examine the effects of IPV-related BI (Likitlersuang et al., 2022; Valera et al., 2019; Valera & Kucyi, 2017). Though the sample sizes were somewhat small, they nonetheless provided yet additional evidence of the potential dangers of BI in these women. Two of the studies showed relationships between BI scores and measures of structural and functional connectivity thought to be important for a range of behaviors (Valera et al., 2019; Valera & Kucyi, 2017). Furthermore, one of the studies showed a relationship between the functional connectivity findings and a woman's ability to learn and remember a list of words (Valera & Kucyi, 2017). Another study showed cortical thickness differences between women who had experienced a traumatic BI from IPV relative to women who had experienced a traumatic BI from other causes (Likitlersuang et al., 2022). Though preliminary, these data underscore the importance of conducting more and larger neuroimaging studies aimed at understanding the neural impacts of BIs sustained specifically from partner violence.

A critically important yet outstanding question is to what degree IPV-related BIs have long lasting, permanent or neurodegenerative effects. At least some studies that assessed time since last abuse and/or BI, have shown high levels of post-concussive or neurobehavioral symptoms, as well as relationships between these symptoms and BI scores, well after a 3-month average (Smirl et al., 2019). Furthermore, two case reports (Danielsen, Hauch, Kelly, & White, 2021; Roberts, Whitwell, Acland, & Bruton, 1990) of women ages 29 and 76, who had sustained severe partner abuse to the head, showed chronic traumatic encephalopathy (CTE) type brain pathology on autopsy. Both of these women also demonstrated so-called "cauliflower ears" or deformities of the outer ears that are often observed in boxers and are indicative of blunt-force trauma. However, there are no studies to date specifically examining the effects of IPV-related BI on cognitive or psychological functioning in late life (e.g., systematically assessing women over the age of 65 or conducting longitudinal studies into late life). However, one study suggesting that there could be longer term effects found that women who reported histories of partner violence with head trauma showed a trend for being more likely to develop dementia than women who had not experienced partner violence with head trauma (Leung, Thompson, & Weaver, 2006). Nonetheless, this study was not able to determine causality, nor could it disentangle the potential effects of IPV itself from IPV-related BI.

In summary, our knowledge pertaining to IPV-related BI is currently in its infancy, but we know enough to understand that it is critical to learn more about this largely overlooked public health issue. Women who experience partner violence have a high likelihood of sustaining repetitive BIs that have negative effects on a range of cognitive, psychological, and somatic behaviors as well as brain structure and function. It is crucial that we increase our understanding of the short- and long-term effects including whether there are negative consequences associated with or interacting with aging.

Summary including sex and gender issues is fundamental to understanding concussions/mTBI since it is now known that sex and gender influence the risk of concussion, symptoms, and outcomes. As well, the mechanisms of injury associated with concussion differ between the sexes. Any hope of improving prevention, treatments, and outcomes after concussions/mTBI will have to consider sex and gender issues.

Key take home messages

- Sex differences have been described in mild traumatic brain injury/concussion.
- Several theories have been proposed to explain the sex differences in mild traumatic brain injury/concussion including sex hormones, brain anatomy and function, and neck muscle strength and head/neck dimensions.
- Intimate partner violence is an often-overlooked cause of mild traumatic brain injury/concussion in women.

References

Aboitiz, F., Rodriguez, E., Olivares, R., & Zaidel, E. (1996). Age-related changes in fibre composition of the human corpus callosum: Sex differences. *Neuroreport*, 7(11), 1761–1764.

Agha, A., & Thompson, C. J. (2006). Anterior pituitary dysfunction following traumatic brain injury (TBI). *Clinical Endocrinology*, *64*(5), 481–488.

Allen, L. S., Richey, M. F., Chai, Y. M., & Gorski, R. A. (1991). Sex differences in the corpus callosum of the living human being. *The Journal of Neuroscience*, *11*(4), 933–942.

American Psychiatric Association. (1996). *Diagnostic and Statistical Manual of Mental Disorders* (p. 2000). Washington, DC: American Psychiatric Association. Text Revision (DSM–IV-TR).

Anderson, T., Heitger, M., & Macleod, A. D. (2006). Concussion and mild head injury. *Practical Neurology*, *6*(6), 342–357.

Anto-Ocrah, M., Tiffany, K., Hasman, L., & van Wijngaarden, E. (2020). Mild traumatic brain injury/concussion and female sexuality, a scoping review of the literature. *Injury Epidemiology*, 7(1), 1–3.

Arosarena, O. A., Fritsch, T. A., Hsueh, Y., Aynehchi, B., & Haug, R. (2009). Maxillofacial injuries and violence against women. *Archives of Facial Plastic Surgery*, *11*(1), 48–52.

Association, A. P., & Association, A. P. (1994). *Diagnostic and statistical manual of mental disorders (DSM)* (pp. 143–147). Washington, DC: American Psychiatric Association.

Bakker, J. (2021). The role of steroid hormones in the sexual differentiation of the human brain. *Journal of Neuroendocrinology*, *8*, e13050.

Barker, T., Russo, S. A., Barker, G., Rice, M. A., Jeffrey, M. G., Broderick, G., et al. (2017). A case matched study examining the reliability of using ImPACT to assess effects of multiple concussions. *BMC Psychology*, *5*(1), 1–9.

Barnes, D. E., Byers, A. L., Gardner, R. C., Seal, K. H., Boscardin, W. J., & Yaffe, K. (2018). Association of mild traumatic brain injury with and without loss of consciousness with dementia in US military veterans. *JAMA Neurology*, *75*(9), 1055–1061.

Barnes, B. C., Cooper, L., Kirkendall, D. T., McDermott, T. P., Jordan, B. D., & Garrett, W. E. (1998). Concussion history in elite male and female soccer players. *The American Journal of Sports Medicine*, *26*(3), 433–438.

Barnes, D. E., Kaup, A., Kirby, K. A., Byers, A. L., Diaz-Arrastia, R., & Yaffe, K. (2014). Traumatic brain injury and risk of dementia in older veterans. *Neurology*, *83*(4), 312–319.

Bauman, S., Ray, M., & Joseph, P. P. (2017). Gender differences in clinical presentation and recovery of sports related concussion. *British Journal of Sports Medicine*, *51*(11), A35.

Bazarian, J. J., Blyth, B., Mookerjee, S., He, H., & McDermott, M. P. (2010). Sex differences in outcome after mild traumatic brain injury. *Journal of Neurotrauma*, *27*(3), 527–539.

Berry, C., Ley, E. J., Tillou, A., Cryer, G., Margulies, D. R., & Salim, A. (2009). The effect of gender on patients with moderate to severe head injuries. *Journal of Trauma and Acute Care Surgery*, *67*(5), 950–953.

Biegon, A. (2021). Considering biological sex in traumatic brain injury. *Frontiers in Neurology*, *39*.

Black, A. M., Sergio, L. E., & Macpherson, A. K. (2017). The epidemiology of concussions: Number and nature of concussions and time to recovery among female and male Canadian varsity athletes 2008 to 2011. *Clinical Journal of Sport Medicine*, *27*(1), 52–56.

Breiding, M. J., Chen, J., & Black, M. C. (2010). *Intimate partner violence in the United States*.

Brickell, T. A., Lippa, S. M., French, L. M., Kennedy, J. E., Bailie, J. M., & Lange, R. T. (2017). Female service members and symptom reporting after combat and non-combat-related mild traumatic brain injury. *Journal of Neurotrauma*, *34*(2), 300–312.

Brook, E. M., Luo, X., Curry, E. J., & Matzkin, E. G. (2016). A heads up on concussions: Are there sex-related differences? *The Physician and Sports Medicine*, *44*(1), 20–28.

Brooks, B. L., Mrazik, M., Barlow, K. M., McKay, C. D., Meeuwisse, W. H., & Emery, C. A. (2014). Absence of differences between male and female adolescents with prior sport concussion. *The Journal of Head Trauma Rehabilitation*, *29*(3), 257–264.

Brooks, B. L., Silverberg, N., Maxwell, B., Mannix, R., Zafonte, R., Berkner, P. D., et al. (2018). Investigating effects of sex differences and prior concussions on symptom reporting and cognition among adolescent soccer players. *The American Journal of Sports Medicine*, *46*(4), 961–968.

Broshek, D. K., Kaushik, T., Freeman, J. R., Erlanger, D., Webbe, F., & Barth, J. T. (2005). Sex differences in outcome following sports-related concussion. *Journal of Neurosurgery*, *102*(5), 856–863.

Brown, D. A., Elsass, J. A., Miller, A. J., Reed, L. E., & Reneker, J. C. (2015). Differences in symptom reporting between males and females at baseline and after a sports-related concussion: A systematic review and meta-analysis. *Sports Medicine*, *45*(7), 1027–1040.

Campbell, J. C., Anderson, J. C., McFadgion, A., Gill, J., Zink, E., Patch, M., et al. (2018). The effects of intimate partner violence and probable traumatic brain injury on central nervous system symptoms. *Journal of Women's Health*, *27*(6), 761–767.

Campbell, J. C., Joseph, A. L. C., Rothman, E., & Valera, E. M. (2022). The prevalence of brain injury among survivors and perpetrators of intimate partner violence and the prevalence of violence victimization and perpetration among people with brain injury: A scoping review. *Current Epidemiology Reports*. In Press.

Campbell, A. J., Spears, G. F. S., & Borrie, M. J. (1990). Examination by logistic regression modeling of the variables which increase the relative risk of elderly women falling compared to elderly men. *Journal of Clinical Epidemiology*, *43*, 1415–1420.

Carroll, L., Cassidy, J. D., Peloso, P., Borg, J., Von Holst, H., Holm, L., et al. (2004). Prognosis for mild traumatic brain injury: Results of the WHO Collaborating Centre Task Force on Mild Traumatic Brain Injury. *Journal of Rehabilitation Medicine*, *36*, 84–105.

Cassidy, J. D., Carroll, L., Peloso, P., Borg, J., Von Holst, H., Holm, L., et al. (2004). Incidence, risk factors and prevention of mild traumatic brain injury: Results of the WHO Collaborating Centre Task Force on Mild Traumatic Brain Injury. *Journal of Rehabilitation Medicine*, *36*, 28–60.

Center, B. I. (2014). Neuroendocrine dysfunction following mild TBI: When to screen for it. *The Journal of Family Practice*, *63*(1).

Centers for Disease Control and Prevention. (2008). Adverse health conditions and health risk behaviors associated with intimate partner violence—United States, 2005. *MMWR: Morbidity and Mortality Weekly Report*, *57*(5), 113–117.

Centers for Disease Control and Prevention. (2014). *Traumatic brain injury in the United States: Fact sheet*. Washington, DC: Centers for Disease Control and Prevention.

Centers for Disease Control and Prevention. (2019). *Surveillance report of traumatic brain injury-related emergency department visits, hospitalizations, and deaths, United States, 2014. Atlanta, GA*.

Chang, V. C., Guerriero, E. N., & Colantonio, A. (2015). Epidemiology of work-related traumatic brain injury: A systematic review. *American Journal of Industrial Medicine*, *58*(4), 353–377.

Choi, J. C., Park, S. K., Kim, Y. H., Shin, Y. W., Kwon, J. S., Kim, J. S., et al. (2006). Different brain activation patterns to pain and pain-related unpleasantness during the menstrual cycle. *The Journal of the American Society of Anesthesiologists*, *105*(1), 120–127.

Clayton, J. A., & Tannenbaum, C. (2017). Sex and gender reporting in research—Reply. *Journal of the American Medical Association*, *317*(9), 975.

Colantonio, A., Mroczek, D., Patel, J., Lewko, J., Fergenbaum, J., & Brison, R. (2010). Examining occupational traumatic brain injury in Ontario. *Canadian Journal of Public Health*, *101*(1), S58–S62.

Connor, T. A., Clark, J. M., Jayamohan, J., Stewart, M., McGoldrick, A., Williams, C., et al. (2019). Do equestrian helmets prevent concussion? A retrospective analysis of head injuries and helmet damage from real-world equestrian accidents. *Sports Medicine*, *5*(1), 1–8.

Corrigan, J. D., Wolfe, M., Mysiw, W. J., Jackson, R. D., & Bogner, J. A. (2003). Early identification of mild traumatic brain injury in female victims of domestic violence. *American Journal of Obstetrics and Gynecology*, *188*(5), S71–S76.

Covassin, T., Elbin, R., Kontos, A., & Larson, E. (2010). Investigating baseline neurocognitive performance between male and female athletes with a history of multiple concussion. *Journal of Neurology, Neurosurgery, and Psychiatry*, *81*(6), 597–601.

Covassin, T., Moran, R., & Elbin, R. J. (2016). Sex differences in reported concussion injury rates and time loss from participation: An update of the National Collegiate Athletic Association Injury Surveillance Program from 2004–2005 through 2008–2009. *Journal of Athletic Training*, *51*(3), 189–194.

Covassin, T., Savage, J. L., Bretzin, A. C., & Fox, M. E. (2018). Sex differences in sport-related concussion long-term outcomes. *International Journal of Psychophysiology*, (132), 9–13.

Covassin, T., Schatz, P., & Swanik, C. B. (2007). Sex differences in neuropsychological function and post-concussion symptoms of concussed collegiate athletes. *Neurosurgery*, *61*(2), 345–351.

Covassin, T., Swanik, C. B., Sachs, M., Kendrick, Z., Schatz, P., Zillmer, E., et al. (2006). Sex differences in baseline neuropsychological function and concussion symptoms of collegiate athletes. *British Journal of Sports Medicine*, *40*(11), 923–927.

Daneshvar, D. H., Nowinski, C. J., McKee, A. C., & Cantu, R. C. (2011). The epidemiology of sport-related concussion. *Clinics in Sports Medicine*, *30*(1), 1–7.

Danielsen, T., Hauch, C., Kelly, L., & White, C. L. (2021). Chronic traumatic encephalopathy (CTE)-type neuropathology in a young victim of domestic abuse. *Journal of Neuropathology and Experimental Neurology*, *80*(6), 624–627.

Davis, D. P., Douglas, D. J., Smith, W., Sise, M. J., Vilke, G. M., Holbrook, T. L., et al. (2006). Traumatic brain injury outcomes in pre-and post-menopausal females versus age-matched males. *Journal of Neurotrauma*, *23*(2), 140–148.

Department of Defense. (2015). *2015 Demographics: Profile of the military community*. Retrieved from http://download.militaryonesource.mil/12038/MOS/Reports/2015-Demographics-Report.

Devries, K. M., Mak, J. Y., Garcia-Moreno, C., Petzold, M., Child, J. C., Falder, G., et al. (2013). The global prevalence of intimate partner violence against women. *Science*, *340*(6140), 1527–1528.

Diviccaro, S., Melcangi, R. C., & Giatti, S. (2021). Neuroactive steroids and neurodevelopment. *Factors Affecting Neurodevelopment*, *1*, 309–320. Academic Press.

Dollé, J. P., Jaye, A., Anderson, S. A., Ahmadzadeh, H., Shenoy, V. B., & Smith, D. H. (2018). Newfound sex differences in axonal structure underlie differential outcomes from in vitro traumatic axonal injury. *Experimental Neurology*, *1*(300), 121–134.

Doroszkiewicz, C., Gold, D., Green, R., Tartaglia, M. C., Ma, J., & Tator, C. H. (2021). Anxiety, depression, and quality of life: A long-term follow-up study of patients with persisting concussion symptoms. *Journal of Neurotrauma*, *38*(4), 493–505.

Duhart, O. (2011). PTSD and women warriors: Causes, controls and a congressional cure. *Cardozo Journal of Law & Gender*, *18*, 327.

Eisenberg, M. A., Andrea, J., Meehan, W., & Mannix, R. (2013). Time interval between concussions and symptom duration. *Pediatrics*, *132*(1), 8–17.

Ellis, A. A., & Trent, R. B. (2001). Do the risks and consequences of hospitalized fall injuries among older adults in California vary by type of fall? *The Journals of Gerontology Series A: Biological Sciences and Medical Sciences*, *56*(11), M686–M692.

Escorial, S., Román, F. J., Martínez, K., Burgaleta, M., Karama, S., & Colom, R. (2015). Sex differences in neocortical structure and cognitive performance: A surface-based morphometry study. *NeuroImage*, *1*(104), 355–365.

Fakhran, S., Yaeger, K., Collins, M., & Alhilali, L. (2014). Sex differences in white matter abnormalities after mild traumatic brain injury: Localization and correlation with outcome. *Radiology*, *272*(3), 815–823.

Fann, J. R., Ribe, A. R., Pedersen, H. S., Fenger-Grøn, M., Christensen, J., Benros, M. E., et al. (2018). Long-term risk of dementia among people with traumatic brain injury in Denmark: A population-based observational cohort study. *The Lancet Psychiatry*, *5*(5), 424–431.

Farace, E., & Alves, W. M. (2000a). Do women fare worse: A metaanalysis of gender differences in traumatic brain injury outcome. *Journal of Neurosurgery*, *93*(4), 539–545.

Farace, E., & Alves, W. M. (2000b). Do women fare worse? A metaanalysis of gender differences in outcome after traumatic brain injury. *Neurosurgical Focus*, *8*(1), 1–8.

Fillingim, R. B., Maixner, W., Girdler, S. S., Light, K. C., Harris, M. B., Sheps, D. S., et al. (1997). Ischemic but not thermal pain sensitivity varies across the menstrual cycle. *Psychosomatic Medicine*, *59*(5), 512–520.

Frommer, L. J., Gurka, K. K., Cross, K. M., Ingersoll, C. D., Comstock, R. D., & Saliba, S. A. (2011). Sex differences in concussion symptoms of high school athletes. *Journal of Athletic Training*, *46*(1), 76–84.

Frost, R. B., Farrer, T. J., Primosch, M., & Hedges, D. W. (2013). Prevalence of traumatic brain injury in the general adult population: A meta-analysis. *Neuroepidemiology*, *40*(3), 154–159.

Fuller, C. W., Junge, A., & Dvorak, J. (2005). A six year prospective study of the incidence and causes of head and neck injuries in international football. *British Journal of Sports Medicine*, *39*(Suppl. 1), i3–i9.

Gallagher, V., Kramer, N., Abbott, K., Alexander, J., Breiter, H., Herrold, A., et al. (2018). The effects of sex differences and hormonal contraception on outcomes after collegiate sports-related concussion. *Journal of Neurotrauma*, *35*(11), 1242–1247.

Gardner, R. C., Burke, J. F., Nettiksimmons, J., Kaup, A., Barnes, D. E., & Yaffe, K. (2014). Dementia risk after traumatic brain injury vs nonbrain trauma: The role of age and severity. *JAMA Neurology*, *71*(12), 1490–1497.

Gaspard, U. J., Romus, M. A., Gillain, D., Duvivier, J., Demey-Ponsart, E., & Franchimont, P. (1983). Plasma hormone levels in women receiving new oral contraceptives containing ethinyl estradiol plus levonorgestrel or desogestrel. *Contraception*, *27*(6), 577–590.

Giordano, C., & Kleiven, S. (2014). Evaluation of axonal strain as a predictor for mild traumatic brain injuries using finite element modeling. *Stapp Car Crash Journal*, *58*(14), 29–61.

Giordano, C., Zappalà, S., & Kleiven, S. (2017). Anisotropic finite element models for brain injury prediction: The sensitivity of axonal strain to white matter tract inter-subject variability. *Biomechanics and Modeling in Mechanobiology*, *16*(4), 1269–1293.

Gulland, A. (2016). Women have higher rates of mental disorders than men, NHS Survey finds. *British Medical Journal*, *354*, i5320.

Gupte, R. P., Brooks, W. M., Vukas, R. R., Pierce, J. D., & Harris, J. L. (2019). Sex differences in traumatic brain injury: what we know and what we should know. *Journal of Neurotrauma*, *36*(22), 3063–3091.

Halstead, M. E., & Walter, K. D. (2010). Sport-related concussion in children and adolescents. *Pediatrics*, *126*(3), 597–615.

Henry, L. C., Elbin, R. J., Collins, M. W., Marchetti, G., & Kontos, A. P. (2016). Examining recovery trajectories after sport-related concussion with a multimodal clinical assessment approach. *Neurosurgery*, *78*(2), 232–241.

Hiploylee, C., Dufort, P. A., Davis, H. S., Wennberg, R. A., Tartaglia, M. C., Mikulis, D., et al. (2017). Longitudinal study of postconcussion syndrome: Not everyone recovers. *Journal of Neurotrauma*, *34*(8), 1511–1523.

Hourani, L., Williams, J., Bray, R., & Kandel, D. (2015). Gender differences in the expression of PTSD symptoms among active duty military personnel. *Journal of Anxiety Disorders*, *1*(29), 101–108.

Ingalhalikar, M., Smith, A., Parker, D., Satterthwaite, T. D., Elliott, M. A., Ruparel, K., et al. (2014). Sex differences in the structural connectome of the human brain. *Proceedings of the National Academy of Sciences of the United States of America*, *111*(2), 823–828.

Iverson, K. M., Dardis, C. M., & Pogoda, T. K. (2017). Traumatic brain injury and PTSD symptoms as a consequence of intimate partner violence. *Comprehensive Psychiatry*, *1*(74), 80–87.

Iverson, K. M., Hendricks, A. M., Kimerling, R., Krengel, M., Meterko, M., Stolzmann, K. L., et al. (2011). Psychiatric diagnoses and neurobehavioral symptom severity among OEF/OIF VA patients with deployment-related traumatic brain injury: A gender comparison. *Women's Health Issues*, *21*(4), S210–S217.

Iverson, K. M., & Pogoda, T. K. (2015). Traumatic brain injury among women veterans: An invisible wound of intimate partner violence. *Medical Care*, *53*(4), S112–S119.

Jackson, H., Philp, E., Nuttall, R. L., & Diller, L. (2002). Traumatic brain injury: A hidden consequence for battered women. *Professional Psychology: Research and Practice*, *33*(1), 39.

Kamins, J., & Giza, C. C. (2016). Concussion-mild traumatic brain injury: Recoverable injury with potential for serious sequelae. *Neurosurgery Clinics of North America*, *27*(4), 441–452.

Kannus, P., Parkkari, J., Koskinen, S., Niemi, S., Palvanen, M., Järvinen, M., et al. (1999). Fall-induced injuries and deaths among older adults. *Journal of the American Medical Association*, *281*(20), 1895–1899.

Kim, H., Colantonio, A., & Chipman, M. (2006). Traumatic brain injury occurring at work. *NeuroRehabilitation*, *21*(4), 269–278.

King, N. (2014a). Permanent post concussion symptoms after mild head injury: A systematic review of age and gender factors. *NeuroRehabilitation*, *34*(4), 741–748.

King, N. S. (2014b). A systematic review of age and gender factors in prolonged post-concussion symptoms after mild head injury. *Brain Injury*, *28*(13–14), 1639–1645.

King, N. S., & Kirwilliam, S. (2013). The nature of permanent post-concussion symptoms after mild traumatic brain injury. *Brain Impairment*, *14*(2), 235–242.

Kontos, A. P., Covassin, T., Elbin, R. J., & Parker, T. (2012). Depression and neurocognitive performance after concussion among male and female high school and collegiate athletes. *Archives of Physical Medicine and Rehabilitation*, *93*(10), 1751–1756.

Langlois, J. A., Marr, A., Mitchko, J., & Johnson, R. L. (2005). Tracking the silent epidemic and educating the public: CDC's traumatic brain injury-associated activities under the TBI act of 1996 and the Children's Health Act of 2000. *The Journal of Head Trauma Rehabilitation*, *20*(3), 196–204.

Lee, Y. K., Hou, S. W., Lee, C. C., Hsu, C. Y., Huang, Y. S., & Su, Y. C. (2013). Increased risk of dementia in patients with mild traumatic brain injury: A nationwide cohort study. *PLoS One*, *8*(5), e62422.

Leung, F. H., Thompson, K., & Weaver, D. F. (2006). Evaluating spousal abuse as a potential risk factor for Alzheimer's disease: Rationale, needs and challenges. *Neuroepidemiology*, *27*(1), 13–16.

Léveillé, E., Guay, S., Blais, C., Scherzer, P., & De Beaumont, L. (2017). Sex-related differences in emotion recognition in multi-concussed athletes. *Journal of the International Neuropsychological Society*, *23*(1), 65–77.

Ley, E. J., Short, S. S., Liou, D. Z., Singer, M. B., Mirocha, J., Melo, N., et al. (2013). Gender impacts mortality after traumatic brain injury in teenagers. *Journal of Trauma and Acute Care Surgery*, *75*(4), 682–686.

Likitlersuang, J., Brown, E. M., Salat, D. H., Iverson, K. M., Werner, K., McGlinchey, R. E., et al. (2022). Neural correlates of traumatic brain injury in women survivors of intimate partner violence: A structural and functional connectivity neuroimaging study. *The Journal of Head Trauma Rehabilitation*, *37*(1), E30–E38.

Lincoln, A. E., Hinton, R. Y., Almquist, J. L., Lager, S. L., & Dick, R. W. (2007). Head, face, and eye injuries in scholastic and collegiate lacrosse: A 4-year prospective study. *The American Journal of Sports Medicine*, *35*(2), 207–215.

Luders, E., Gaser, C., Narr, K. L., & Toga, A. W. (2009). Why sex matters: Brain size independent differences in gray matter distributions between men and women. *The Journal of Neuroscience*, *29*(45), 14265–14270.

Luders, E., Narr, K. L., Zaidel, E., Thompson, P. M., Jancke, L., & Toga, A. W. (2006). Parasagittal asymmetries of the corpus callosum. *Cerebral Cortex*, *16*(3), 346–354.

Luders, E., Toga, A. W., & Thompson, P. M. (2014). Why size matters: Differences in brain volume account for apparent sex differences in callosal anatomy: The sexual dimorphism of the corpus callosum. *NeuroImage*, *1*(84), 820–824.

Macpherson, A., Fridman, L., Scolnik, M., Corallo, A., & Guttmann, A. (2014). A population-based study of paediatric emergency department and office visits for concussions from 2003 to 2010. *Paediatrics & Child Health*, *19*(10), 543–546.

Makdissi, M., Davis, G., Jordan, B., Patricios, J., Purcell, L., & Putukian, M. (2013). Revisiting the modifiers: How should the evaluation and management of acute concussions differ in specific groups? *British Journal of Sports Medicine*, *47*(5), 314–320.

Maldonado-Rodriguez, N., Crocker, C. V., Taylor, E., Jones, K. E., Rothlander, K., Smirl, J., et al. (2021). Canadian Traumatic Brain Injury Research Consortium. Characterization of cognitive-motor function in women who have experienced intimate partner violence-related brain injury. *Journal of Neurotrauma*, *38*(19), 2723–2730.

Mansell, J., Tierney, R. T., Sitler, M. R., Swanik, K. A., & Stearne, D. (2005). Resistance training and head-neck segment dynamic stabilization in male and female collegiate soccer players. *Journal of Athletic Training*, *40*(4), 310.

Marrocco, J., & McEwen, B. S. (2022). Sex in the brain: hormones and sex differences. *Dialogues in Clinical Neuroscience*, *1*.

Mattocks, K. M., Skanderson, M., Goulet, J. L., Brandt, C., Womack, J., Krebs, E., et al. (2010). Pregnancy and mental health among women veterans returning from Iraq and Afghanistan. *Journal of Women's Health*, *19*(12), 2159–2166.

McAllister, T. W., Ford, J. C., Ji, S., Beckwith, J. G., Flashman, L. A., Paulsen, K., et al. (2012). Maximum principal strain and strain rate associated with concussion diagnosis correlates with changes in corpus callosum white matter indices. *Annals of Biomedical Engineering*, *40*(1), 127–140.

McCrea, M., Guskiewicz, K., Randolph, C., Barr, W. B., Hammeke, T. A., Marshall, S. W., et al. (2013). Incidence, clinical course, and predictors of prolonged recovery time following sport-related concussion in high school and college athletes. *Journal of the International Neuropsychological Society*, (1), 22–33.

McCrory, P., Meeuwisse, W. H., Aubry, M., et al. (2013). Consensus statement on concussion in sport: The 4th International Conference on Concussion in Sport, Zurich, November 2012. *Journal of Athletic Training*, *48*(4), 554–575.

McEwen, B. S., Alves, S. E., Bulloch, K., & Weiland, N. G. (1998). Clinically relevant basic science studies of gender differences and sex hormone effects. *Psychopharmacology Bulletin*, *34*(3), 251–259.

McGraw, K. (2016). Gender differences among military combatants: Does social support, ostracism, and pain perception influence psychological health? *Military Medicine*, *181*(suppl_1), 80–85.

McManus, S., Bebbington, P. E., Jenkins, R., & Brugha, T. (2016). Mental health and wellbeing in England: The adult psychiatric morbidity survey 2014. *NHS Digital*.

Menzler, K., Belke, M., Wehrmann, E., Krakow, K., Lengler, U., Jansen, A., et al. (2011). Men and women are different: Diffusion tensor imaging reveals sexual dimorphism in the microstructure of the thalamus, corpus callosum and cingulum. *NeuroImage, 54*(4), 2557–2562.

Merritt, V. C., Padgett, C. R., & Jak, A. J. (2019). A systematic review of sex differences in concussion outcome: What do we know? *The Clinical Neuropsychologist, 33*(6), 1016–1043.

Mihalik, J. P., Ondrak, K. S., Guskiewicz, K. M., & McMurray, R. G. (2009). The effects of menstrual cycle phase on clinical measures of concussion in healthy college-aged females. *Journal of Science and Medicine in Sport, 12*(3), 383–387.

Mikolic, A., Groeniger, J. O., Zeldovich, M., Wilson, L., van Lennep, J. R., van Klaveren, D., et al. (2021). Collaborative European NeuroTrauma Effectiveness Research in Traumatic Brain Injury (CENTER-TBI) Participants and Investigators. Explaining outcome differences between men and women following mild traumatic brain injury. *Journal of Neurotrauma, 38*(23), 3315–3331.

Miller, J. H., Gill, C., Kuhn, E. N., Rocque, B. G., Menendez, J. Y., O'Neill, J. A., et al. (2016). Predictors of delayed recovery following pediatric sports-related concussion: A case-control study. *Journal of Neurosurgery. Pediatrics, 17*(4), 491–496.

Mishell, D. R., Jr., Thorneycroft, I. H., Nakamura, R. M., Nagata, Y., & Stone, S. C. (1972). Serum estradiol in women ingesting combination oral contraceptive steroids. *American Journal of Obstetrics and Gynecology, 114*(7), 923–928.

Mollayeva, T., El-Khechen-Richandi, G., & Colantonio, A. (2018). Sex & gender considerations in concussion research. *Concussion, 3*(1), CNC51.

Monahan, K., Bannon, S., & Dams-O'Connor, K. (2020). Nonfatal strangulation (NFS) and intimate partner violence: A brief overview. *Journal of Family Violence, 13*, 1–2.

Monahan, K., Goldfine, A., & Biegon, A. (2017). Traumatic brain injuries in victims of intimate partner violence: An underappreciated source of neurological morbidity. *Future Neurology*, (4), 189–191.

Monahan, K., & O'Leary, K. D. (1999). Head injury and battered women: An initial inquiry. *Health & Social Work, 24*(4), 269–278.

Moser, R. S., Olek, L., & Schatz, P. (2019). Gender differences in symptom reporting on baseline sport concussion testing across the youth age span. *Archives of Clinical Neuropsychology, 34*(1), 50–59.

Musumeci, G., Ravalli, S., Amorini, A. M., & Lazzarino, G. (2019). Concussion in sports. *Journal of Functional Morphology and Kinesiology, 4*(2), 37.

National Center for Injury Prevention and Control (US). (2003). Report to Congress on mild traumatic brain injury in the United States: Steps to prevent a serious public health problem. *Centers for Disease Control and Prevention.*

Ng, L. J., Volman, V., Gibbons, M. M., Phohomsiri, P., Cui, J., Swenson, D. J., et al. (2017). A mechanistic end-to-end concussion model that translates head kinematics to neurologic injury. *Frontiers in Neurology, 15*(8), 269.

Nordström, P., Michaëlsson, K., Gustafson, Y., & Nordström, A. (2014). Traumatic brain injury and young onset dementia: A nationwide cohort study. *Annals of Neurology, 75*(3), 374–381.

Nordström, A., & Nordström, P. (2018). Traumatic brain injury and the risk of dementia diagnosis: A nationwide cohort study. *PLoS Medicine, 15*(1), e1002496.

O'Connor, K. L., Baker, M. M., Dalton, S. L., Dompier, T. P., Broglio, S. P., & Kerr, Z. Y. (2017). Epidemiology of sport-related concussions in high school athletes: National Athletic Treatment, Injury and Outcomes Network (NATION), 2011–2012 through 2013–2014. *Journal of Athletic Training*, (3), 175–185.

Pakkenberg, B., & Gundersen, H. J. (1997). Neocortical neuron number in humans: Effect of sex and age. *The Journal of Comparative Neurology, 384*(2), 312–320.

Peel, N. M., Kassulke, D. J., & McClure, R. J. (2002). Population based study of hospitalised fall related injuries in older people. *Injury Prevention, 8*(4), 280–283.

Preiss-Farzanegan, S. J., Chapman, B., Wong, T. M., Wu, J., & Bazarian, J. J. (2009). The relationship between gender and postconcussion symptoms after sport-related mild traumatic brain injury. *PM & R: The Journal of Injury, Function, and Rehabilitation, 1*(3), 245–253.

Queen, R. M., Weinhold, P. S., Kirkendall, D. T., & Yu, B. (2003). Theoretical study of the effect of ball properties on impact force in soccer heading. *Medicine and Science in Sports and Exercise, 35*(12), 2069–2076.

Ramage-Morin, P. L. (2008). Motor vehicle accident deaths, 1979 to 2004. *Health Reports, 19*(3), 45.

Riley, J. L., III, Robinson, M. E., Wise, E. A., Myers, C. D., & Fillingim, R. B. (1998). Sex differences in the perception of noxious experimental stimuli: A meta-analysis. *Pain, 74*(2–3), 181–187.

Riley, J. L., III, Robinson, M. E., Wise, E. A., & Price, D. (1999). A meta-analytic review of pain perception across the menstrual cycle. *Pain, 81*(3), 225–235.

Ring, C., van Zanten, J. J., & Kavussanu, M. (2009). Effects of sex, phase of the menstrual cycle and gonadal hormones on pain in healthy humans. *Biological Psychology, 81*(3), 189–191.

Ripley, D. L., Harrison-Felix, C., Sendroy-Terrill, M., Cusick, C. P., Dannels-McClure, A., & Morey, C. (2008). The impact of female reproductive function on outcomes after traumatic brain injury. *Archives of Physical Medicine and Rehabilitation, 89*(6), 1090–1096.

Roberts, G. W., Whitwell, H. L., Acland, P. R., & Bruton, C. J. (1990). Dementia in a punch-drunk wife. *The Lancet, 335*(8694), 918–919.

Rueda, M. A., Halley, W. L., & Gilchrist, M. D. (2010). Fall and injury incidence rates of jockeys while racing in Ireland, France and Britain. *Injury, 41*(5), 533–539.

Ruigrok, A. N., Salimi-Khorshidi, G., Lai, M. C., Baron-Cohen, S., Lombardo, M. V., Tait, R. J., et al. (2014). A meta-analysis of sex differences in human brain structure. *Neuroscience and Biobehavioral Reviews, 1*(39), 34–50.

Schatz, P., & Moser, R. S. (2011). Current issues in pediatric sports concussion. *The Clinical Neuropsychologist, 25*(6), 1042–1057.

Schneider, K. J., Emery, C. A., Kang, J., Schneider, G. M., & Meeuwisse, W. H. (2010). Examining sport concussion assessment tool ratings for male and female youth hockey players with and without a history of concussion. *British Journal of Sports Medicine, 44*(15), 1112–1117.

Schneider, H. J., Kreitschmann-Andermahr, I., Ghigo, E., Stalla, G. K., & Agha, A. (2007). Hypothalamopituitary dysfunction following traumatic brain injury and aneurysmal subarachnoid hemorrhage: A systematic review. *Journal of the American Medical Association, 298*(12), 1429–1438.

Schneider, A. L., Selvin, E., Latour, L., Turtzo, L. C., Coresh, J., Mosley, T., et al. (2021). Head injury and 25-year risk of dementia. *Alzheimer's & Dementia, 17*(9), 1432–1441.

Seiger, A., Goldwater, E., & Deibert, E. (2015). Does mechanism of injury play a role in recovery from concussion? *The Journal of Head Trauma Rehabilitation., 30*(3), E52–E56.

Shafi, R., Crawley, A. P., Tartaglia, M. C., Tator, C. H., Green, R. E., Mikulis, D. J., et al. (2020). Sex-specific differences in resting-state functional connectivity of large-scale networks in postconcussion syndrome. *Scientific Reports, 10*(1), 1–2.

Shah, K. N., Ba, E. J., & Daniels, A. H. (2020). Concussion: mechanisms of injury and trends from 1997 to 2019. *Rhode Island Medical Journal, 103*(7), 71–75.

Shehata, N., Wiley, J. P., Richea, S., Benson, B. W., Duits, L., & Meeuwisse, W. H. (2009). Sport concussion assessment tool: Baseline values for varsity collision sport athletes. *British Journal of Sports Medicine, 43*(10), 730–734.

Sherman, J. J., & LeResche, L. (2006). Does experimental pain response vary across the menstrual cycle? A methodological review. *American Journal of Physiology. Regulatory, Integrative and Comparative Physiology*, *291*(2), R245–R256.

Shin, Y. W., Kim, D. J., Hyon, T., Park, H. J., Moon, W. J., Chung, E. C., et al. (2005). Sex differences in the human corpus callosum: Diffusion tensor imaging study. *Neuroreport*, *16*(8), 795–798.

Smirl, J. D., Jones, K. E., Copeland, P., Khatra, O., Taylor, E. H., & Van Donkelaar, P. (2019). Characterizing symptoms of traumatic brain injury in survivors of intimate partner violence. *Brain Injury*, *33*(12), 1529–1538.

Smith, S. G., Basile, K. C., Gilbert, L. K., Merrick, M. T., Patel, N., Walling, M., et al. (2017). *National intimate partner and sexual violence survey (NISVS): 2010–2012 state report*. Atlanta, GA: National Center for Injury Prevention and Control, Centers for Disease Control and Prevention.

Solomito, M. J., Reuman, H., & Wang, D. H. (2019). Sex differences in concussion: A review of brain anatomy, function, and biomechanical response to impact. *Brain Injury*, *33*(2), 105–110.

Stortz, M., Presman, D. M., Pecci, A., & Levi, V. (2021). Phasing the intranuclear organization of steroid hormone receptors. *The Biochemical Journal*, *478*(2), 443–461.

Tanriverdi, F., Unluhizarci, K., & Kelestimur, F. (2010). Pituitary function in subjects with mild traumatic brain injury: A review of literature and proposal of a screening strategy. *Pituitary*, *13*(2), 146–153.

Tator, C. H., & Davis, H. (2014). The postconcussion syndrome in sports and recreation: Clinical features and demography in 138 athletes. *Neurosurgery*, *75*(Suppl_4), S106–S112.

Tator, C. H., Davis, H. S., Dufort, P. A., Tartaglia, M. C., Davis, K. D., Ebraheem, A., et al. (2016). Postconcussion syndrome: Demographics and predictors in 221 patients. *Journal of Neurosurgery*, *125*(5), 1206–1216.

Thompson, H. J., McCormick, W. C., & Kagan, S. H. (2006). Traumatic brain injury in older adults: Epidemiology, outcomes, and future implications. *Journal of the American Geriatrics Society*, *54*(10), 1590–1595.

Tierney, R. T., Higgins, M., Caswell, S. V., Brady, J., McHardy, K., Driban, J. B., et al. (2008). Sex differences in head acceleration during heading while wearing soccer headgear. *Journal of Athletic Training*, *43*(6), 578–584.

Tierney, R. T., Sitler, M. R., Swanik, C. B., Swanik, K. A., Higgins, M. I., & Torg, J. (2005). Gender differences in head-neck segment dynamic stabilization during head acceleration. *Medicine and Science in Sports and Exercise*, *37*(2), 272–279.

Tsushima, W. T., Lum, M., & Geling, O. (2009). Sex differences in the long-term neuropsychological outcome of mild traumatic brain injury. *Brain Injury*, *23*(10), 809–814.

Valera, E. M., & Berenbaum, H. (2003). Brain injury in battered women. *Journal of Consulting and Clinical Psychology*, *71*(4), 797.

Valera, E. M., Cao, A., Pasternak, O., Shenton, M. E., Kubicki, M., Makris, N., et al. (2019). White matter correlates of mild traumatic brain injuries in women subjected to intimate-partner violence: A preliminary study. *Journal of Neurotrauma*, *36*(5), 661–668.

Valera, E., & Kucyi, A. (2017). Brain injury in women experiencing intimate partner-violence: Neural mechanistic evidence of an "invisible" trauma. *Brain Imaging and Behavior*, *11*(6), 1664–1677.

Varriano, B., Tomlinson, G., Tarazi, A., Wennberg, R., Tator, C., & Tartaglia, M. C. (2018). Age, gender and mechanism of injury interactions in post-concussion syndrome. *The Canadian Journal of Neurological Sciences*, *45*(6), 636–642.

Vehicle RSaM, Regulation TC. (2011). *Road Safety in Canada*. [cited 2017 March 16, 2017]. Available at: https://www.tc.gc.ca/eng/motorvehiclesafety/tp-tp15145-1201.htm#s35.

Wagner, A. K., Franzese, K., Weppner, J. L., Kwasnica, C., Galang, G. N., Edinger, J., et al. (2021). Traumatic brain injury. *Braddom's Physical Medicine and Rehabilitation*, 916–953. Elsevier.

World Health Organization. (2015). *WHO global disability action plan 2014-2021. Better health for all people with disability. Report*. Geneva: World Health Organization (WHO). Available at www.who.int/disabilities/actionplan/en.

Wright, D. W., Kellermann, A. L., Hertzberg, V. S., Clark, P. L., Frankel, M., Goldstein, F. C., et al. (2007). ProTECT: A randomized clinical trial of progesterone for acute traumatic brain injury. *Annals of Emergency Medicine*, *49*(4), 391–402.

WSIB. *Workplace Safety and Insurance Board*. www.wsib.on.ca/WSIBPortal/faces/WSIBDetailPage?cGUID=WSIB066165&rDef=WSIBRDARTICLE.

Wu, V., Huff, H., & Bhandari, M. (2010). Pattern of physical injury associated with intimate partner violence in women presenting to the emergency department: A systematic review and meta-analysis. *Trauma, Violence & Abuse*, *11*(2), 71–82.

Wunderle, M. K., Hoeger, K. M., Wasserman, M. E., & Bazarian, J. J. (2014). Menstrual phase as predictor of outcome after mild traumatic brain injury in women. *The Journal of Head Trauma Rehabilitation*, *29*(5), E1.

Xiao, G., Wei, J., Yan, W., Wang, W., & Lu, Z. (2008). Improved outcomes from the administration of progesterone for patients with acute severe traumatic brain injury: A randomized controlled trial. *Critical Care*, *12*(2). 1-0.

Yaffe, K., Lwi, S. J., Hoang, T. D., Xia, F., Barnes, D. E., Maguen, S., et al. (2019). Military-related risk factors in female veterans and risk of dementia. *Neurology*, *92*(3), e205–e211.

Zaidi, Z. F. (2010). Gender differences in human brain: A review. *The Open Anatomy Journal*, *2*(1).

Zemek, R., Barrowman, N., Freedman, S. B., Gravel, J., Gagnon, I., McGahern, C., et al. (2016). Clinical risk score for persistent postconcussion symptoms among children with acute concussion in the ED. *Journal of the American Medical Association*, *315*(10), 1014–1025.

Zuckerman, S. L., Apple, R. P., Odom, M. J., Lee, Y. M., Solomon, G. S., & Sills, A. K. (2014). Effect of sex on symptoms and return to baseline in sport-related concussion. *Journal of Neurosurgery. Pediatrics*, *13*(1), 72–81.

Zuckerman, S. L., Solomon, G. S., Forbes, J. A., Haase, R. F., Sills, A. K., & Lovell, M. R. (2012). Response to acute concussive injury in soccer players: Is gender a modifying factor? *Journal of Neurosurgery. Pediatrics*, *10*(6), 504–510.

Kolibri 360 GmbH
Ferdinand-Jühlke-Straße 7
99095 Erfurt, Deutschland
produktsicherheit@kolibri360.de